METHODS IN MOLECULAR BIOLOGY

Series Editor
John M. Walker
School of Life and Medical Sciences
University of Hertfordshire
Hatfield, Hertfordshire, UK

For further volumes:
http://www.springer.com/series/7651

For over 35 years, biological scientists have come to rely on the research protocols and methodologies in the critically acclaimed *Methods in Molecular Biology* series. The series was the first to introduce the step-by-step protocols approach that has become the standard in all biomedical protocol publishing. Each protocol is provided in readily-reproducible step-by-step fashion, opening with an introductory overview, a list of the materials and reagents needed to complete the experiment, and followed by a detailed procedure that is supported with a helpful notes section offering tips and tricks of the trade as well as troubleshooting advice. These hallmark features were introduced by series editor Dr. John Walker and constitute the key ingredient in each and every volume of the *Methods in Molecular Biology* series. Tested and trusted, comprehensive and reliable, all protocols from the series are indexed in PubMed.

Modeling Transcriptional Regulation

Methods and Protocols

Edited by

Shahid Mukhtar

Department of Biology, University of Alabama, Birmingham, AL, USA

Humana Press

Editor
Shahid Mukhtar
Department of Biology
University of Alabama
Birmingham, AL, USA

ISSN 1064-3745 ISSN 1940-6029 (electronic)
Methods in Molecular Biology
ISBN 978-1-0716-1533-1 ISBN 978-1-0716-1534-8 (eBook)
https://doi.org/10.1007/978-1-0716-1534-8

This Humana imprint is published by the registered company Springer Science+Business Media, LLC, part of Springer Nature.
The registered company address is: 1 New York Plaza, New York, NY 10004, U.S.A.

Preface

This book provides a broad and systematic overview of new methods and cutting-edge techniques used in the construction of global transcriptional regulatory networks, various layers of gene regulation, and mathematical as well as computational modeling of transcriptional gene regulation in diverse systems. It is comprehensive and accessible, which makes it suitable as a graduate textbook and a standard laboratory reference. Moreover, it is also targeted to a specialized audience for an in-depth understanding of new approaches and tools to study transcriptional gene regulation.

Birmingham, AL, USA *Shahid Mukhtar*

Contents

Contributors

SYNAN F. ABUQAMAR • *Department of Biology, College of Science, United Arab Emirates University, Al Ain, United Arab Emirates*

FATIMAH ALJEDAANI • *King Abdullah University of Science and Technology (KAUST), Biological and Environmental Sciences and Engineering (BESE), Thuwal, Saudi Arabia*

JOSÉ M. ALVAREZ • *ANID-Millennium Science Initiative Program- Millenium Institute for Integrative Biology (iBio), Santiago, Chile; Centro de Genómica y Bioinformática, Facultad de Ciencias, Universidad Mayor, Santiago, Chile*

MOHAMMAD ATHAR • *Department of Dermatology, School of Medicine, University of Alabama at Birmingham, Birmingham, AL, USA*

IKRAM BLILOU • *King Abdullah University of Science and Technology (KAUST), Biological and Environmental Sciences and Engineering (BESE), Thuwal, Saudi Arabia*

YOGENDRA BORDIYA • *Department of Biology, Texas State University, San Marcos, TX, USA; Department of Molecular Biosciences, Institute for Cellular & Molecular Biology, The University of Texas at Austin, Austin, TX, USA*

JAKE Y. CHEN • *The University of Alabama at Birmingham, Birmingham, AL, USA*

THOMAS DANDEKAR • *Department of Bioinformatics, Biocenter, University of Wuerzburg, Am Hubland, Wuerzburg, Germany*

KHALED A. EL-TARABILY • *Department of Biology, College of Science, United Arab Emirates University, Al Ain, United Arab Emirates; Harry Butler Institute, Murdoch University, Murdoch, WA, Australia*

AMANDA CARVALHO GARCIA • *Endocrinology and Metabolism Service of the University Hospital of the University of Paraná, Paraná, Brazil; PhD student in Internal Medicine and Health Sciences at the Federal University of Paraná, Paraná, Brazil*

MARIAH GOBBLE • *Plant and Microbial Biology Department, North Carolina State University, Raleigh, NC, USA*

GUIMU GUO • *The University of Alabama at Birmingham, Birmingham, AL, USA*

RODRIGO A. GUTIÉRREZ • *ANID-Millennium Science Initiative Program- Millenium Institute for Integrative Biology (iBio), Santiago, Chile; Departamento de Genética Molecular y Microbiología, Facultad de Ciencias Biológicas, Pontificia Universidad Católica de Chile, Santiago, Chile; FONDAP Center for Genome Regulation, Santiago, Chile*

JOANNA E. HANDZLIK • *Department of Biology, University of North Dakota, Grand Forks, ND, USA*

JIMING JIANG • *Department of Plant Biology, Michigan State University, East Lansing, MI, USA; Department of Horticulture, Michigan State University, East Lansing, MI, USA; Michigan State University AgBioResearch, East Lansing, MI, USA*

HONG-GU KANG • *Department of Biology, Texas State University, San Marcos, TX, USA*

SUNG IL KIM • *Department of Biology, Texas State University, San Marcos, TX, USA*

NILESH KUMAR • *Department of Biology, University of Alabama at Birmingham, Birmingham, AL, USA*

ALICIA T. LAMERE • *Mathematics Department, Bryant University, Smithfield, RI, USA*

JIYOUNG LEE • *Ph.D. program in Genetics, Bioinformatics and Computational Biology, Virginia Polytechnic Institute and State University, Blacksburg, VA, USA*

SONG LI • *School of Plant and Environmental Sciences, Virginia Polytechnic Institute and State University, Blacksburg, VA, USA*

ZIWEN LI • *Zhongzhi International Institute of Agricultural Biosciences, Biology and Agriculture Research Center, University of Science and Technology Beijing, Beijing, China; Beijing Engineering Laboratory of Main Crop Bio-Tech Breeding, Beijing International Science and Technology Cooperation Base of Bio-Tech Breeding, Beijing Solidwill Sci-Tech Co. Ltd., Beijing, China*

YUAN LIN • *Department of Plant Biology, Michigan State University, East Lansing, MI, USA*

JINBAO LIU • *Department of Biology, University of Alabama at Birmingham, Birmingham, AL, USA*

YEN LEE LOH • *Department of Physics and Astrophysics, University of North Dakota, Grand Forks, ND, USA*

MANU • *Department of Biology, University of North Dakota, Grand Forks, ND, USA*

MEGAN L. MATTHEWS • *Department of Civil and Environmental Engineering, University of Illinois at Urbana-Champaign, Urbana, IL, USA; Institute for Sustainability, Energy, and Environment, University of Illinois at Urbana-Champaign, Urbana, IL, USA; Carl R. Woese Institute for Genomic Biology, University of Illinois at Urbana–Champaign, Urbana, IL, USA*

JOSÉ MAYORGA • *Department of Biology, Texas State University, San Marcos, TX, USA*

BHARAT MISHRA • *Department of Biology, University of Alabama at Birmingham, Birmingham, AL, USA*

TOMÁS C. MOYANO • *ANID-Millennium Science Initiative Program- Millenium Institute for Integrative Biology (iBio), Santiago, Chile; Departamento de Genética Molecular y Microbiología, Facultad de Ciencias Biològicas, Pontificia Universidad Católica de Chile, Santiago, Chile; FONDAP Center for Genome Regulation, Santiago, Chile*

SHAHID MUKHTAR • *Department of Biology, University of Alabama at Birmingham, Birmingham, AL, USA*

VIJAYKUMAR YOGESH MULEY • *Instituto de Neurobiología, Universidad Nacional Autónoma de México, Querétaro, Mexico*

JI-CHUL NAM • *Department of Biology, Texas State University, San Marcos, TX, USA*

MUHAMMAD NASEEM • *Department of Bioinformatics, Biocenter, University of Wuerzburg, Am Hubland, Wuerzburg, Germany; Department of Life and Environmental Sciences, College of Natural and Health Sciences, Zayed University, Abu Dhabi, UAE*

ÖZGE OSMANOGLU • *Department of Bioinformatics, Biocenter, University of Wuerzburg, Am Hubland, Wuerzburg, Germany*

KAROLINA M. PAJEROWSKA-MUKHTAR • *Department of Biology, University of Alabama at Birmingham, Birmingham, AL, USA*

NAGANAND RAYAPURAM • *King Abdullah University of Science and Technology (KAUST), Biological and Environmental Sciences and Engineering (BESE), Thuwal, Saudi Arabia*

MICHAEL SCHWARTZ • *Plant and Microbial Biology Department, North Carolina State University, Raleigh, NC, USA*

ARJUN SHAM • *Department of Biology, College of Science, United Arab Emirates University, Al Ain, United Arab Emirates*

SHABANA SHAMS • *Department of Animal Sciences, Faculty of Biological Sciences, Quaid-i-Azam University Islamabad, Islamabad, Pakistan*

ROSANGELA SOZZANI • *Plant and Microbial Biology Department, North Carolina State University, Raleigh, NC, USA*

RYAN SPURNEY • *Electrical and Computer Engineering Department, North Carolina State University, Raleigh, NC, USA*
LISA VAN DEN BROECK • *Plant and Microbial Biology Department, North Carolina State University, Raleigh, NC, USA*
XIANGYUAN WAN • *Zhongzhi International Institute of Agricultural Biosciences, Biology and Agriculture Research Center, University of Science and Technology Beijing, Beijing, China; Beijing Engineering Laboratory of Main Crop Bio-Tech Breeding, Beijing International Science and Technology Cooperation Base of Bio-Tech Breeding, Beijing Solidwill Sci-Tech Co. Ltd., Beijing, China*
WEI WANG • *State Key Laboratory for Protein and Plant Gene Research, School of Life Sciences, Peking University, Beijing, China; Center for Life Sciences, Beijing, China*
XINGWEI WANG • *State Key Laboratory for Protein and Plant Gene Research, School of Life Sciences, Peking University, Beijing, China; Center for Life Sciences, Beijing, China*
CRANOS M. WILLIAMS • *Department of Electrical and Computer Engineering, North Carolina State University, Raleigh, NC, USA*
YUFENG XU • *College of Life Sciences, Capital Normal University, Beijing, China*
DA YAN • *The University of Alabama at Birmingham, Birmingham, AL, USA*
HAIDONG YAN • *School of Plant and Environmental Sciences, Virginia Polytechnic Institute and State University, Blacksburg, VA, USA*
ZONGLIANG YUE • *The University of Alabama at Birmingham, Birmingham, AL, USA*
MIAN ZHOU • *College of Life Sciences, Capital Normal University, Beijing, China*

Chapter 1

Co-expression Networks in Predicting Transcriptional Gene Regulation

Synan F. AbuQamar, Khaled A. El-Tarabily, and Arjun Sham

Abstract

Recent progress in transcriptomics and co-expression networks have enabled us to predict the inference of the biological functions of genes with the associated environmental stress. Microarrays and RNA sequencing (RNA-seq) are the most commonly used high-throughput gene expression platforms for detecting differentially expressed genes between two (or more) phenotypes. Gene co-expression networks (GCNs) are a systems biology method for capturing transcriptional patterns and predicting gene interactions into functional and regulatory relationships. Here, we describe the procedures and tools used to construct and analyze GCN and investigate the integration of transcriptional data with GCN to provide reliable information about the underlying biological mechanism.

Key words Biological networks, Co-expression networks, Network analysis, Systems biology, Target gene identification, Transcriptomics

1 Introduction

Systems biology research has recently gained great attention, mainly in the recent developments of biological, medical, and environmental fields. This biology-based interdisciplinary approach integrates mathematical analysis and computational modeling to better understand complex biological systems. To study biological or ecological effects, this type of analysis can be performed with the help of high-throughput data within organisms or between organisms sharing a common environment. Omics techniques, such as transcriptomics, metabolomics, and proteomics, are now being used for the identification of target genes/proteins and prediction of unknown gene/protein functions [1–3].

Network biology has been widely used to study in-depth knowledge and comprehensive understanding of a system within an organism. This can be achieved with the predicted/experimental biological networks of the organism [4]. Interactomes may also be described as biological networks, which refer to protein–protein

Shahid Mukhtar (ed.), *Modeling Transcriptional Regulation: Methods and Protocols*, Methods in Molecular Biology, vol. 2328, https://doi.org/10.1007/978-1-0716-1534-8_1,

interaction (PPI), transcription-regulatory, gene co-expression networks (GCNs), metabolic networks, signaling networks, and others. All these networks are constructed using experimentally proven or predicted interactome research data. Typically, a network is a set of nodes (genes/proteins) inter-connected using edges (connections between nodes), based on their interaction with other nodes to form a complex structure [4–6]. Among the biotechnology approaches, microarrays and RNA sequencing (RNA-seq) can produce huge amounts of transcriptomic data that correspond to the transcript levels within an organism (Fig. 1) [7–9]. The bioinformatics can reduce the complexity of big data. There are many existing tools nowadays to visually explore the biological networks in the format of matrices, which are backed by the support of network analyses and functional-pathway predictions [10]. Expression network data may also contain a gene/protein interacting with others (gene regulation) of which the direction of interaction can be single-sided; or multiple genes/proteins involved in the regulation of which the edges/directions of interaction can be found in both sides forming a sub-network within the main interaction network (Fig. 2). The complexity of a network is scaled by the presence of substantial sub-units or motifs within the network. These motifs form the basic units of co-expressing genes/proteins in the network. In other words, network motifs are sub-graphs that repeat themselves in a specific network or among various networks. Each of the sub-graphs, defined by a specific pattern of interactions between vertices, may reflect a framework in which functions are efficiently achieved. Biological network motifs are highly important because they reflect the functional properties of the biological system [11]. Nowadays, motifs are considered a useful concept to uncover structural design principles of complex networks. Each type of network seems to display its own set of characteristic motifs. For example, ecological networks have different motifs than gene regulation networks [12]. Directed networks are the type of networks in which the edges or connections between each node have a specific direction (e.g., transcriptional regulatory network), whereas undirected networks are those networks in which the edges between the nodes can be in both directions (e.g., PPI network) [13, 14].

In this chapter, we will introduce the bioinformatic tools/packages to perform a comprehensive co-expression network analysis. As mentioned above, the co-expression network is a set of nodes, which are connected using undirected edges, in which the co-expressed genes are controlled by the same transcriptional regulation (e.g., treatments at different time points). Network analysis generates different node and edge attributes such as predicted data for the multiple gene/protein interactions, functional relation of genes/proteins within a network, biological process/cellular function (gene annotation) of the gene/protein of interest and

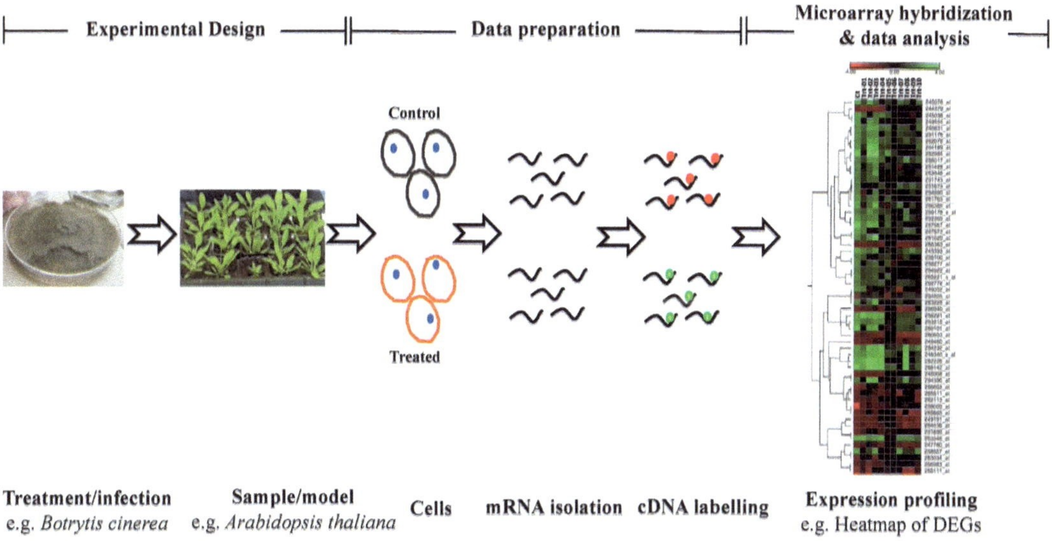

Fig. 1 Overview of steps for obtaining data of differentially expressed genes (DEGs) in a typical gene expression microarray experiment. Workflow of (1) experimental design, (2) experimental procedure, and (3) microarray hybridization and data analysis. An example of a treatment with a stress or pathogen infection (treatment/infection), i.e., *Botrytis cinerea* on *Arabidopsis* plants (sample/model). Total RNA or mRNA (mRNA) will be extracted, followed by cDNA synthesis and labeling for microarray hybridization and data analysis purposes. This displays different colors, which correspond to the up-/downregulation of genes. Heatmap analysis shows part of data analysis to determine the expression profiling of genes or DEGs. To obtain proper results, Arabidopsis non-treated plants will serve as a control treatment

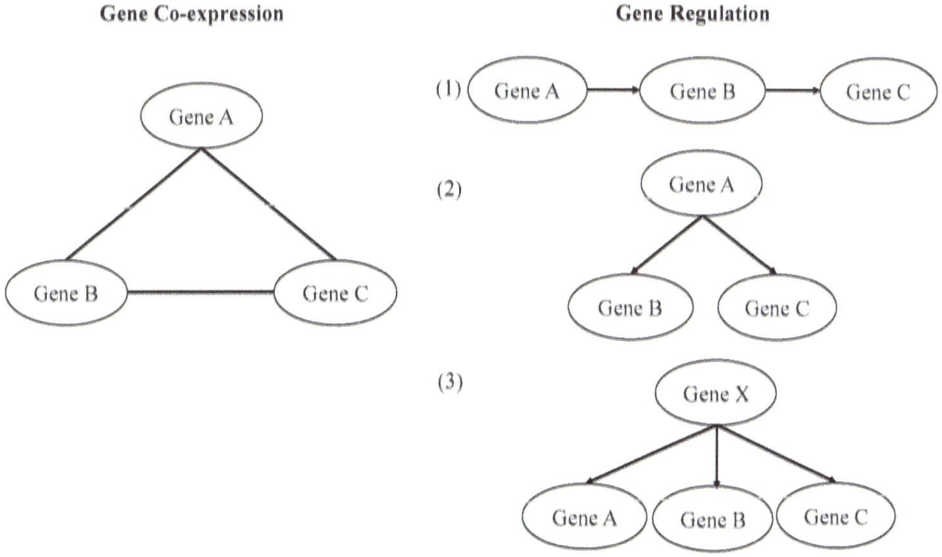

Fig. 2 Gene co-expression networks (GCNs) versus gene regulation. The GCNs are undirected networks (the edges between genes do not indicate the direction of the interaction). This, however, is not determined between the three genes A, B, and C that are co-expressed. Scenarios of gene regulation can be: (1) A activates B and B activates C; (2) A activates B and C; or (3) another gene (X) activates A, B, and C

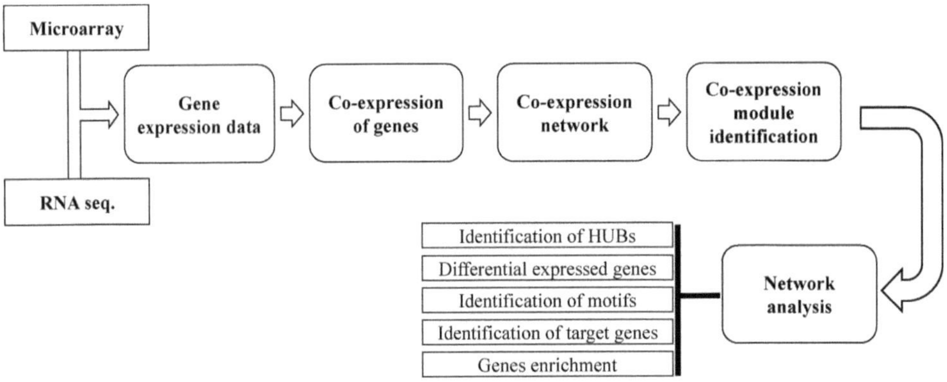

Fig. 3 Major steps describing the integration of biological networks and gene expression data. Expression data obtained from high-throughput methods (microarray or RNA-seq) are validated using bioinformatic tools to find out the co-expression pattern of genes and to plot the gene co-expression network (GCNs). The GCN analysis using specific software/tools helps in identifying the key factors of a network such as hubs, differentially expressed genes, motif patterns, the role of target genes or genes-of-interest and further gene enrichment studies and others

eventually the topology, degree distribution, and centrality of the nodes within the network (Fig. 3) [15]. The analysis of GCNs helps understand the transcriptional activities of the gene(s), and hence the topological parameters of the network which opens a window to the pathways and processes involved by the gene(s) [16, 17]. Complex interaction networks, i.e., human brain, can be examined by the structural and functional changes [18]. Structural networks describe the anatomical connections linking a set of neural elements of the brain based on relatively stable and short time intervals and may change with long exposures (e.g., neural imaging) [19]. Functional networks, however, are derived from time series of measurements/observations and describe statistical patterns of neural elements among anatomically separated regions. The analysis of these parameters will explore the condition-specific transcriptional changes or comparative transcriptomic studies. Here, the network of the model plant *Arabidopsis thaliana* will be used as an example. We will explain the methods and tools used to analyze and predict the correlation of *Arabidopsis* genes within the network in response to pathogen infections and/or other types of environmental stresses.

To study the application of co-expression networks in predicting the gene transcriptional regulation, we will use co-expression datasets and the generated data for the network analysis from the recently published article [20] as an illustration.

2 *Arabidopsis* Transcriptional Data for Multiple Stress Conditions

The transcription data for *Arabidopsis* plants under different stresses are obtained from NASCArrays at the BAR (http://bar.utoronto.ca/NASCArrays/index.php) [21]. For GCN analysis purposes, the previously published article containing five biotic stresses, two abiotic stresses, and four hormonal treatments [20] will be analyzed and compared with *Arabidopsis* plants infected with *B. cinerea* dataset (NASCArray-167).

3 Transcriptome Analysis Console (TAC)

The transcriptome datasets (.CEL files) are normalized using Expression Console software from Transcriptome Analysis Console (TAC) software package suit released by ThermoFisher Scientific (www.affymetrix.com).

4 *Arabidopsis* Protein–Protein Interaction (PPI) Network

Arabidopsis PPI dataset can be downloaded from *A. thaliana* Protein Interaction Network (AtPIN), which contains both experimentally proven and predicted interactions [22].

5 MDraw: Network Motif Analysis Tool

Motifs present in the network are analyzed using MFinder application in the MDraw tool (http://www.weizmann.ac.il/mcb/UriAlon/download/network-motif-software) [23] and the corresponding network motifs forming 3-, 4-, and 5-nodes are identified.

6 Cytoscape: Network Visualization and Analysis Software

The *Arabidopsis* PPI network is visualized and analyzed using Cytoscape software version 3.7.0 (https://cytoscape.org/) [24]. For further analysis and modification on the network, other Cytoscape associated applications are used. Such useful applications are: CentiScaPe for centrality measures [25]; BINGO for Gene Ontology (GO) annotations [26]; and String for associated pathways [27].

7 Methods

7.1 Normalizing Arabidopsis Microarray Data

1. Download all the microarray data from the NASCArray database (http://affymetrix.arabidopsis.info).
2. Open TAC. From the software package, also open the Expression Console Suit and set the configuration panel.
3. Specify the "user profile" and set the "library path" to the folder which has the *.CEL and *.CHP format files (*see* **Note 1**).
4. Download and save the "library" and "annotation" files. In the example, it is considered as ATH1-121501.
5. Specify the "report controls" based on the microarray data used for the analysis. For ATH1-121501 data, Spike Controls (AFFX-BioB, AFFX-BioC, AFFX-CreX, etc.) and Housekeeping Controls (AFFX-r2-At-Actin, AFFX-r2-At-GAPDH, and AFFX-r2-At-Ubq) are used. Optional "report thresholds" can also be set according to the data used (*see* **Note 2**).
6. After setting all the configurations, open the "study" tab and click "create a new study/open existing study." Add the intensity files (*.CEL) followed by adding the summarization files (*.CHP).
7. Run analysis using the preferred method of normalization, Robust Multichip Average (RMA), Probe Logarithmic Intensity Error Estimation (PLIER) Workflow or MAS5 [28–30]. Nowadays, the most commonly used one is the RMA normalization due to its specificity and low errors [31].
8. Click the "reports" tab to generate the normalized data, which can be then exported to excel files for further analysis.

7.2 Principle Component Analysis (PCA)

Once the data are normalized, the next step is to reduce the complexity of the data. One of the most effective way to do this is to perform Principle Component Analysis (PCA). It helps reduce the dimensionality or the variables in that data, thereby making it easier to visualize large datasets [32].

1. The PCA can be done manually by calculating the eigenvalues and eigenvectors for the corresponding covariance matrix or by simply using a software (e.g., Expression Console Suit).
2. Select the list of data which needs to be sorted, click "graph" menu and select PCA. Alternatively, select PCA from the "QC Array Comparisons" tab.

7.3 Identification of Differentially Expressed Genes (DEGs)

1. The normalized microarray data can be sorted out based on the probe detection calls: Present (P), Absent (A), and Medium (M). The (A) and (M) detection calls are removed and only (P) is considered. For the signal replicates, an average signal value can be used for further analysis.
2. In case of RNA-seq data, an additional quality control (QC) check is recommended to curate the data and to produce the best results. One of the widely used software is RNA-SeQC (https://software.broadinstitute.org/cancer/cga/rna-seqc), which calculates yield, alignment and duplication rates, GC bias, rRNA content, regions of alignment (exon, intron and intragenic), continuity of coverage, 3′/5′ bias and count of detectable transcripts, and others [33].
3. As the first step of identifying DEGs, the individual treated data are compared with a control (non-treated/non-infected) microarray data. Fold change (FC) for each gene is calculated by dividing the gene's signal intensity (SI) in the treated data by the same gene's SI in control/non-treated data. A gene is considered upregulated if the FC obtained is ≥ 2 and is considered downregulated if the FC of that gene is ≤ 0.5 (or ≥ -2). The FC values obtained are log transformed prior to the plotting of DEGs.
4. To compare the different datasets and to plot the DEGs, Morpheus (https://software.broadinstitute.org/morpheus/) which is an online tool can be used to plot the heatmap of the multiple datasets. In this case, a graphical representation of the DEGs is generated.

7.4 Identification of Co-expressing Genes and Network Motifs Across the Datasets

1. Once DEGs are identified, the multiple datasets are compared to check for any co-expressing genes.
2. The PPI data can be opened in MDraw software (http://www.weizmann.ac.il/mcb/UriAlon/download/network-motif-software) to visualize the whole PPI network. This can be followed by opening the "Tools" menu and run "MFinder" to select the unique motifs present in the network. Set the parameters for running the "MFinder" tool. Note that motif size must be mentioned. The program searches simultaneously for 2 sizes (e.g., 3- and 4-node motifs).
3. Select either the directed or undirected "network type," the "input file format" of the network data (e.g., Source to Target or Target to Source) and the number of "sampling" are required to run the data (normally ranges 100–1000). After completing the parameters, click "run" to execute the program. Once finished, the result page will show the unique motif IDs for the corresponding 3-, 4-, and/or 5-node motifs and their nodes forming the motifs (*see* **Note 3**).

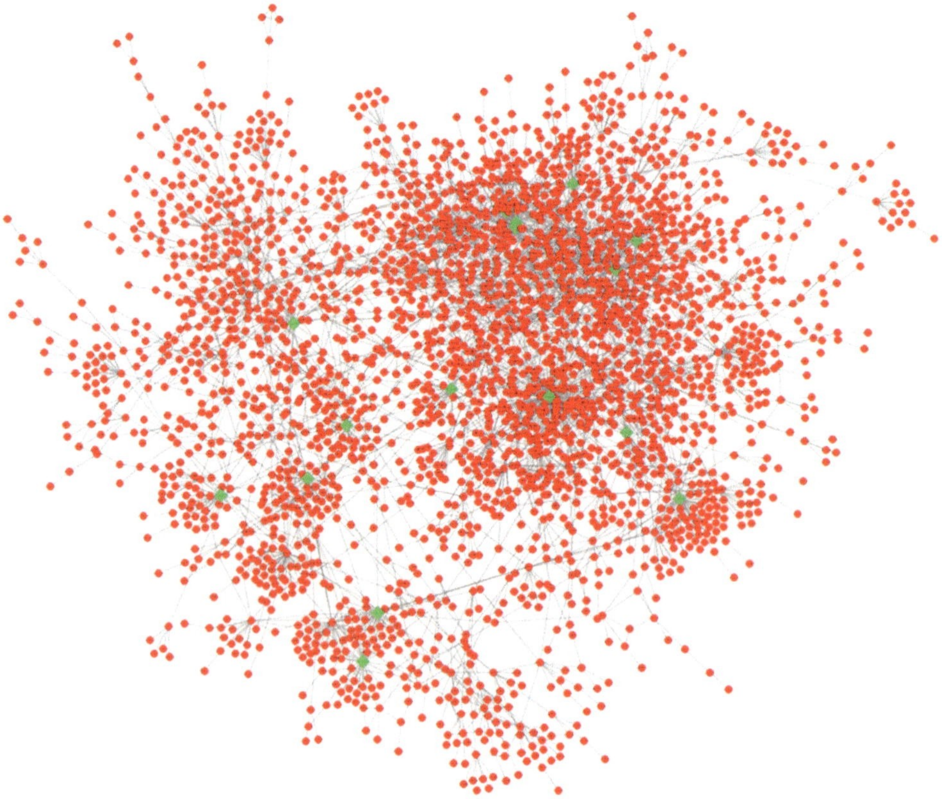

Fig. 4 Gene co-expression network (GCN) obtained from *Arabidopsis thaliana* Protein Interaction Network (AtPIN) post inoculation with *Botrytis cinerea*. The GCN shows the genes involved in response to *B. cinerea* infection in *Arabidopsis*. Green and red colored nodes represent the hubs and the interacting genes, respectively

7.5 Visualization and Analysis of Co-expression Network

1. Open the PPI network file with Cytoscape and select "import>network>file" from the main "file" menu. Select the network file; and from the pop-up window, select the "source node," "target node," and "edge attribute," if any. The network will be opened in a new window (Fig. 4).
2. Once the network is opened, select the nodes from the identified co-expression analysis data to highlight the GCN formed by the selected genes.
3. Cytoscape has a variety of applications and plugins incorporated with the software for network analyses. Here, use the CentiScaPe app [25] to pull out the hierarchy of the network by analyzing the degree (number of interactions) of the nodes that identifies the hubs of the network. Other results from the CentiScaPe may include, but not limited to, centrality, betweenness, and closeness. This reveals the importance of the node in the network, and hence identifies the target genes/proteins.

4. The GO enrichment plays a crucial role in understanding the functional characteristics, biological processes, and cellular locations of the genes in the network [26]. BINGO, a plugin in Cytoscape, is used to obtain the over- and under-repressed GO of the selected genes in the network (*see* **Note 4**). This will help in predicting the transcriptional pathways in which the gene is involved.
5. Another application in Cytoscape known as STRING helps in understanding the pathways and transcriptional regulations of the genes/proteins of the selected network [27]. In general, the information can be pulled out from the STRING database (https://string-db.org/) and annotate it with the selected network to identify the function of the proteins and much more by using the functional enrichment analysis.
6. Using these network analysis methods in Cytoscape, a better understanding of the regulation and correlation of the genes/proteins can be visualized.

8 Notes

1. While normalizing the data, make sure to add the exact *.CHP/summarization file along with the *.CEL/intensity file. Any error in mislabeling may turn out to totally different signal values for the genes.
2. Controls of the data for normalization have to be verified with the microarray data used for the study. These must be the same ones entered in the Expression Console.
3. For network motif analysis, most motifs are identified for up to five-node motifs. If we go higher, there is a chance of a combination of lower-numbered motifs to return as a bigger-numbered motif. For example, two 3-node motifs combined together can be obtained while searching for a 6-node motif will lead to the loss of the uniqueness of the motif data.
4. The parameters for the gene enrichment analysis using BINGO must match with the network type used, and the organism and the annotation file selected for the enrichment. For example, *Arabidopsis* network has to be analyzed using the ATH1 annotation file and the organism should be selected as *A. thaliana* per se. Any error in this step may turn non-specific results or end up crashing the Cytoscape.

Acknowledgement

This work was supported by Khalifa Center for Biotechnology and Genetic Engineering-UAEU grant 12R028 to S.AQ.

References

1. AbuQamar SF, Moustafa K, Tran L-SP (2016) 'Omics' and plant responses to *Botrytis cinerea*. Front Plant Sci 7:1658. https://doi.org/10.3389/fpls.2016.01658
2. Fey WKD, Ryan CJ, Tavassoly I et al (2018) Systems biology primer: the basic methods and approaches. Essays Biochem 62(4):487–500. https://doi.org/10.1042/EBC20180003
3. Breitling R (2010) What is systems biology? Front Physiol 1:9. https://doi.org/10.3389/fphys.2010.00009
4. Proulx SR, Promislow DE, Phillips PC (2005) Network thinking in ecology and evolution. Trends Ecol Evol 20(6):345–353. https://doi.org/10.1016/j.tree.2005.04.004
5. Barabasi A, Oltvai Z (2004) Network biology: understanding the cell's functional organization. Nat Rev Genet 5:101–113. https://doi.org/10.1038/nrg1272
6. Ideker T, Nussinov R (2017) Network approaches and applications in biology. PLoS Comput Biol 13(10):e1005771. https://doi.org/10.1371/journal.pcbi.1005771
7. Sham A, Al-Azzawi A, Al-Ameri S et al (2014) Transcriptome analysis reveals genes commonly induced by *Botrytis cinerea* infection, cold, drought and oxidative stresses in *Arabidopsis*. PLoS One 9(11):e113718. https://doi.org/10.1371/journal.pone.0113718
8. Sham A, Moustafa K, Al-Ameri S et al (2015) Identification of *Arabidopsis* candidate genes in response to biotic and abiotic stresses using comparative microarrays. PLoS One 10(5):e0125666. https://doi.org/10.1371/journal.pone.0125666
9. Sham A, Moustafa K, Al-Shamisi S et al (2017) Microarray analysis of *Arabidopsis WRKY33* mutants in response to the necrotrophic fungus *Botrytis cinerea*. PLoS One 12(2):e0172343. https://doi.org/10.1371/journal.pone.0172343
10. Kusonmano K (2016) Gene expression analysis through network biology: bioinformatics approaches. In: Nookaew I (ed) Network biology. Advances in biochemical engineering/biotechnology, vol 160. Springer, Cham, pp 15–32
11. Milo R, Shen-Orr S, Itzkovitz S et al (2002) Network motifs: simple building blocks of complex networks. Science 298:824–827
12. Wernicke S (2006) Efficient detection of network motifs. IEEE/ACM Trans Comput Biol Bioinform 3(4):347–359. https://doi.org/10.1109/TCBB.2006.51
13. Alon U (2007) Network motifs: theory and experimental approaches. Nat Rev Genet 8:450–461. https://doi.org/10.1038/nrg2102
14. Itzkovitz S, Alon U (2005) Subgraphs and network motifs in geometric networks. Phys Rev E 71:026117. https://doi.org/10.1103/PhysRevE.71.026117
15. van Dam S, Võsa U, Van-der-Graaf A et al (2018) Gene co-expression analysis for functional classification and gene–disease predictions. Brief Bioinform 19(4):575–592. https://doi.org/10.1093/bib/bbw139
16. Des-Marais DL, Guerrero RF, Lasky JR et al (2017) Topological features of a gene co-expression network predict patterns of natural diversity in environmental response. Proc Biol Sci 284:20170914. https://doi.org/10.1098/rspb.2017.0914
17. Schäpe P, Kwon MJ, Baumann B (2019) Updating genome annotation for the microbial cell factory *Aspergillus niger* using gene co-expression networks. Nucleic Acids Res 47(2):559–569. https://doi.org/10.1093/nar/gky1183
18. Yao Z, Hu B, Xie Y et al (2015) A review of structural and functional brain networks: small world and atlas. Brain Inform 2:45–52. https://doi.org/10.1007/s40708-015-0009-z
19. Sporns O (2013) Structure and function of complex brain networks. Dialogues Clin Neurosci 15(3):247–262
20. Sham A, Al-Ashram H, Whitley K et al (2019) Metatranscriptomic analysis of multiple environmental stresses identifies *RAP2.4* gene associated with *Arabidopsis* immunity to *Botrytis cinerea*. Sci Rep 9:17010. https://doi.org/10.1038/s41598-019-53694-1
21. Toufighi K, Brady SM, Austin R et al (2005) The botany array resource, e-northerns,

expression angling, and promoter analyses. Plant J 43:153–163. https://doi.org/10.1111/j.1365-313X.2005.02437.x

22. Brandao MM, Dantas LL, Silva-Filho MC (2009) AtPIN, *Arabidopsis thaliana* protein interaction network. BMC Bioinformatics 10:454. https://doi.org/10.1186/1471-2105-10-454
23. Kashtan N, Itzkovitz S, Milo R et al (2004) Efficient sampling algorithm for estimating sub-graph concentrations and detecting network motifs. Bioinformatics 20:1746–1758. https://doi.org/10.1093/bioinformatics/bth163
24. Shannon P, Markiel A, Ozier O et al (2003) Cytoscape: a software environment for integrated models of biomolecular interaction networks. Genome Res 13:2498–2504. https://doi.org/10.1101/gr.1239303
25. Scardoni G, Petterlini M, Laudanna C (2009) Analyzing biological network parameters with CentiScaPe. Bioinformatics 25 (21):2857–2859. https://doi.org/10.1093/bioinformatics/btp517
26. Maere S, Heymans K, Kuiper M (2005) BiNGO: a Cytoscape plugin to assess overrepresentation of GO categories in biological networks. Bioinformatics 21(16):3448–3449. https://doi.org/10.1093/bioinformatics/bti551
27. Doncheva NT, Morris JH, Gorodkin J et al (2019) Cytoscape StringApp: network analysis and visualization of proteomics data. J Proteome Res 18(2):623–632
28. Irizarry RA, Hobbs B, Collin F et al (2003) Exploration, normalization and summaries of high-density oligonucleotide array probe level data. Biostatistics 4(2):249–264. https://doi.org/10.1093/biostatistics/4.2.249
29. Terry M, Therneau BKV (2008) What does PLIER really do? Cancer Inform 6:423–431
30. Pepper SD, Saunders EK, Edwards LE et al (2007) The utility of MAS5 expression summary and detection call algorithms. BMC Bioinformatics 8:273. https://doi.org/10.1186/1471-2105-8-273
31. Parrish RS, Spencer HJ (2004) Effect of normalization on significance testing for oligonucleotide microarrays. J Biopharm Stat 14 (3):575–589. https://doi.org/10.1081/BIP-200025650
32. Yeung KY, Ruzzo WL (2001) Principal component analysis for clustering gene expression data. Bioinformatics 17(9):763–774. https://doi.org/10.1093/bioinformatics/17.9.763
33. DeLuca DS, Levin JZ, Sivachenko A et al (2012) RNA-SeQC: RNA-seq metrics for quality control and process optimization. Bioinformatics 28(11):1530–1532. https://doi.org/10.1093/bioinformatics/bts196

Chapter 2

Inference of Gene Coexpression Networks from Bulk-Based RNA-Sequencing Data

Alicia T. Lamere

Abstract

Gene coexpression networks (GCNs) are useful tools for inferring gene functions and understanding biological processes when properly constructed. Traditional microarray analysis is being more frequently replaced by bulk-based RNA-sequencing as a method for quantifying gene expression. This new technology requires improved statistical methods for generating GCNs. This chapter explores several popular methods for constructing GCNs using bulk-based RNA-Seq data, such as distribution-based methods and normalization techniques, implemented using the statistical programming language R.

Key words Gene coexpression network, Gene regulatory network, Bulk-based RNA-Seq, Correlation coefficient, Count data

1 Introduction

The "guilt-by-association" heuristic has led to the wide use of Gene Coexpression Networks (GCNs) when performing transcriptome analysis. The belief is that, if genes display coexpression, they are likely involved with the same cellular processes [1]. Therefore, if we construct a GCN and discover simultaneous expression/silence of two or more genes, and the function of one of the coexpressing genes is previously known, we can infer that the others are also somehow involved with that function. Many researchers have demonstrated the validity of the associations identified by GCNs, particularly in the realm of cancer research [2–5]. When constructed properly, as illustrated in Fig. 1, GCNs can be a valuable tool for better understanding the molecular mechanisms underlying biological processes and predicting unknown gene functions.

The data generated by RNA-Seq consist of the number of times each gene was observed expressing in a sample. This is accomplished by isolating the messenger RNA, or "reads," from a given sample before passing them through Illumina Sequencing, which outputs the coded "TCGA" sequences. Reads are then mapped to

Shahid Mukhtar (ed.), *Modeling Transcriptional Regulation: Methods and Protocols*, Methods in Molecular Biology, vol. 2328,
https://doi.org/10.1007/978-1-0716-1534-8_2,

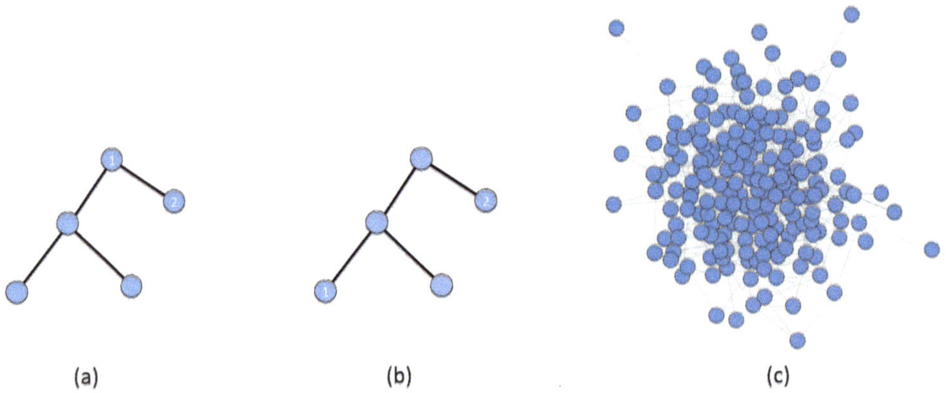

Fig. 1 Examples of GCNs. Nodes represent genes and edges represent coexpression between a pair of genes, (**a**) genes 1 and 2 in the network have an edge between them, meaning they have been observed coexpressing, (**b**) genes 1 and 2 do not share an edge, so there is no evidence of coexpression, (**c**) an example of an actual GCN generated from RNA-Seq data. To construct a GCN, we must first obtain a measure of gene expression. In recent years, RNA-Sequencing (RNA-Seq) has replaced microarrays as the technological standard for measuring high-throughput gene expression. RNA-Seq is perceived to be not only more efficient at discovering new genes or isoforms, but also allows for a much larger dynamic range for measuring each gene's expression [6]. However, the non-Gaussian nature of the data generated requires an adjustment to the statistical tools used to analyze this data

the genome of the organism being studied [6]. We can then "count" the number of reads that are mapped to each gene.

These counts consist of nonnegative integer values, and hence should not be directly modeled using a Gaussian distribution. Instead, the data must be either normalized or distributions such as the Poisson or negative binomial should be used. There are many different types of GCNs that can be constructed. Allen et al. [7] classified these methods into four categories: correlation-based, probabilistic network-based, partial-correlation-based, and information-theory-based. Correlation-based methods remain the simplest and fastest methods to implement, and therefore will be the focus of this chapter.

2 Materials

1. Gene count data.
2. R software for statistical computing.
3. Standard laptop computer should suffice for data that has been reduced to approximately 1000 genes.

    R Packages:
4. edgeR (available through Bioconductor).
5. DiPhiSeq [8] (available through CRAN).
6. Weighted Gene Coexpression Network Analysis (WGCNA) [9] (available through CRAN).

3 Methods

This section discusses the steps necessary to properly prepare count data, followed by explanations of several widely used methods for GCN construction. The assumption of this chapter is that counts have already been generated for the sample of interest (*see* **Note 1**).

An important consideration when working with RNA-Seq data is the choice of which distribution to use to model the data. Generally, we can model our data as m RNA-Seq experiments measuring the expression of n genes. Let x_{ij} represent the expression of gene i in experiment j, where $i = 1, \ldots, n$ and $j = 1, \ldots, m$. Then this expression x_{ij} is a nonnegative integer, for which one of the most commonly used distributions is the negative binomial. Alternatively, some researchers choose to transform x_{ij} to normalize it in such a way that methods developed for microarray data can be applied.

3.1 Sequencing Depth

An important component of any model is the sequencing depth (*see* **Note 2**). RNA-Seq experiments inherently tend to have a large variation in the total counts, or depth, observed (Table 1).

Best practices are to estimate this depth for each experiment using all genes so that each experiment can then be scaled to allow for a fairer comparison across experiments [10–12]. This can be easily accomplished with R using code similar to the following:

1. Create a vector `dep` of length m to store the depths, where m is the number of experiments.

```
> dep = rep(0,m)
```

2. Let `data` represent the n x m matrix of observed counts. Calculate the totals for each column.

```
> for (i in (1:m)){dep[i] = sum(data[,i])}
```

3.2 Identify Highly Expressed Genes

Once sequencing depths are found, it is important to determine the genes with enough expression and variation to provide accurate and informative edges within our GCN. Genes with low observed expression could either have been poorly captured by the experiment or only represent noise in our data, and therefore may not reflect true coexpression in the organism of interest. Again, we can

Table 1
Example observed counts for RNA-Seq data across different experimental conditions, demonstrating the impact of sequencing depth. Consider Gene 4. Naively, it appears to be most highly expressed under condition 3, with observed counts of 200, and the least expressed under condition 1, with only 10 observed counts. However, relative to the other observed counts in condition 3, 200 is actually a low level of expression for this condition, while 10 is relatively the highest expression of all genes under condition 1

Gene	Condition 1	Condition 2	Condition 3
Gene 1	5	40	1700
Gene 2	2	30	1200
Gene 3	7	50	900
Gene 4	10	130	200
Gene 5	1	260	1400
Gene 6	3	110	2000

identify genes appropriate for analysis easily through R using code such as:

1. First, standardize the observed counts using the sequencing depths. It's good practice to store them as a new object, `stand_data`.

```
> stand_data = data
> for (i in (1:m)){stand_data[,i]=data[,i]/dep[i]}
```

2. Next, find the mean and inter-quartile range of the standardized counts for each gene.

```
> mean_val = apply(stand_data, 1, mean)
> iqr_vals = apply(stand_data, 1, IQR)
```

3. Now scale the inter-quartile range for each gene with its mean:

```
> iqr_scale = iqr_vals/mean_val
```

4. Finally, keep only those genes with sufficiently high mean and variation in expression. These can usually be found by retaining those with a minimum mean of 0.1 and a minimum scaled IQR of 0.5. These minimums should be adjusted as necessary for each dataset (*see* **Note 3**). Again, it is good practice to store

these identified genes as a new data object, here `filter_data`.

```
> keep = (mean_val >= 0.1 & iqr_scale >= 0.5)
> filter_data = stand_data[keep,]
```

3.3 Negative Binomial Model

When using a negative binomial model for count data, the model is as follows:

$$x_{ij}\tilde{N}B(d_j\mu_i, \phi_i)$$

where d_j represents the sequencing depth of experiment j, μ_i is the mean expression for gene i, and ϕ_i is a dispersion parameter capturing the disparity in variance from what we'd expect to observe for a Poisson distribution. Hence, $d_j\mu_i$ represents our mean for the distribution, $d_j\mu_i + \phi_i(d_j\mu_i)^2$ is the variance, and a Poisson distribution is represented when $\phi_i = 0$. This is useful, as some RNA-Seq data does not demonstrate overdispersion, so our model will still allow for these instances. To implement any model-based methods, these parameters must be estimated for our distributions.

3.3.1 iCC

Distribution-inversed and Gaussian-transformed Correlation Coefficient (iCC) is a GCN construction method developed directly for use on RNA-Seq data [13]. It works with the count data directly, allowing for the use of whatever distribution, negative binomially or otherwise, that is desired. This, and the direct incorporation of the sequencing depths, increases the power of iCC over normalization-dependent methods. Though not provided as a package, we can implement it through R fairly easily:

1. Let `filter_data` be the filtered dataset of the most highly expressed genes. Use this data to find the transformed sequencing depth for each experiment:

```
> d = colMeans(filter_data)
> depth = exp(log(d) - mean(log(d)))
```

2. Next, we need to find the values of the other parameters defining the distribution—these need to be found for each gene using the data across all experiments. If using a negative binomial model, this can be done with the DiPhiSeq R package [8] using the following code for genes i and j:

```
> library(DiPhiSeq)
> results_i = robnb(filter_data[i,], depth)
> results_j = robnb(filter_data[j,], depth)
```

Table 2
Example adjacency matrix for six genes containing correlations. High absolute correlation values (close to 1) such as those between Gene 1 and Gene 4 and between Gene 4 and Gene 5 are strong indications of a coexpression relationship (*see* Note 5)

Genes	Gene 1	Gene 2	Gene 3	Gene 4	Gene 5	Gene 6
Gene 1	1	0.21	−0.34	−0.87	−0.11	−0.42
Gene 2		1	0.33	0.53	−0.62	0.25
Gene 3			1	−0.17	0.31	0.23
Gene 4				1	0.84	−0.51
Gene 5					1	0.32
Gene 6						1

3. Then, for each observation for the genes i and j, calculate the associated probability of observing a given value k using the parameters found by DiPhiSeq:

```
> for (k in (1: length(depth)){
                    pval_i[k] = pnbinom(filter_data[j,k],
                    size=1/results_i$phi,
                    mu=exp(results_i$beta)*depth[k])
                    pval_j[k] = pnbinom(filter_data[j,k],
                    size=1/results_j$phi,
                    mu=exp(results_j$beta)*depth[k])}
```

4. Now, these probabilities can be used to find the associated values from a standard Gaussian distribution.

```
> norm_i = qnorm(pval_i, 0, 1)
> norm_j = qnorm(pval_j, 0, 1)
```

5. Finally, for the gene pair, use the standard Gaussian-distributed values to estimate their correlation.

```
> corr_ij = cor(norm_i, norm_j)
```

These correlations define the adjacency matrix that describes the GCN (*see* Table 2). Typically, a cutoff should be chosen for the correlations to construct the network (*see* **Notes 4** and **5**).

3.4 Normalization-Based Models

An alternative to using distribution-based methods, the count data can be normalized (*see* **Note 6**) and then traditional methods designed for microarray data can be applied to construct GCNs. This normalization must account for differences in means and variances between samples.

3.4.1 Variance Stabilizing Transformation

A widely used method is VST [14], which attempts to take both these mean and variance differences into account. The transformation is simple and can easily be implemented with R:

1. For a pair of genes *i* and *j*, create vectors to store the normalized counts for each experiment.

```
> norm.i = rep(0,length(dep))
> norm.j = rep(0,length(dep))
```

2. Using the sequencing depth `dep` calculated in 3.1, find the VST normalized counts.

```
> for (k in (1:length(dep))){
              x.norm[k] = sqrt(x[k]/dep[k])
              y.norm[k] = sqrt(y[k]/dep[k])}
```

It is important to note that this VST transformation was designed to accommodate a Poisson distribution, and consequentially does not perform as well as distribution-based methods when the data is truly negative binomially distributed [13]. There is no VST for the more appropriate negative binomial distribution, as it is impossible in that case to stabilize both the mean and variance simultaneously due to the dispersion's association with the mean (*see* model description in Subheading 3.3).

3.4.2 Other Normalization Methods

In some cases, RNA-Seq data have been demonstrated to contain a significant amount of technical noise. Normalization methods such as relative log expression [11], upper quartile [15], or trimmed mean of M-values [16] have been developed to counter this concern (*see* **Note 7**). The package edgeR can be used to perform any of these normalizations:

1. Use the function `calcNormFactors` on your original raw count dataset with your desired normalization method selected. Here we use relative log expression:

```
> norm_factors = calcNormFactors(data, method="RLE")
```

2. Once the factors are found, we must use them to generate our normalized data with the `cpm()` function:

```
> norm_data = cpm(norm_factors, normalized.lib.sizes = TRUE)
```

Be sure that the `normalized.lib.sizes = TRUE` option is selected.

3.4.3 Weighted Gene Coexpression Network Analysis

WGCNA [9] is arguably the most widely used algorithm for developing GCNs using microarray data. Due to its popularity and effectiveness, many choose to continue to implement it with RNA-Seq data. This can safely be done once counts have been normalized using methods such as described above, although methods directly incorporating non-Gaussian distributions are shown to be more effective [13]. Similar to the method employed by iCC, once the data is normalized the Pearson's correlation coefficient is calculated to generate the adjacency matrix used to create the GCN. To create this matrix, we can use the WGCNA package created for R, and use code similar to what is provided by the package's authors in their tutorial:

1. First, it is useful to enable the use of multi-threading to speed up the calculations, particularly when using a larger data set:

```
> enableWGCNAThreads()
```

2. One benefit of using WGCNA is the ability to incorporate a soft-threshold when determining the final GCN through the use of weighted edges. In this way, users can view the ranked edges and use WGCNA's algorithm to help decide on a threshold. This can be done with the following code. Note that `filter_data` should be your normalized dataset, filtered to keep the most highly expressed genes.

```
> Powers = c(c(1:10), seq(from=12, to=20, by=2))
> soft = pickSoftThreshold(filter_data, powerVector =
 Powers, verbose = 5)
> R_sqr= -sign(soft$fitIndices[,3]) *soft$fitIndices[,2]
> plot(soft$fitIndices[,1], R_sqr)
```

The resulting plot can be used to identify a threshold based on the lowest power for which the curve flattens after reaching a high value for the signed R square.

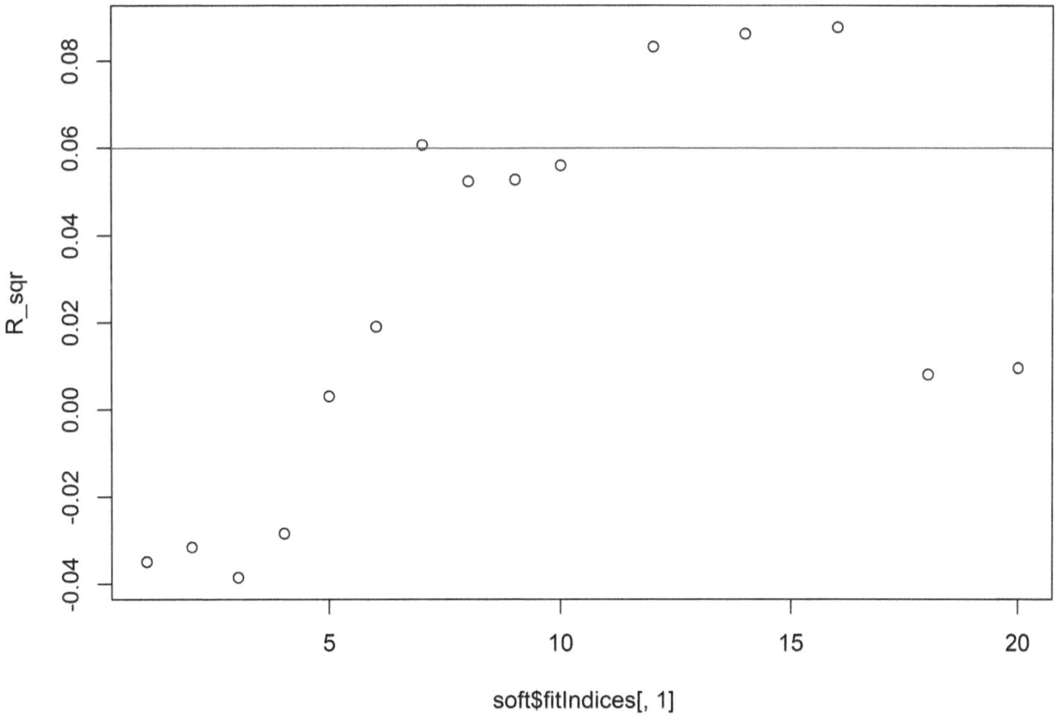

Fig. 2 The resulting plot from WGCNA for identifying power threshold. The red line has been added to indicate where the curve flattens after reaching a high value of approximately 0.06. This begins at the 7th index, indicating that a power of 7 would be appropriate for this data

3. Looking at Fig. 2, it appears that an appropriate choice for the soft threshold of `power = 7`. Our network can then be constructed using the following code:

```
> adj = adjacency(filter_data, selectCols = NULL,
type = "unsigned", power=7, corFnc = "cor", corOptions
= list(use = "p"), weights = NULL, distFnc = "dist",
distOptions = "method = 'euclidean'")
```

This code will create the adjacency matrix for a basic, undirected GCN. Users are encouraged to review WGCNA's documentation for information on how to create weighted or signed GCNs.

4 Notes

1. **Processing RNA-Seq data**. For readers working with raw read files, these can be processed using tools such as Tophat2 [17], Bowtie2 [18], and HTSeq 2.0 [19] to obtain the observed

counts. It is important to pay attention to the alignment of your read files and whether they are paired-end, as this will determine certain options when employing these methods.

2. **Checking the quality of data**. Although there are no exact requirements when working with RNA-Seq data, and every experimental design will inherently have its own restrictions, it is generally accepted that a minimum of around 1 million reads per experimental condition is enough sequencing depth to identify meaningful signal when working with bulk-based data. However, when the goal is to study genes that are inherently more lowly expressed, a depth closer to 100 million reads per experiment may be required [20].

3. **Choosing cutoff for identifying highly expressed genes**: The choice of cutoffs for mean expression and scaled interquartile range should be determined based on the RNA-Seq dataset, and in practice must often be tweaked to identify the most informative and practical set of genes. Using the `summary()` function in R can be a helpful way to better understand the range of values of mean and IQR for the genes in a given dataset. Another consideration to keep in mind is that most GCN construction methods will require a reduction in the number of genes being considered to a computationally feasible size. In practice, one-thousand genes or less work well for most methods on a typical laptop computer.

4. **Adjacency cutoffs**. It's generally recommended that absolute values of 0.7 or more be used, while not going below 0.5 as these edges have a greater chance of being the result of noise in the data. Additionally, correlation values closer to 0 require a greater number of experiments (usually at least 50) to be statistically significant.

5. **Diagonal of the adjacency matrix**. Before analyzing the resulting network, it is important to "zero-out" this diagonal, otherwise these self-associations will be included. Methods vary in their handling of the diagonal of the adjacency matrix, some do this automatically.

6. **Filtered data**. These normalizations should be performed with the original count data files. Once normalized, the filtering step should be performed before estimating the network.

7. **Additional transformation**. In some cases, after applying these transformations the data may still appear skewed due to extremely high counts. This can often be remedied by applying an additional $\log(x + 1)$ transformation, where 1 is added before taking the log of the values to avoid the issue presented by 0 observations. This is particularly helpful when the experiment is concerned with more lowly expressed genes, as they may be eliminated otherwise.

References

1. Wolfe C, Kohane I, Butte A (2005) Systematic survey reveals general applicability of "guilt-by-association" within gene coexpression networks. BMC Bioinformatics 6(1):227
2. Yang Y et al (2014) Gene co-expression network analysis reveals common system-level properties of prognostic genes across cancer types. Nat Commun 5(1):1–9
3. Liu Y, Zhao M (2016) lnCaNet: pan-cancer co-expression network for human lncRNA and cancer genes. Bioinformatics 32 (10):1595–1597
4. Zhai X et al (2017) Colon cancer recurrence-associated genes revealed by WGCNA co-expression network analysis. Mol Med Rep 16(5):6499–6505
5. Liu R et al (2015) Identification and validation of gene module associated with lung cancer through coexpression network analysis. Gene 563(1):56–62
6. Wilhelm B, Landry JR (2009) RNA-Seq—quantitative measurement of expression through massively parallel RNA-sequencing. Methods 48(3):249–257
7. Allen JD et al (2012) Comparing statistical methods for constructing large scale gene networks. PLoS One 7(1):e29348
8. Li J, Lamere AT (2019) DiPhiSeq: robust comparison of expression levels on RNA-Seq data with large sample sizes. Bioinformatics 35 (13):2235–2242
9. Langfelder P, Horvath S (2008) WGCNA: an R package for weighted correlation network analysis. BMC Bioinformatics 9(1):559
10. Robinson MD, McCarthy DJ, Smyth GK (2010) edgeR: a Bioconductor package for differential expression analysis of digital gene expression data. Bioinformatics 26 (1):139–140
11. Anders S, Huber W (2010) Differential expression analysis for sequence count data. Nat Precedings 1(1)
12. Li J et al (2012) Normalization, testing, and false discovery rate estimation for RNA-sequencing data. Biostatistics 13 (3):523–538
13. Specht AT, Li J (2015) Estimation of gene co-expression from rna-seq count data. Stat Interface 8(4):507–515
14. Giorgi FM, Del Fabbro C, Licausi F (2013) Comparative study of rna-seq-and microarray-derived coexpression networks in *arabidopsis thaliana*. Bioinformatics 29(6):717–724
15. Bullard JH et al (2010) Evaluation of statistical methods for normalization and differential expression in mrna-seq experiments. BMC Bioinformatics 11(1):94
16. Robinson MD, Oshlack A (2010) A scaling normalization method for differential expression analysis of RNA-seq data. Genome Biol 11 (3):R25
17. Kim D et al (2013) TopHat2: accurate alignment of transcriptomes in the presence of insertions, deletions and gene fusions. Genome Biol 14(4):R36
18. Langmead B, Salzberg SL (2012) Fast gapped-read alignment with bowtie 2. Nat Methods 9 (4):357
19. Zanini F et al (2020) HTSeq 2.0: Efficient manipulation of high-throughput sequencing data with long genomes. In preparation
20. Conesa A, Madrigal P, Tarazona S et al (2016) A survey of best practices for RNA-seq data analysis. Genome Biol 17(1):181

Chapter 3

Genomic Footprinting Analyses from DNase-seq Data to Construct Gene Regulatory Networks

Tomás C. Moyano, Rodrigo A. Gutiérrez, and José M. Alvarez

Abstract

Chromatin accessibility is directly linked with transcription in eukaryotes. Accessible regions associated with regulatory proteins are highly sensitive to DNase I digestion and are termed DNase I hypersensitive sites (DHSs). DHSs can be identified by DNase I digestion, followed by high-throughput DNA sequencing (DNase-seq). The single-base-pair resolution digestion patterns from DNase-seq allows identifying transcription factor (TF) footprints of local DNA protection that predict TF–DNA binding. The identification of differential footprinting between two conditions allows mapping relevant TF regulatory interactions. Here, we provide step-by-step instructions to build gene regulatory networks from DNase-seq data. Our pipeline includes steps for DHSs calling, identification of differential TF footprints between treatment and control conditions, and construction of gene regulatory networks. Even though the data we used in this example was obtained from *Arabidopsis thaliana*, the workflow developed in this guide can be adapted to work with DNase-seq data from any organism with a sequenced genome.

Key words DNase-seq, Chromatin, Genomic Footprinting, Transcription, Gene Regulatory Networks

1 Introduction

Eukaryotic genomes are tightly packed, wrapping short fragments of DNA around nucleosomes in a repeating unit that forms the structural basis of chromatin [1]. Chromatin is classified as either heterochromatin or euchromatin based on its transcriptional capability, compaction, and accessibility [2]. The distribution of nucleosomes along the chromosome provides different levels of accessibility of transcriptional machinery to cis-regulatory elements such as promoters and enhancers [3, 4]. These cis-regulatory regions are targeted by regulatory proteins such as transcription factors (TFs), which play central roles in regulating gene expression [5]. Therefore, the identification of cis-regulatory sequences in their native chromatin environment is important for understanding

Shahid Mukhtar (ed.), *Modeling Transcriptional Regulation: Methods and Protocols*, Methods in Molecular Biology, vol. 2328, https://doi.org/10.1007/978-1-0716-1534-8_3,

how gene expression is regulated in response to internal and environmental signals.

Genomic regions associated with regulatory proteins are generally depleted of nucleosomes and represent "open" chromatin. One common characteristic of these genomic regions is their pronounced hypersensitivity to DNase I cleavage. These genomic regions are termed DNase I hypersensitive sites (DHSs) [5]. DHSs can be identified by the DNase-seq technique in which DNase I cleaves at open chromatin, releasing low molecular weight DNA fragments, which can be purified, sequenced, and mapped back to the genome. DHSs sizes range from 200 bp to 1 kb or larger [5, 6]. Within DHSs, the single-base-pair resolution digestion patterns from DNase-seq allow to identifying footprints of local DNA protection due to TF–DNA binding [5, 7–10]. In genomic footprinting, the presence of DNAse I protection at a binding motif is associated with TF occupancy of that site and protection from attack by steric blockage of the DNAse I [5].

The application of genomic footprinting has yielded valuable insights into the structure, function, and evolution of TF occupancy patterns across different cell types, differentiation states, and environmental conditions [5, 6, 8, 11–15]. Genomic footprinting has also been combined with databases containing defined TF recognition sequences to enable the construction of gene regulatory networks [7, 10].

Conceptually, approaches to analyze genomic footprinting data have centered on footprints or TF binding sites. The first approach focuses on de novo detection and annotation of DNase I footprints, whereas the second strategy attempts to determine TF binding at defined genomic locations. Several algorithms for de novo annotation of DNase I footprints have been developed [16]. A major difficulty inherent to de novo footprint detection is that the basic parameters defining a TF–DNA interaction are unknown in advance and must be learned simultaneously with the footprint identification process. For approaches focusing on TF binding sites, genomic matches to position weight matrices (PWMs) for hundreds of known TF binding motifs can be achieved with algorithms such as Find Individual Motif Occurrences (FIMO) [17]. At each TF binding motif scanned in the genome, the numbers of DNase I cut at each nucleotide are counted and regions with low numbers of DNA I cuts embedded in high-cut peaks are identified as footprints. This approach is considerably more straightforward than de novo footprinting because using predefined motif matches (such as PWMs) as a prior allows to delimitate basic parameters, such as the genomic location and the expected width of the TF–DNA interaction [16]. The CENTIPEDE algorithm has been widely employed to accurately infer TF binding genome-wide using information of known TF binding motifs [7, 16, 18, 19]. C ENTIPEDE predicts footprints and label binding sites for a set of desired TFs by integrating both DNase-seq data and PWMs [9].

Hundreds of PWMs have been identified by determining DNA sequence preferences for >1000 TFs encompassing 54 different TF classes from 131 diverse eukaryotes using protein binding microarrays [20]. By exploiting the conservation of DNA binding domains, these data were used for the inference of motifs for ~57.000 TFs [20]. These data represent a comprehensive resource of PWMs across eukaryotes.

Although open chromatin is linked with transcription in eukaryotes, rapid transcriptional changes do not necessarily involve overall changes in chromatin accessibility. Instead, local variations in chromatin accessibility caused by TF binding within DHSs—that can be detected by genomic footprinting—correlate with transcriptional activation in response to environmental stimuli [7, 10, 21, 22]. Thus, large chromatin changes would be required for major changes at the cellular level, for instance cell differentiation during development, but not for rapid changes in response to an environmental stimulus. Although a quick and timely transcriptional response is of critical importance to all organism's survival, it is particularly key for plants that must execute immediate transcriptional reprogramming to cope with a changing environment from which they cannot escape. In plants, local changes in TF binding within DHSs correlate with gene expression changes in response to light, darkness, heat, cold, salt, drought, and nitrate [7, 10, 22, 23]. Identification of differential TF footprints between two conditions allows mapping relevant TF regulatory interactions driving gene expression changes.

This chapter provides step-by-step instructions to analyze DNase-seq data and identify differential TF footprints in treatments versus control conditions. We provide a framework to construct gene regulatory networks where TF regulatory interactions within open chromatin regions are integrated with differential transcriptome changes (Fig. 1). We present a straightforward pipeline for users with basic bioinformatics skills. This chapter can be used as a starting point to identify key TFs in a gene regulatory network from available or newly generated DNase-seq data. The examples provided herein use DNAse-seq data from KNO_3 or KCl treatments in Arabidopsis root [7], but the protocol is applicable to any organism or experimental condition for which similar data is available.

2 Materials

Personal computer or server with Internet access. Computer requirements vary depending on the amount of data to be analyzed. In this guide, we use a server with 2× Intel(R) Xeon(R) CPU X5670 @ 2.93 GHz (24 threads), 64 Gb RAM, and 1 TB of free disk space with Ubuntu 18.04 distribution.

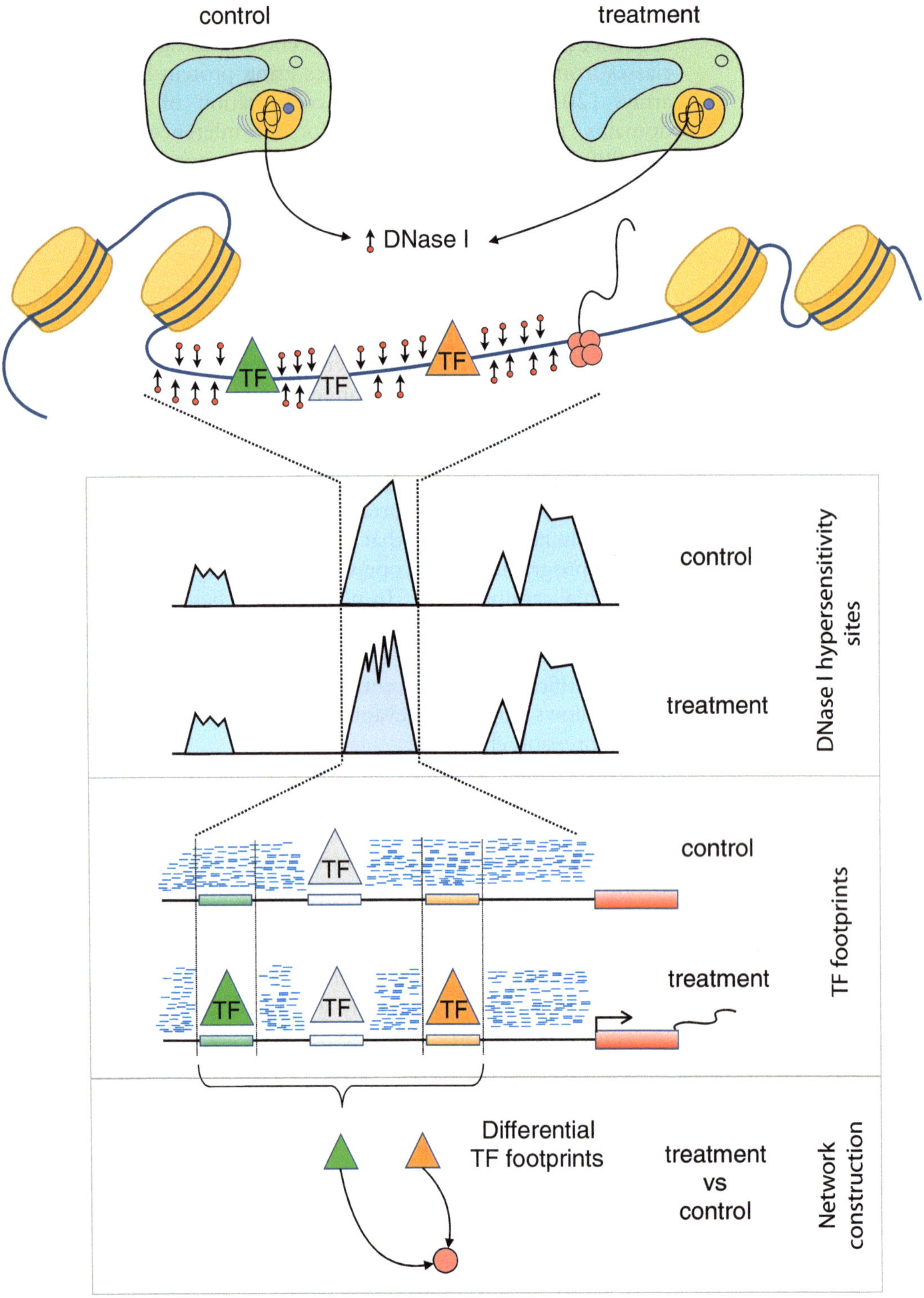

Fig. 1 Analysis of DHSs and genomic footprinting from DNase-seq data for network construction. In the DNase-seq experiment, nuclei are harvested from tissues and treated with the endonuclease DNase I. Open chromatin regions are hypersensitive to cleavage by DNase I. TF-bound regions within DNase I hypersensitive sites are protected from DNase I cleavage leaving detectable "footprints." Differential TF footprints can be integrated with gene expression data to generate gene regulatory networks

To run this pipeline, the following programs need to be installed:

1. Conda 4.8.3 (https://anaconda.org/anaconda/conda) with the bioconda channel (https://bioconda.github.io/).
2. JAVA OpenJDK version "1.8.0_112".
3. Cytoscape 3.8.0 from https://cytoscape.org/ [24].
4. R version 3.6.1.

Other tools will be described in the text, with the respective instructions on how to obtain them.

In this chapter, specific programs have been developed to facilitate the workflow. These programs are indicated through the chapter and can be downloaded from http://virtualplant.bio.puc.cl/cgi-bin/Lab/DNAse.cgi

The DNAse-seq data used herein can be obtained from PRJNA563066 BioProject (https://www.ncbi.nlm.nih.gov/bioproject/PRJNA563066/). The codes for the libraries used in this study are: DH_012_KCl_replicate1 (SRR10051093), DH_012_KCl_replicate2 (SRR10051092), DH_012_KNO3_replicate1 (SRR10051080) and DH_012_KNO3_replicate2 (SRR10051104).

3 Methods

The following sections describe a simple pipeline to generate a gene regulatory network starting from DNase-seq data. The next steps will guide the user from obtaining data for selected DNase-seq experiments to the identification of DHSs, scanning of TF motifs, footprinting analysis, and building a gene regulatory network (Fig. 1).

The instructions of this chapter were designed and tested in a Linux and R environment. A basic level understanding of both Linux and R is recommended. There are different tools available to analyze DNAse-seq data and for the identification of DHSs [25]. In this chapter, we will use a standardized procedure described in the ENCODE project (https://github.com/ENCODE-DCC/dnase_pipeline). The ENCODE procedure contains ready-to-use programs in a pipeline designed for the human genome. However, in this work, we adapted this pipeline to make it compatible with any species of interest. The different tools necessary for the pipeline are present in Bioconda (https://bioconda.github.io/) [26]. We recommend to install this program since it resolves most necessary dependencies for executing the instructions below. For nomenclature, we used a "$" character to indicate beginning of a new line in the UNIX environment. We indicate a new line in the R environment with the ">" character.

3.1 Identification of DHSs

3.1.1 Preparing the Genome-Specific Files

In this first section, we will identify open regions of the chromatin (DHSs) using the DNAse-seq data. To do this, it is necessary to prepare the working folder with the programs and files necessary for DHSs identification. First, in the console, we will create or select a working directory and set the pathway in an environmental variable.

```
$ export WD=$(pwd)
```

The first element we need is a file in Fasta format with the genome of the organism. In this chapter, we will use the *Arabidopsis thaliana* genome (T10.fa) as an example. This file can be downloaded from http://virtualplant.bio.puc.cl/cgi-bin/Lab/DNAse.cgi. This file is a modified version from TAIR10_Chr.all.fa genome file from www.arabidopsis.org, in which we eliminated the spaces in the identifiers (spaces in identifiers can cause problems in the pipeline).

```
$ wget http://virtualplant.bio.puc.cl/share/DNAse/T10.fa
```

The ENCODE pipeline for the analysis of DNAse-seq must be downloaded and unzipped with the following instructions:

```
$ wget https://github.com/ENCODE-DCC/dnase_pipeline/archive/master.zip -O dnase_pipeline.zip

$ unzip dnase_pipeline.zip
```

Depending on the genome composition and the size of DNA-seq reads, it is necessary to know the mappable regions in the genome (regions uniquely mappable given the length of the reads). Depending on the DNA-seq protocol used, read length can vary, with 20 and 36 nucleotides being common sizes [10, 15]. For some organisms, these files are available at https://bismap.hoffmanlab.org/.

If the file is not available for the required read size, it can be created using the "enumerateUniquelyMappableSpace.pl" tool available at https://github.com/rthurman/hotspot/tree/master/hotspot-distr and the alignment tool bowtie. This chapter will use the Arabidopsis genome, and DNAse-seq reads with a size of 20 nucleotides.

The HOTSPOT program will be used for DHSs identification and can be downloaded and unzipped with the following instructions:

```
wget https://github.com/rthurman/hotspot/archive/master.zip -O hotspot-master.zip

$ unzip hotspot-master.zip
```

In this and subsequent steps, the programs will be included in the PATH as an environment tool.

```
$ export PATH=$WD/hotspot-master/hotspot-distr/hotspot-de-
ploy/bin/:$PATH
$ export PATH=$WD/dnase_pipeline-master/dnanexus/dnase-index-
bwa/resources/usr/bin/:$PATH
```

We will execute the enumerateUniquelyMappableSpace.pl script to create the file with the mappable regions in the genome.

Then, the genome needs to be indexed for the bowtie tool. To continue, we recommend you to install bowtie with conda. To do this, follow the next instructions:

```
$ conda install bowtie
```

Index the reference genome for bowtie alignment tool:

```
$ bowtie-build -f T10.fa T10.bowtie
```

Once this file is obtained, the mappable regions are generated for a given DNAse-seq read length. In this case, we used 20 bp:

```
$ $WD/hotspot-master/hotspot-distr/hotspot-deploy/bin/enumer-
ateUniquelyMappableSpace.pl 20 T10.bowtie T10.fa | sort-bed -
| bedops -m - > T10.bowtie.read_length20.mappable_only.bed
```

The genome.read_length20.mappable_only.bed file contains the mappable coordinates of the genome. Now, we have all the files required for the next steps to do the mapping and DHSs calling. The following steps align the DNAse-seq reads of a library of interest to the previously indexed genome. This step's output is a ".bam" file, which is the compressed binary version of a SAM file and contains the mapping coordinates of each read in the genome, discarding the unmappable regions. The instructions provided consider a single-end DNAse-seq library.

```
$ export PATH=$WD/dnase_pipeline-master/dnanexus/dnase-index-
bwa/resources/usr/bin/:$PATH
```

First, a genome index is generated, excluding the non-mappable regions corresponding to read length.

```
$ $WD/dnase_pipeline-master/dnanexus/dnase-index-bwa/re-
sources/usr/bin/dnase_index_bwa.sh T10 T10.fa false T10.bow-
tie.read_length20.mappable_only.bed
```

This step's output is a BWA index of the genome. Up to this point, each step should be done only once per genome and once per read size if it is necessary.

3.1.2 Aligning DNAse-Seq Reads to the Genome

Next, the reads from DNAse-seq libraries are mapped to the genome. The following step may take a long time. The following steps will be an example for one of the libraries. In this case, we will use the DH_012_KCl_replicate1.fastq.gz library, which can be downloaded from http://virtualplant.bio.puc.cl/cgi-bin/Lab/DNAse.cgi . The following steps must be repeated with the other DNAse-seq libraries available at the same webpage.

```
$ wget http://virtualplant.bio.puc.cl/share/DNAse/DH_012_KCl_replicate1.fastq.gz
```

You can choose the number of processors to use in the alignment process. Here we use 6 processors.

```
$ export PATH=$WD/dnase_pipeline-master/dnanexus/dnase-align-bwa-se/resources/usr/bin/:$PATH
```

3.1.3 Filtering the Alignment Files

```
$ $WD/dnase_pipeline-master/dnanexus/dnase-align-bwa-se/resources/usr/bin/dnase_align_bwa_se.sh T10_bwa_index.tgz DH_012_KCl_replicate1.fastq.gz 6 DH_012_KCl_replicate1
```

This step generates an alignment file with a ".bam" extension. Also, this step generates a file with *.flagstat.txt and *edwBamStats.txt suffix, which includes alignment statistics.

The next instructions allow filtering the .bam file to discard low-quality alignments and duplicated reads. This step requires the picard.jar file in the working directory, which can be downloaded from https://github.com/broadinstitute/picard/releases/tag/2.23.2 (use version 2.23.2). After picard.jar is saved in the working directory, run the following commands:

```
$ export PATH=$WD/dnase_pipeline-master/dnanexus/dnase-filter-se/resources/usr/bin/:$PATH
```

```
$ $WD/dnase_pipeline-master/dnanexus/dnase-filter-se/resources/usr/bin/dnase_filter_se.sh DH_012_KCl_replicate1.bam 10 60 DH_012_KCl_replicate1.filter
```

3.1.4 Identification of Open Chromatin Regions

Once we obtain the filtered alignments, we proceed to identify the open regions of the chromatin. For this example, we will use the HOTSPOT tool. The HOTSPOT program identifies open genomic regions based on statistically significant cleavage activity on DHSs as compared to the surrounding genomic context. For the DHSs identification, we call the program hotspot2.sh, which executes a series of steps to identify open chromatin regions in .starch format as output. These steps include identification of the cleavage sites in the mappable regions and comparison of small windows of genomic sequences against larger windows (used as background) for identification of DHSs or hotspots. The inputs for this tool are a chromosome size file in bed format, the mappable sites for the corresponding genome and read length, the filtered bam file, and the output directory.

```
$ export PATH=$WD/dnase_pipeline-master/dnanexus/dnase-call-hotspots/resources/usr/bin/:$PATH
```

```
$ nohup hotspot2.sh -c chrom_sizes.bed -C center_sites.starch -M T10.bowtie.read_length20.mappable_only.bed -P DH_012_KCl_replicate1.filter.bam DH_012_KCl_replicate1.DH >DH_012_KCl_replicate1.DH.nohup
```

The output directory at this stage should contain ten files with information of the DHSs or hotspots (for details, *see* https://github.com/Altius/hotspot2). The data used in the next step is the starch file that contains the coordinates of the DHSs passing the adjusted *p*-values filters, the narrow peak file, and the Signal Portion of Tags (SPOT) score. SPOT measures signal-to-noise as the fraction of cuts within DHS in the DNase-seq library. Thus, the SPOT score is a measure of the library's quality. The SPOT value is indicated in the file whose name ends with SPOT.fdr0.05.txt. The value corresponds to the number of cleavages observed in the DHSs divided by the total number of cleavages sites in mappable regions, so the value is between 0 and 1. According to the ENCODE criteria, if the value is below 0.25, it is recommended to discard the library, over 0.4 is considered a high-quality DNase-seq library. Next, the starch file needs to be converted to bed format for further analysis.

```
$ unstarch DH_012_KCl_replicate1.DH/DH_012_KCl_replicate1.filter.hotspots.fdr0.05.starch >DH_012_KCl_replicate1.filter.hotspots.fdr0.05.bed
```

```
$ unstarch DH_012_KCl_replicate1.DH/DH_012_KCl_replicate1.filter.peaks.narrowpeaks.starch > DH_012_KCl_replicate1.filter.peaks.narrowpeaks.bed
```

3.2 Scanning for TF Binding Motifs within DHSs

3.2.1 Collecting and Formatting TF Binding Sites

Once we have the open chromatin regions, we proceed to the identification of sites protected by DNA-binding proteins (such as TFs) which are known as "footprint." The following part of the pipeline is based on the CENTIPEDE tutorial for footprinting analysis (https://slowkow.github.io/CENTIPEDE.tutorial/). In order to predict whether a TF is bound to DNA, to count on known TF binding sites for the organism of interest is necessary.

One of the most comprehensive source of TF binding sites available as Position Weight Matrix (PWM) files come from the CISBP database (http://cisbp.ccbr.utoronto.ca/) [20]. The PWMs can be downloaded from the website by selecting the binding motifs for the species of interest. Herein, we will use the TF binding sites associated with *Arabidopsis thaliana*. In the Bulk Download page (http://cisbp.ccbr.utoronto.ca/bulk.php), select *Arabidopsis thaliana* and download the PWMs and TF info (Download or copy the URL). The filename changes according to the download date. Correct the link for your downloading instance.

```
$ wget http://cisbp.ccbr.utoronto.ca/tmp/Arabidopsis_thaliana_2020_06_24_5:54_pm.zip -O Arabidopsis_thaliana_pwm.zip
```

Then the file is unzipped:

```
$ unzip Arabidopsis_thaliana_pwm.zip
```

These files need to be converted to MEME format. To do this, we need to download and install the meme-suite program.

```
$ wget http://meme-suite.org/meme-software/5.1.1/meme-5.1.1.tar.gz
$ tar -xzvf meme-5.1.1.tar.gz
$ cd meme-5.1.1/
$ ./configure
$ make
$ make install
$ cd ..
```

The next instruction will be displayed for a single PWM file. The way to automate all available PWM files will be shown below. First, to transform the file from PWM to MEME, we need to enter to the folder where the PWM files are located and execute the following instruction:

```
$ cd pwms_all_motifs
$ $WD/meme-5.1.1/scripts/matrix2meme < <(tail -n+2 M06794_2.00.txt | cut -f2-) > M06794_2.00.txt.meme
```

This can be automatized for all available PWM files with the following instruction (the user can select the number of processors in this example "-P 2":

```
$ echo '$WD/meme-5.1.1/scripts/matrix2meme < <(tail -n+2 $1 | cut -f2-) > $1.meme' > script.pwm2meme.sh

$ ls M*.txt | xargs -n1 -P 2 bash ./script.pwm2meme.sh
```

Then, returns to the working directory with:

```
$ cd $WD
```

3.2.2 Extracting DNA Sequence from Open Chromatin Regions

We need to obtain the nucleotide sequence of each DHS from the . bed file that contains all DHSs called using hotspot to scan TF binding sites. We will extract the sequence in Fasta format. To do this, we will use the bedtools program. We extract the sequences in Fasta format from the .bed coordinates from the genome (T10.fa) using the following instruction:

```
$ bedtools getfasta -fi T10.fa -bed DH_012_KCl_replicate1.
filter.peaks.narrowpeaks.bed -fo DH_012_KCl_replicate1.fil-
ter.peaks.narrowpeaks.bed.fasta
```

3.2.3 Scanning for TF Binding Motifs Within Open Chromatin Regions

The next step scans the TF binding site in .meme format in the extracted Fasta sequences of the DHSs. This is done using the FIMO program that we recently installed. Thus, for each .meme file, we search for potential TF binding sites in the DHSs. We can automate this step for each .pwm file.

```
$ echo '$WD/meme-5.1.1/src/fimo --text --parse-genomic-coord
$1 DH_012_KCl_replicate1.filter.peaks.narrowpeaks.bed.fasta |
gzip > $1__DH_012_KCl_replicate1.narrowpeaks_fimo.gz ' >
script.meme_fimo.sh
```

Run the script

```
$ ls pwms_all_motifs/*.meme |xargs -n1 bash script.meme_fimo.sh
```

Each FIMO file contains the information of the predicted TF binding site in the genome. Each line includes the coordinates, sequence, score, and p-value for each motif as shown in Table 1.

3.3 Genomic Footprinting

3.3.1 Installation and Execution of CENTIPEDE

After obtaining the putative TF binding sites within the DHS, the footprints will be identified using the alignment pattern in the regions surrounding the binding site. For this example, we will use the CENTIPEDE program, which integrates the DNase-seq data .bam files and the information obtained from the FIMO program. The CENTIPEDE algorithm uses the R program for its operation, so it must be installed before execution. To apply and install CENTIPEDE on R, you need to install some additional libraries. Therefore, in R, the following instructions should be executed:

```
> if (!requireNamespace("BiocManager", quietly =
TRUE)) install.packages("BiocManager")
> BiocManager::install("Rsamtools")
> install.packages("CENTIPEDE", repos="http://R-
Forge.R-project.org")
> library(Rsamtools)
> library(CENTIPEDE)
```

Table 1
First 4 lines of the meme/fimo output. This file contains the genome coordinate of each predicted binding site and the statistics associated

Motif_id	Motif_alt_id	Sequence_name	Start	Stop	Strand	Score	*p*-Value	Matched_sequence
1	DRTGACGTCAKCDDH	Chr1	3601	3615	–	9.47959	5.52E-05	AGTGAAGTCAGCGTT
1	DRTGACGTCAKCDDH	Chr1	55471	55485	–	9.2449	6.07E-05	TATGACGCCACCCTT
1	DRTGACGTCAKCDDH	Chr1	68642	68656	+	12.0102	1.88E-05	CATCACGTCACCATC

The following function will determine whether a TF is likely bound to the putative TF binding site. We will use the default parameters shown in the CENTIPEDE tutorial, which stands that the reads that map in a flanking area of 100 nucleotides of the predicted TF binding site will be counted. The identified footprints showing a P-value lower than 1×10^{-4} will be kept for further analysis.

After the libraries are loaded in R, the footprints will be searched. The necessary functions are available in the CENTIPEDE GitHub tutorial (https://slowkow.github.io/CENTIPEDE.tuto rial/). For this pipeline, the functions present in the CENTIPEDE tutorial have adaptations that can be downloaded from http:// virtualplant.bio.puc.cl/share/DNAse/DNAse-functions.R.

First, the functions are loaded in R.

```
> source("DNAse-functions.R") .
```

The FIMO files obtained in the previous step are listed in the "fimos" object with the function "dir."

```
> allfimos<-dir("pwms_all_motifs/",pattern="fimo")
```

The CENTIPEDE pipeline contains two main functions, the centipede_data that combines the information from FIMO with the .bam file and the fitCentipede that performs footprint calling and tests whether a TF is bound to the predicted TF binding site. For each putative TF binding site, a window of 100 bp is selected upstream and downstream to quantify read counts in the corresponding .bam file. With this information, the fitCentipede function calculates the probability that a TF is bound to each predicted TF binding site. The search_protectedSite function was created for this chapter to combine the two tasks mentioned above.

```
> conditions<-"DH_012_KCl_replicate1"

> search_protectedSite(conditions, allfimos, FLANK_size=100,
log10p=4)
```

As a result, this function generates three files for each PWM file; the "protected.txt" is a bed file with the positions of the identified footprints. The "regions.txt" is a file with one row for each motif evaluated, and the 100 bp flanking windows. The "mat.txt" file contains the read counts at each position of the motif and the100 bp flanking windows.

The first lines of the "protected.txt" file for the M11491_2.00 motif and DH_012_KCl_replicate1.filter.bam file are shown in Table 2.

Table 2
First 3 lines of the protected.txt file. This file contains the genome coordinate and information of each footprint identified. The original file does not contain header

Chr	Start	Stop	Sequence	Library	Strand	Motif id
Chr1	295177	295191	CTATATAAATGCCCT	DH_012_KCl_replicate1	+	M11491_2.00
Chr1	518218	518232	CTATAAATAACCAAC	DH_012_KCl_replicate1	+	M11491_2.00
Chr1	701430	701444	CTTTAAAAGGCCGTT	DH_012_KCl_replicate1	–	M11491_2.00

Up to this point in the pipeline, the steps must be repeated for each library analyzed. The following steps will be performed using four DNAse-seq libraries. We will use two biological replicates of KNO_3 treatments and two biological replicates of KCl treatments as the control in Arabidopsis roots. As mentioned above, these libraries can be downloaded from PRJNA563066 BioProject (https://www.ncbi.nlm.nih.gov/bioproject/PRJNA563066/) or http://virtualplant.bio.puc.cl/cgi-bin/Lab/DNAse.cgi. All libraries are processed and available in the http://virtualplant.bio.puc.cl/share/DNAse/protected.tgz file to facilitate the analysis with the example proposed herein.

3.3.2 TF/Motif Assignation

Next, we need to assign a TF to each of the identified footprints. First, in R, we need to read the file that contains the list of identified footprints and the TF_information.txt file that contains the TF information necessary to associate a TF to each footprint downloaded previously.

First, the files obtained for each motif are concatenated in a single file with the following command in UNIX:

```
$ cat *.protected.txt >protected.total
```

The protected.total file is read in R with the following command:

```
> protected.total<-read.table("pwms_all_motifs/protected.to-
tal")
```

The TF_information.txt is read in R with the following instruction:

```
> TF_info<-read.table("TF_Information.txt",header=T,sep="\t")
```

Then, the protected.total and TF_info file are merged and a column with the TF gene identifier is added:

```
> protected.TF<-merge(protected.total,TF_info[,c(4,6,9,10)],
by.x=7,by.y=1)
```

The protected.TF object contains the information necessary to continue our analysis to build gene regulatory networks. The columns of the file need to be reordered for further analysis.

```
> protected.bed<-(protected.TF[,c(2:4,1,8,7,9,10,5,6)])
```

The protected.TF object is saved as a file by the following steps:

```
> write.table(protected.bed,"protected.TF.sites.txt",
sep="\t",row.names=F,col.names=F,quote=F)
```

For simplification purposes, if a footprint shares the same coordinates, the motif and the associated TF can be saved on a single line. The .bed file will be sorted using the command bedtools sort. Besides, up to this point, we have identified TF footprints for the four DNase-seq libraries. We will use the collapse.pl to add the library's name in which the footprint was detected. The collapse.pl program is available at http://virtualplant.bio.puc.cl/cgi-bin/Lab/DNAse.cgi. Thus, all information will be present in a single line with the library information in the last column.

```
$ bedtools sort -i protected.TF.sites.txt |perl collapse.pl
>protected.TF.sites.conditions.bed
```

3.3.3 Assigning Footprints to Genes

The following steps are necessary to assign the footprints to the promoter of genes. Herein, a promoter will be defined as the region spanning 1000 bp upstream of the TSS of each gene. The file containing the promoter region for Arabidopsis genes (TAIR10.1000pb5p.bed) can be downloaded from http://virtualplant.bio.puc.cl/cgi-bin/Lab/DNAse.cgi.

```
$ bedtools intersect -wo -a protected.TF.sites.conditions.bed
-b TAIR10.1000pb5p.bed >protected.TF.sites.conditions.1000_5p.bed
```

The file protected.TF.sites.conditions.1000_5p.bed contains the information for each footprint, the motif that was used for the prediction, the TF associated with each motif, the target gene considering 1000 bp upstream of TSS as the promoter, and the DNase-seq library for which the footprint was detected.

Below, we show a general summary of the results. This table shows the number of identified footprints for each combination of DNase-seq libraries. We identified 242,378 footprints that are detected in all libraries. This result suggests that an important number of TFs are bound regardless of the experimental conditions. We define differential footprints when they are detected in both DNase-seq replicates treatments, and none of the DNase-seq replicates of control treatments and vice versa. Thus, we identified 21,339 differential footprints for KNO_3 treatments (DH_012_KNO3_replicate1|DH_012_KNO3_replicate2) and 17,504 differential footprints for KCl treatments (DH_012_KCl_replicate1|DH_012_KCl_replicate2).

3.4 Network Visualization

3.4.1 Selection of Genes and Network Preparation

The protected.TF.sites.conditions.1000_5p.bed file contains the identified footprints for all Arabidopsis TFs, for which a PWM is available at CISBP, at the promoter of all Arabidopsis genes. To build the gene regulatory network, an edge was created when a footprinted motif of a source TF overlapped a target gene, including 1000 bp upstream of the target gene's TSS. As a first step to visualize the resulting network:

In R, the file with the information of footprinting and the associated TFs needs to be loaded:

```
> protected.sites<-read.table("protected.TF.sites.condi-
tions.1000_5p.bed",sep="\t")
```

For this example, we will add the gene expression data as an attribute to the target genes. This step allows us to limit our network to genes that are regulated at 60 min of KNO_3 treatments in roots in further steps. The 60 min time point was selected based on the high overlap with genes for which differential footprinting by KNO_3 was detected on their promoter [7]. The list of regulated genes at 60 min by KNO_3 can be downloaded from http://virtualplant.bio.puc.cl/share/DNAse/ and can be loaded using the following command:

```
> response_to_nitrate060min<-as.matrix(read.table("response_-
to_nitrate060min.txt"))
```

The selected targets are extracted, and the columns of interest are selected to build the network.

```
> network<-merge(response_to_nitrate060min , protected.sites,
by.x=1,by.y=14)
 > network<-network[,c(1:11)]
```

The names can be assigned to each column to be used as identifiers in the network.

```
> colnames(network)<-c("TARGET","Chr","start","end","moti-
f_id","TF","strand","TF_assign","TF_family","motif_sequen-
ce","DH_pattern")
```

The network object is saved as a .txt file to be used in Cytoscape.

```
> write.table(network,"network.txt",sep="\t",col.names=NA,
quote=F)
```

Finally, the list of TFs in the network can be obtained to use it as an attribute with the following command:

```
> TFs.att<-cbind(unique(as.matrix(network$TF)),"TF")

> colnames(TFs.att)<-c("agi","TF")

> write.table(TFs.att,"TFs.att",sep="\t",row.names=F,quote=F)
```

3.4.2 Network Visualization in Cytoscape

Cytoscape is a JAVA based tool that allows to visualize the network. The newly created file will be imported by selecting the following options in the toolbar: File, Import, Import Network from File. The network.txt file is selected, and the TF column is defined as Source and the TARGET column as target.

These instructions load to Cytoscape the entire network with all possible footprints identified in the previous steps. Since different footprints for the same TF may be identified in the same promoter, a TF can have multiple interactions with the same target gene. We recommend adding tables with attributes to the nodes to facilitate the network visualization. A detailed pipeline for network visualization and analysis using Cytoscape is described in [27]. The list of TFs created above (TFs.att) is an example of an attribute that can be added to the network. This can be done with the following steps in the toolbar by clicking "File," "Import," "Table from File," select the TF.att file, and press click in OK. The list of TFs is now added as a node attribute and allows distinguishing TFs from targets.

The shape and color of the nodes can be changed in the "Style" tab. In this example, the target genes are circles, and TFs are triangles (Fig. 2). The generated network is very large since it contains all possible footprints. In this example, we will focus on differential footprints defined as the ones that change in response to KNO_3 treatments, as described above. To do this, we will use the Select tab of the control panel and select "Column Filter," "Edge: DH_pattern," and on the right side, select "matches regex." Write "DH_012_KNO3_replicate1\|DH_012_KNO3_replicate2" which selects the differential footprints by KNO_3 treatments. Then, with the "+" sign on the left side, an additional filter with differential footprints by KCl treatments can be added by selecting "DH_012_KCl_replicate1\|DH_012_KCl_replicate2." With the selected edges, a new network can be created by selecting "File,"

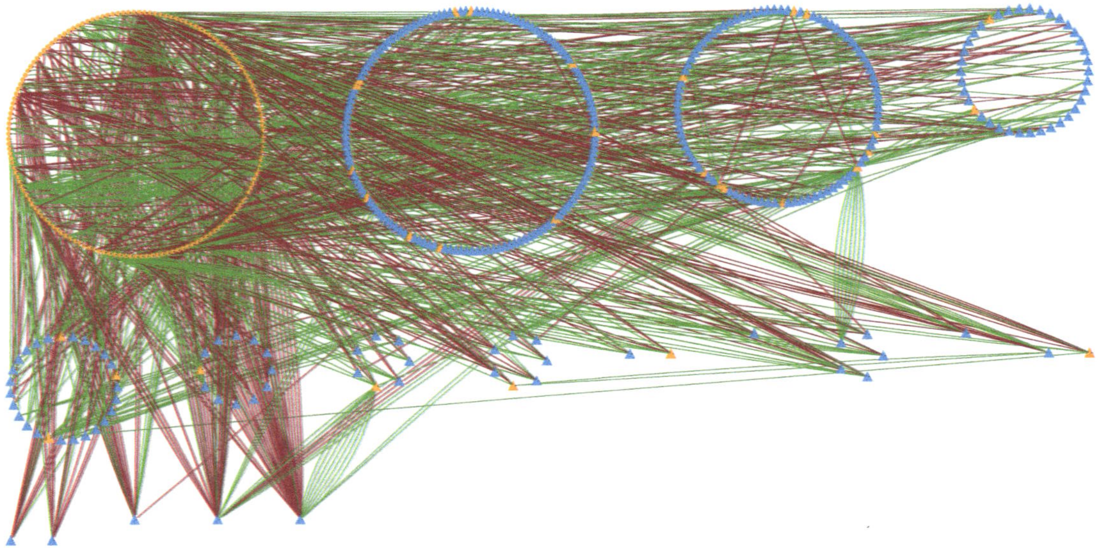

Fig. 2 Gene regulatory network integrates differential footprinting and gene expression data. To build the network, an edge was created when a differential footprint of a source TF (triangles) was detected in the promoter of the target gene (circles). Orange nodes represent genes for which transcript levels change in response to KNO_3 treatments at 60 min [7]. Blue triangles are TFs for which transcript levels do not change by KNO_3 treatments. Green edges represent protein–DNA interactions from TFs that are bound to the target gene in response to KNO_3 (footprint detected in both DNase-seq replicates of KNO_3 treatments and none of the DNase-seq replicates of KCl treatments). Blue edges represent protein–DNA interactions from TFs that are detected in response to KCl treatments (footprint detected in both DNase-seq replicates of KCl treatments and none of the DNase-seq replicates of KNO_3 treatments). To design the network, we used the "Attribute Circle Layout" available in Cytoscape tools to arrange the TFs according to outdegree (number of target genes) from high (lower layers) to low (higher layers)

"New Network," "From selected Nodes, Selected edges" option. The edges can be colored according to the footprint type. To do this, in the "Style" tab of "Control Panel" select "edges" and check "edge color to arrows" and the user can change the edge color. In this example, we selected green for "DH_012_KNO3_replicate1\|DH_012_KNO3_replicate2" footprints and red for "DH_012_KCl_replicate1\|DH_012_KCl_replicate2" footprints. For the network layout, we used the "Attribute Circle Layout" in the "Layout" option. The "Outdegree" attribute was selected for the final design. This analysis connected 357 TFs and 140 targets. The TFs are arranged into different tiers according to the number of target genes (outdegree) (Fig. 2). Our analysis allows us to identify TF and targets whose expression is regulated by KNO_3 at 60 min (orange triangles) and TFs that are not regulated by KNO_3 (blue triangles).

Our analysis allows us to integrate DNase-seq with gene expression data to construct gene regulatory networks. This type of analysis generates rich gene networks that help biologists build a testable hypothesis. For example, influential TFs based on network connectivity and their hierarchical position can be tested experimentally.

4 Notes

1. The user should be aware that the program versions may change in the future resulting in different available options or software incompatibility.
2. The user should notice that data, methods, and parameters used at each step are intended as a simple first guideline for DNAse-seq data and footprinting analyses. Each software tool utilized has parameters that can be optimized, and new tools come out regularly in the literature. Indeed, since this is an active field of research, the developers' tools downloaded in the chapter are permanently improved. Thus, the changes may alter these programs' output, and it will be necessary to modify the script.
3. The tools that were used in this chapter will be backed up on the webpage http://virtualplant.bio.puc.cl/share/DNAse/, but it is recommended to download the tool from the developer's page.
4. The operative system must have the libssl.so.1.0.0 shared libraries installed.
5. Before you start using the pipeline, ensure that the library is of good quality, and remove the adapters. We recommend the use of Fastqc https://www.bioinformatics.babraham.ac.uk/

projects/fastqc/ and Trimmomatic http://www.usadellab.org/cms/?page=trimmomatic for trimming and checking the quality.

6. In the step of .bam file filtering, a temporary file is created in the working directory called "sorted.bam". If you run the filtering program in parallel, you should modify the script to prevent the file from being overwritten during execution.
7. By default, the hotspot2.sh tool creates files in the /tmp folder in the operating system, which must have enough space. Once the execution is finished, these files should be deleted.
8. The pipeline uses the picard.jar version 2.23.2 file (https://github.com/broadinstitute/picard/releases/tag/2.23.2). Other picard.jar versions were tried, but the pipeline failed.

Acknowledgements

Research in R.A.G.'s laboratory is funded by FONDECYT 1180759, ANID/FONDAP/15090007, ANID – Millennium Science Initiative Program- Millennium Institute for Integrative Biology (iBio) ICN17_022 and EvoNet project DE-SC0014377. Research in J.M.A.'s laboratory is funded by ANID – Millennium Science Initiative Program- Millennium Institute for Integrative Biology (iBio) ICN17_022 and ANID FONDECYT 1210389.

References

1. Kornberg RD, Lorch Y (1999) Twenty-five years of the nucleosome, fundamental particle of the eukaryote chromosome. Cell 98 (3):285–294
2. Schones DE, Cui K, Cuddapah S, Roh T-Y, Barski A, Wang Z, Wei G, Zhao K (2008) Dynamic regulation of nucleosome positioning in the human genome. Cell 132(5):887–898
3. Orphanides G, Reinberg D (2002) A unified theory of gene expression. Cell 108 (4):439–451. https://doi.org/10.1016/s0092-8674(02)00655-4
4. He HH, Meyer CA, Shin H, Bailey ST, Wei G, Wang Q, Zhang Y, Xu K, Ni M, Lupien M (2010) Nucleosome dynamics define transcriptional enhancers. Nat Genet 42(4):343
5. Boyle AP, Davis S, Shulha HP, Meltzer P, Margulies EH, Weng Z, Furey TS, Crawford GE (2008) High-resolution mapping and characterization of open chromatin across the genome. Cell 132(2):311–322. https://doi.org/10.1016/j.cell.2007.12.014
6. Thurman RE, Rynes E, Humbert R, Vierstra J, Maurano MT, Haugen E, Sheffield NC, Stergachis AB, Wang H, Vernot B, Garg K, John S, Sandstrom R, Bates D, Boatman L, Canfield TK, Diegel M, Dunn D, Ebersol AK, Frum T, Giste E, Johnson AK, Johnson EM, Kutyavin T, Lajoie B, Lee BK, Lee K, London D, Lotakis D, Neph S, Neri F, Nguyen ED, Qu H, Reynolds AP, Roach V, Safi A, Sanchez ME, Sanyal A, Shafer A, Simon JM, Song L, Vong S, Weaver M, Yan Y, Zhang Z, Zhang Z, Lenhard B, Tewari M, Dorschner MO, Hansen RS, Navas PA, Stamatoyannopoulos G, Iyer VR, Lieb JD, Sunyaev SR, Akey JM, Sabo PJ, Kaul R, Furey TS, Dekker J, Crawford GE, Stamatoyannopoulos JA (2012) The accessible chromatin landscape of the human genome. Nature 489 (7414):75–82. https://doi.org/10.1038/nature11232
7. Alvarez JM, Moyano TC, Zhang T, Gras DE, Herrera FJ, Araus V, O'Brien JA, Carrillo L, Medina J, Vicente-Carbajosa J, Jiang J, Gutierrez RA (2019) Local changes in chromatin

accessibility and transcriptional networks underlying the nitrate response in Arabidopsis roots. Mol Plant 12(12):1545–1560. https://doi.org/10.1016/j.molp.2019.09.002

8. Hesselberth JR, Chen X, Zhang Z, Sabo PJ, Sandstrom R, Reynolds AP, Thurman RE, Neph S, Kuehn MS, Noble WS, Fields S, Stamatoyannopoulos JA (2009) Global mapping of protein-DNA interactions in vivo by digital genomic footprinting. Nat Methods 6 (4):283–289. https://doi.org/10.1038/nmeth.1313
9. Pique-Regi R, Degner JF, Pai AA, Gaffney DJ, Gilad Y, Pritchard JK (2011) Accurate inference of transcription factor binding from DNA sequence and chromatin accessibility data. Genome Res 21(3):447–455. https://doi.org/10.1101/gr.112623.110
10. Sullivan AM, Arsovski AA, Lempe J, Bubb KL, Weirauch MT, Sabo PJ, Sandstrom R, Thurman RE, Neph S, Reynolds AP, Stergachis AB, Vernot B, Johnson AK, Haugen E, Sullivan ST, Thompson A, Neri FV 3rd, Weaver M, Diegel M, Mnaimneh S, Yang A, Hughes TR, Nemhauser JL, Queitsch C, Stamatoyannopoulos JA (2014) Mapping and dynamics of regulatory DNA and transcription factor networks in *A. thaliana*. Cell Rep 8 (6):2015–2030. https://doi.org/10.1016/j.celrep.2014.08.019
11. Stamatoyannopoulos JA, Snyder M, Hardison R, Ren B, Gingeras T, Gilbert DM, Groudine M, Bender M, Kaul R, Canfield T (2012) An encyclopedia of mouse DNA elements (mouse ENCODE). Genome Biol 13 (8):1–5
12. Yue F, Cheng Y, Breschi A, Vierstra J, Wu W, Ryba T, Sandstrom R, Ma Z, Davis C, Pope BD (2014) A comparative encyclopedia of DNA elements in the mouse genome. Nature 515(7527):355–364
13. Kharchenko PV, Alekseyenko AA, Schwartz YB, Minoda A, Riddle NC, Ernst J, Sabo PJ, Larschan E, Gorchakov AA, Gu T (2011) Comprehensive analysis of the chromatin landscape in *Drosophila melanogaster*. Nature 471 (7339):480–485
14. Neph S, Vierstra J, Stergachis AB, Reynolds AP, Haugen E, Vernot B, Thurman RE, John S, Sandstrom R, Johnson AK, Maurano MT, Humbert R, Rynes E, Wang H, Vong S, Lee K, Bates D, Diegel M, Roach V, Dunn D, Neri J, Schafer A, Hansen RS, Kutyavin T, Giste E, Weaver M, Canfield T, Sabo P, Zhang M, Balasundaram G, Byron R, MacCoss MJ, Akey JM, Bender MA, Groudine M, Kaul R, Stamatoyannopoulos JA (2012) An expansive human regulatory lexicon encoded in transcription factor footprints. Nature 489 (7414):83–90. https://doi.org/10.1038/nature11212
15. Zhang W, Zhang T, Wu Y, Jiang J (2012) Genome-wide identification of regulatory DNA elements and protein-binding footprints using signatures of open chromatin in Arabidopsis. Plant Cell 24(7):2719–2731. https://doi.org/10.1105/tpc.112.098061
16. Vierstra J, Stamatoyannopoulos JA (2016) Genomic footprinting. Nat Methods 13 (3):213–221
17. Grant CE, Bailey TL, Noble WS (2011) FIMO: scanning for occurrences of a given motif. Bioinformatics 27(7):1017–1018
18. Buenrostro JD, Giresi PG, Zaba LC, Chang HY, Greenleaf WJ (2013) Transposition of native chromatin for fast and sensitive epigenomic profiling of open chromatin, DNA-binding proteins and nucleosome position. Nat Methods 10(12):1213
19. Sherwood RI, Hashimoto T, O'donnell CW, Lewis S, Barkal AA, Van Hoff JP, Karun V, Jaakkola T, Gifford DK (2014) Discovery of directional and nondirectional pioneer transcription factors by modeling DNase profile magnitude and shape. Nat Biotechnol 32 (2):171–178
20. Weirauch MT, Yang A, Albu M, Cote AG, Montenegro-Montero A, Drewe P, Najafabadi HS, Lambert SA, Mann I, Cook K (2014) Determination and inference of eukaryotic transcription factor sequence specificity. Cell 158(6):1431–1443
21. John S, Sabo PJ, Thurman RE, Sung M-H, Biddie SC, Johnson TA, Hager GL, Stamatoyannopoulos JA (2011) Chromatin accessibility pre-determines glucocorticoid receptor binding patterns. Nat Genet 43(3):264–268
22. Liu Y, Zhang W, Zhang K, You Q, Yan H, Jiao Y, Jiang J, Xu W, Su Z (2017) Genome-wide mapping of DNase I hypersensitive sites reveals chromatin accessibility changes in Arabidopsis euchromatin and heterochromatin regions under extended darkness. Sci Rep 7 (1):4093. https://doi.org/10.1038/s41598-017-04524-9
23. Raxwal VK, Ghosh S, Singh S, Agarwal SK, Goel S, Jagannath A, Kumar A, Scaria V, Agarwal M (2020) Abiotic stress mediated modulation of chromatin landscape in *Arabidopsis thaliana*. J Exp Bot 71(17):5280–5293
24. Shannon P, Markiel A, Ozier O, Baliga NS, Wang JT, Ramage D, Amin N, Schwikowski B, Ideker T (2003) Cytoscape: a software environment for integrated models of

biomolecular interaction networks. Genome Res 13(11):2498–2504

25. Madrigal P, Krajewski P (2012) Current bioinformatic approaches to identify DNase I hypersensitive sites and genomic footprints from DNase-seq data. Front Genet 3:230
26. Grüning B, Dale R, Sjödin A, Chapman BA, Rowe J, Tomkins-Tinch CH, Valieris R, Köster J (2018) Bioconda: sustainable and comprehensive software distribution for the life sciences. Nat Methods 15(7):475–476
27. Moyano TC, Vidal EA, Contreras-López O, Gutiérrez RA (2015) Constructing simple biological networks for understanding complex high-throughput data in plants. In: Plant functional genomics. Springer, Berlin, pp 503–526

Chapter 4

Spatiotemporal Gene Expression Profiling and Network Inference: A Roadmap for Analysis, Visualization, and Key Gene Identification

Ryan Spurney, Michael Schwartz, Mariah Gobble, Rosangela Sozzani, and Lisa Van den Broeck

Abstract

Gene expression data analysis and the prediction of causal relationships within gene regulatory networks (GRNs) have guided the identification of key regulatory factors and unraveled the dynamic properties of biological systems. However, drawing accurate and unbiased conclusions requires a comprehensive understanding of relevant tools, computational methods, and their workflows. The topics covered in this chapter encompass the entire workflow for GRN inference including: (1) experimental design; (2) RNA sequencing data processing; (3) differentially expressed gene (DEG) selection; (4) clustering prior to inference; (5) network inference techniques; and (6) network visualization and analysis. Moreover, this chapter aims to present a workflow feasible and accessible for plant biologists without a bioinformatics or computer science background. To address this need, TuxNet, a user-friendly graphical user interface that integrates RNA sequencing data analysis with GRN inference, is chosen for the purpose of providing a detailed tutorial.

Key words Gene regulatory network inference, RNA sequencing, Bioinformatics, Network visualization

1 Introduction

Plant growth and developmental as well as environmental responses are complex traits governed by multiple inputs, including regulation of gene expression via transcription factors (TFs). The importance of transcriptional regulation is shown, for example, when the genetic modification of these regulatory genes greatly influences a plethora of phenotypic traits, most likely via further regulation of many downstream genes [1]. Multiple transcriptional regulators form interconnected networks rather than

Ryan Spurney and Michael Schwartz contributed equally to this work.

Shahid Mukhtar (ed.), *Modeling Transcriptional Regulation: Methods and Protocols*, Methods in Molecular Biology, vol. 2328, https://doi.org/10.1007/978-1-0716-1534-8_4, © Springer Science+Business Media, LLC, part of Springer Nature 2021

linear pathways to control plant traits [2–4]. These networks are often highly complex as a result of the scale of the network and the feedback/feedforward mechanisms implemented to overcome over-activation or to restrict activity over time. It is therefore necessary to study genetic networks as one entity in addition to studying the role of their individual components to gain insights into the arising phenotype. To unravel these complex gene regulatory networks (GRNs) and explore changes in gene expression in gain- or loss-of-function plant lines or in response to system perturbations such as treatments, RNA sequencing (RNA-seq) experiments and analysis are commonly used for the identification of differentially expressed genes (DEGs). Moreover, RNA-seq analysis coupled with GRN inference can be used to explore the dynamic nature of a system as gene expression is constantly fluctuating depending on the time of day, duration of the exposure to stress, and/or the presence or absence of signaling molecules, such as phytohormones.

A GRN consists of nodes and edges, which represent genes and causal regulatory interactions between genes, respectively [5]. By understanding incoming and outgoing interactions of key genes/nodes and their roles in regulatory circuits, verifiable predictions can be made about the mechanisms driving key biological processes. Conclusions can be drawn both globally and locally in the context of GRNs. Globally, major hub genes with high numbers of outgoing regulations and overall network robustness can be identified for further experimental exploration. Locally, the abundance and statistical significance of sub-networks (e.g., feedforward loops, feedback loops, bifans) known as network motifs can be quantified and scored.

Many tools currently exist to analyze RNA-seq data and identify differentially expressed genes. For example, EdgeR and DESeq2 are both used to great extent, but require advanced knowledge in bioinformatics and programming [6, 7]. Additionally, there is a lack of tools that integrate RNA-seq processing, DEG identification, and GRN inference in an automated and easy-to-use package. TuxNet, an open-source software, incorporates a user-friendly RNA-seq pipeline and two downstream computational pipelines, GEne regulatory Network Inference from SpatioTemporal data (GENIST) [8, 9], and Regression Tree Pipeline for Spatial, Temporal, and Replicate data (RTP-STAR) [9, 10], to infer causal relationships among DEGs. GENIST uses a dynamic Bayesian network (DBN) inference approach, while RTP-STAR applies a regression tree machine learning approach to predict the causal relationships between a set of input genes.

In this chapter, a workflow is provided that directs the reader through the experimental design, RNA-seq analysis, DEG selection, network inference, network visualization, and candidate selection for further experiments.

2 Materials

2.1 TuxNet Architecture

TuxNet is downloadable via GitHub (https://github.com/rspurney/TuxNet) and includes three subsections: TUX, GENIST, and RTP-STAR. TuxNet is written in MATLAB App Designer and calls the MATLAB scripts running each task of the workflow.

2.2 Software Versions

1. MATLAB: R2017b or later.
2. MATLAB Bioinformatics Toolbox: 4.9 or later.
3. ea-utils (fastq-mcf): 1.1.2.
4. Hisat2: 2.1.0.
5. SAMtools: 1.2.
6. Cufflinks (Cufflinks, Cuffdiff, Cuffmerge): 2.1.1.
7. Cytoscape: 3.8.0.

2.3 TuxNet Website

Written and video instructions for the installation and use of TuxNet are available online at https://rspurney.github.io/TuxNet/.

3 Methods

3.1 Designing Experiments to Address Specific Hypotheses

Several key variables should be considered when designing RNA-seq experiments (*see* **Note 1**) to ensure relevance and accuracy of downstream analyses:

1. **Type of data**. There are two subdivisions of RNA-seq data, temporal and steady-state data, both of which can provide information on global changes in gene expression (Fig. 1). The preference to obtain temporal or steady-state data depends on the type of perturbation. For example, changes in gene expression can occur rapidly after the application of various chemicals or phytohormones. In this case, acquire temporal data at specific short-term time points after the application of a chemical or phytohormone. In contrast, obtain steady-state data when using a static knockout mutant compared to a wild-type (WT) reference [11].
2. **Prior knowledge**. While investigating the function of a specific gene or group of genes, consider whether a knockout mutant

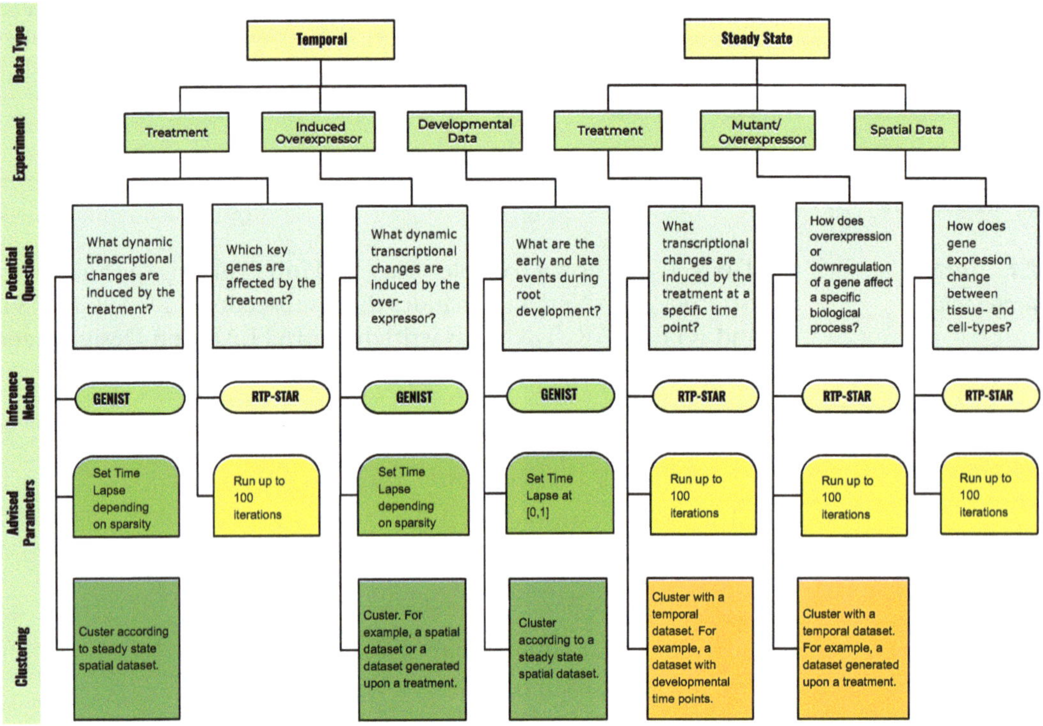

Fig. 1 Workflow to infer gene regulatory networks (GRNs) from temporal or steady-state gene expression data

or overexpression line will be more valuable for network inference (*see* **Note 2**) (Fig. 1).

3. **Data source**. The type of organ or tissue of interest will have implications for the experimental design: for example, when studying root development, samples can be taken at different stages of development along the longitudinal axis of the root starting from the root tip, whereas studying leaf or flower development would require sampling at various time points [12] (*see* **Note 3**). Use GRN inference to show which TFs act as hubs when new cell types or organs are formed.
4. **Timescale of the data points**. Take into account the timescale of the selected datasets when designing an experiment. Depending on the timescale of the biological process of interest, choose experimental time points accordingly. For example, to address the transcriptional changes caused by overexpression, choose short-term inductions (using an inducible transgene), since long-term changes in gene expression can mask the stage-specific function of the gene [11].

3.2 Analyzing Raw Data to Prepare for Network Inference

After sequencing, data from an RNA-seq experiment is often returned in the form of .fastq or .fq files that contain lists of raw reads. The TUX module of TuxNet, which runs on either Mac or Linux, transforms these raw reads into useful expression values and comparisons by applying a series of processing steps (*see* **Note 4**). The TUX module workflow is as follows:

1. Organize the raw read (.fastq or .fq) files for each condition into a separate folder. The condition can indicate a time point, treatment, plant line, etc. Biological replicates of the same condition should be added into the same folder.
2. Open TuxNet and navigate to the TUX tab.
3. Select the Inputs Folder, Reference Genome file (*see* **Note 5**), and FASTA File (*see* **Note 5**) to indicate the main folder containing the subfolders with the raw read files, the annotated genome file (.gft or .gff), and species reference genome (.fasta), respectively.
4. Press Run. Depending on processing power and dataset size, runtimes range between several hours and several days (*see* **Note 6**).
5. Ideally each sample is run onto the same lane during sequencing. If this is not the case, please go to **Note** 7.

3.3 Selecting and Assessing DEGs

RNA-seq analysis and network inference rely on the identification of genes that exhibit differential expression across experimental conditions. Compare expression values of two or more experimental conditions (i.e., WT vs. mutant, early time point vs. later time point, treatment vs. control, etc.) to identify DEGs from RNA-seq data. Determining the comparison criteria for DEG selection is critical for meaningful downstream analysis. Although the genes used for downstream analysis such as network inference can be hand selected based on prior knowledge, statistical selection criteria remove the bias that can cause important, non-described genes to be overlooked. Assess two criteria to categorize a gene as differentially expressed in a given comparison: fold change (FC) and a significance threshold, such as false discovery rate (FDR) or q-value (corrected p-value) (Box 1). After setting a FDR threshold, DEGs can be filtered by setting a FC threshold to reduce the total number, which is important for improving the accuracy of downstream analyses such as network inference (Box 1).

Box 1: TuxNet includes the TuxOP section of the TUX tab to select DEGs based on user-provided values of FC and FDR. TuxOP selects DEGs in either pairwise (i.e., one sample vs. another) or combinatorial (i.e., one or more samples vs. one or more samples) comparisons. Combinatorial DEG selection enables higher specificity: for example, in an experiment with three time points, selecting only genes that are simultaneously differentially expressed in both time point 0 vs. time point 1 and time point 0 vs. time point 2 would capture genes that have changed expression over the entire time course. Basing DEG selection on one or more specific biological questions produces meaningful and directed results. For example, in an experiment aiming to determine how two different types of treatment elicit different responses, first selecting two groups of DEGs by performing a comparison of each treatment to the control and then performing a comparison between the treatment datasets identifies common responses (*see* **Note 8**). By selecting the overlap between the two treatments, a list of only the genes that change in response to treatment *and* change between the treatment types is obtained. The TuxOP workflow is as follows:

1. Select the Inputs Folder to indicate where the processed RNA-seq data is located (this will be the same location as selected for the TUX tab with the initial raw data).
2. Choose Sample Groups 1 and 2 as the experimental conditions for comparison. The name and formatting should be exactly the same as the raw data input subfolders. If performing a combinatorial comparison, sample groups should be separated with a comma. For example, given sample groups T0, T1, T2, and T3 and the goal to compare T0 to T1 and T2, input T0 for Sample Group 1 and T1,T2 for Sample Group 2 (or vice versa).
3. Optionally, select a TF File and Gene Name File (*see* **Note 9**). With a TF File included, TuxOP will provide a second sheet in the output worksheet with only differentially expressed TFs. With a Gene Name File included, TuxOP will label each gene with its locus ID as well as its provided name.
4. Press Run. Depending on processing power and dataset size, runtimes range between several minutes to an hour.

This DEG list is directly used as the Gene Input File for network inference with TuxNet (see below). A DEG list with thousands of genes is refined through an iterative selection process. For example, applying more stringent FDR and FC thresholds, performing an overlap with existing datasets (*see* **Note 8**), and/or focusing on the TFs only.

Box 1: FC quantifies how much a value changes between two compared measurements and is defined as the ratio of the two measurements. The FC of measurement *Y* with respect to measurement *X* is given by Y/X. FDR quantifies the expected proportion of null hypothesis testing type I errors, also known as false positives, and corrects for multiple testing [26]. FDR and FC thresholds are data-dependent and can be assessed by plotting $\log_2$ mean expression against $\log_2$ FC. As p-values are calculated based on differences in expression between two groups and the variation within each group, a gene can have a significant p-value but a small FC, while a gene with a non-significant p-value can have a large FC as a result of high variation between replicates.

FDR and FC thresholds should be strategically chosen to maximize both network size and accuracy, thus providing the most informative possible network. The accuracy of network inference techniques, such as the ones described here, relies on the ratio between the number of genes in the network and the number of experimental conditions to draw inferences from. For example, GENIST, available as part of TuxNet, can be applied to predict causal gene interactions in the range of hundreds of genes [9, 27] and select the most biologically important genes with high accuracy.

3.4 Selecting Clustering Datasets and Methods

Both GENIST and RTP-STAR can optionally perform clustering on an additional dataset to improve GRN inference (*see* **Note 10**). The type of dataset is not restricted and could be, for example, time course, control vs. treatment, or tissue-type specific. However, use spatial and time course datasets for GENIST and RTP-STAR, respectively (Fig. 1). Using a clustering dataset, GENIST and RTP-STAR assume that genes that are not coexpressed (i.e., expressed within the same tissue when using a spatial dataset) cannot regulate each other.

RTP-STAR applies Dynamic Time Warping (DTW) clustering to an additional expression dataset before network inference (*see* **Note 11**), while GENIST applies linkage clustering, a form of agglomerative hierarchical clustering (*see* **Note 11**). Both of these clustering methods enable faster and more accurate GRN inference, assuming a biologically relevant clustering dataset. For this reason, applying clustering is highly recommended when using TuxNet for GRN inference if a relevant dataset is available (Fig. 1). Both in house and publicly available datasets can be used to cluster or infer regulatory relations.

3.5 Selecting a Network Inference Technique

TuxNet offers two options for inferring GRNs: GENIST, a dynamic Bayesian network (DBN) inference approach (Fig. 2a) (*see* **Note 12**), and RTP-STAR, a regression tree machine learning approach (Fig. 2b) (*see* **Note 13**).

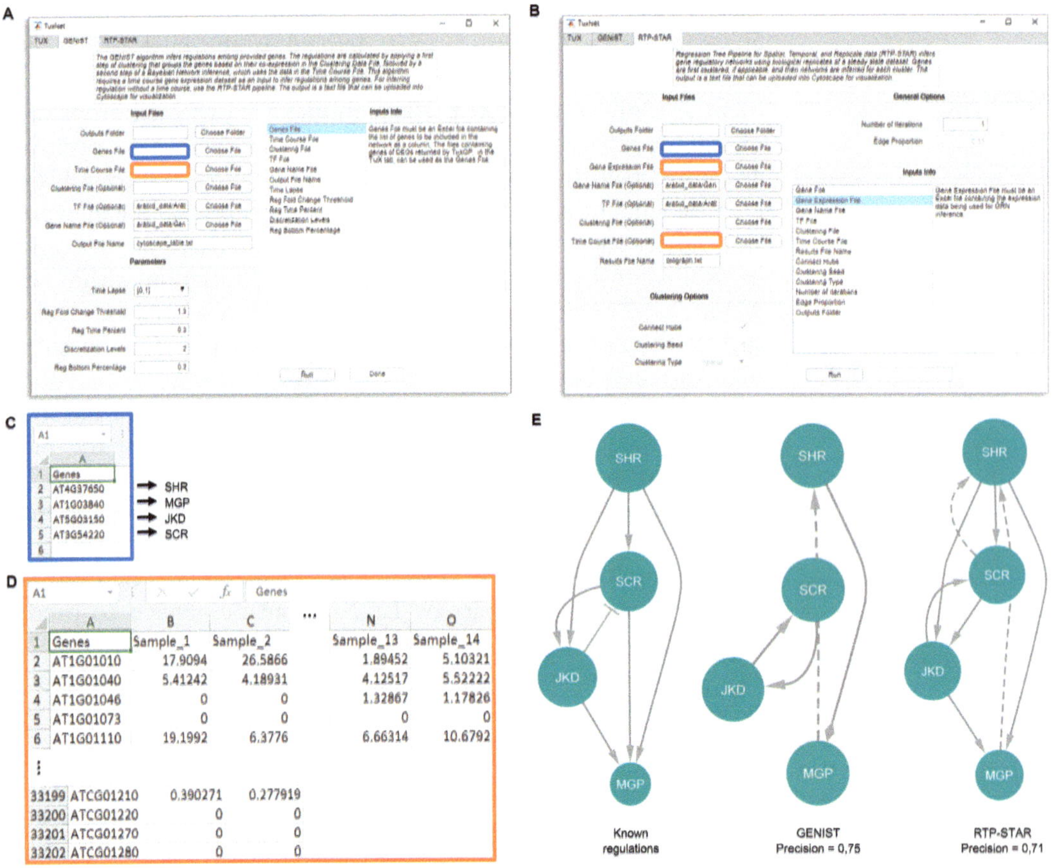

Fig. 2 Example network inference using the GENIST and RTP-STAR tabs of the TuxNet software. (**a**) The GENIST tab of TuxNet. (**b**) The RTP-STAR tab of TuxNet. (**c**) The genes list used to infer the networks. (**d**) The gene expression data used to infer the networks. (**e**) From left to right: experimentally proven regulations, the network inferred using GENIST, and the network inferred using RTP-STAR. Dashed lines indicate the regulations that differ from the experimentally proven regulations

GENIST applies a DBN inference approach to infer causal regulations between selected genes using average expression values from a time course dataset. Several options are available in GENIST to tailor GRN inference (Fig. 2a) (*see* **Note 12**):

1. GENIST includes the option to choose whether a TF can regulate genes within the same time point (preferred for a time course dataset with long intervals between each time point, e.g., a time course over several days) or can regulate genes only in the next time point (preferred with short time intervals, e.g., every 30 min), or both (Parameters: Time Lapse). Selecting both is preferred for a time course dataset with sparse intervals between samples, such as a mixture of short and long intervals (*see* **Note 14**) (Fig. 1).

2. GENIST has the default setting that a regulator needs to elicit a fold change of 1.3 (Parameters: Reg Fold Change Threshold) within the expression of the target gene. Change this parameter to be more or less stringent depending on the context.
3. GENIST has the default setting that a regulator and a target need to show changes of expression in at least 30% (0.3) of the time points for the predicted interaction to be considered true (Parameters: Reg Time Percent). Change this parameter to be more or less stringent depending on the context.
4. GENIST scales the expression values to 0 and 1 or 0, 0.5, and 1 for two or three discretization levels, respectively (Parameters: Discretization Levels). Generally, use a value of either two or three.
5. GENIST has the default setting to remove the bottom 20% (0.2) weakest inferred edges (Parameters: Reg Bottom Percentage). This option is provided as a proportion of the total inferred interactions.

RTP-STAR applies a regression tree algorithm (GENIE3) to expression values of biological replicates to infer regulations between selected genes [13] (*see* **Note 13**). A limitation of RTP-STAR is that without the provision of a TF File, its performance decreases [13]. Thus, include a TF File when possible. To determine the directionality of the inferred regulatory interactions, RTP-STAR has the option to include a time course file. This file will not contribute to the inference of the regulations, but will predict whether an inferred interaction is activating or repressing. Several other parameters are available in RTP-STAR to tailor GRN inference (Fig. 2b):

1. Although recommended to use a time course dataset prior to network interference, RTP STAR can use spatial datasets to cluster as well (Clustering Options: Clustering Type).
2. If using clustering, a seed to generate the pseudorandom starting clusters (Clustering Options: Clustering Seed) can be set. Using the default setting (0) will generate a new random seed each run.
3. If prior knowledge provides that the genes placed in different clusters cannot regulate each other, deselect the Connect Hubs box (Clustering Options: Connect Hubs). Otherwise, connect the hubs of each cluster, where hubs are defined as the node (s) with the most output edges in each cluster.
4. Since RTP-STAR generates regression trees pseudorandomly, the option is included to run the algorithm more than once (General Options: Number of Iterations) and combine the results, removing edges that do not appear in enough runs

(General Options: Edge Proportion). When using spatial clustering data, use 100 or more iterations for accurate results. When using temporal data, use 10 or more iterations (Fig. 1).

To illustrate the use of TuxNet, causal regulations between four well-characterized genes, *SHORTROOT*(*SHR*) [14, 15], *SCARECROW* (*SCR*) [15, 16], *MAGPIE* (*MGP*) [15, 17], and *JACKDAW* (*JKD*) [15, 17], were predicted with GENIST and RTP-STAR (Fig. 2c). These four genes have been shown to regulate stem cell identity and are differentially regulated at different stages of root development [8, 15]. As such, a time course dataset of a stem cell-enriched population collected every 8 h from 4 day to 6 day old plants was chosen (Fig. 2d) [10]. Moreover, because the regulatory interactions between *SHR*, *SCR*, *MGP*, and *JKD* are experimentally identified, the precision of the inferred network can be calculated (*see* **Note 15**) [15, 18, 19]. The precision of the inferred networks were scored against the known experimentally validated regulations and was greater than 70% for both GENIST and RTP-STAR (Fig. 2e).

3.6 Visualizing and Assessing Inferred Networks to Draw Conclusions

The outputs of GENIST and RTP-STAR are formatted to be easily imported into Cytoscape [20–23], an open-source network visualization software tailored to bioinformatics applications. Once imported, network styling options (e.g., size, color, and shape of nodes and edges) can be fully customized (Fig. 3). Further, additional network analysis applications are available to install and run within Cytoscape. To import a network into Cytoscape:

1. File Import Network from File and then choose the network file.
2. Select Advanced Options to choose a delimiter (SPACE for RTP-STAR or TAB for GENIST). If importing a network from RTP-STAR, deselect *Use first line as column names.*
3. Select the source node (green circle), interaction type (purple triangle), and target node columns (red target). These columns will be preselected for a GENIST file. For an RTP-STAR file, they will correspond to columns 1, 2, and 3, respectively.
4. Press OK to finish importing.

Cytoscape can provide topological information for a network such as connectivity, neighborhood connectivity, outdegree, and indegree (*see* **Note 16**) [24]. These and other topological values can help to establish an overview of the network and to identify key genes. To analyze the network for this information:

1. Tools Analyze Network.
2. Check Analyze as Directed Graph.
3. Press OK.

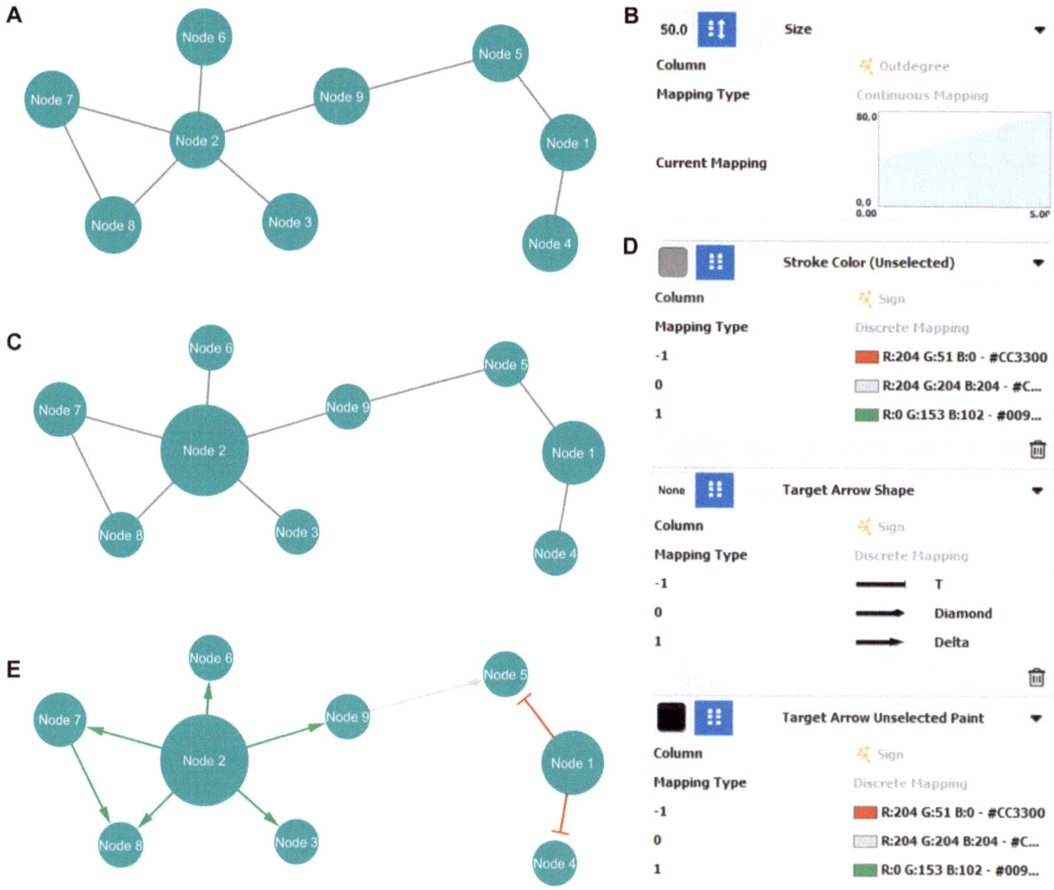

Fig. 3 Network visualization in Cytoscape. (**a**, **c**, **e**) Visualization of an example network with nodes and undirected edges (**a**), nodes sized according to outdegree (**c**), and nodes sized according to outdegree and directional color-coded edges (**e**). (**b**) Settings in Cytoscape to adjust node size continuously according to outdegree. (**d**) Settings in Cytoscape to color-code edges according to the regulation type. Repression (sign = −1), activation (sign = 1), and unknown (sign = 0) regulations are colored red, green, and grey and the tip of the arrow is shaped T, delta, and diamond, respectively

The calculated attributes are used to further customize the network style. For example, a common approach is to continuously map node size to outdegree so that each node scales in size with its number of outgoing regulations (Fig. 3b, c).

Each interaction inferred by both GENIST and RTP-STAR is labeled with one of three edge directions: activation, inhibition, and unknown. For an imported RTP-STAR network file, the interaction column in the edge table will contain this information as either activates, inhibits, or regulates, respectively. For a GENIST network file, 1, −1, and 0 values are used, respectively, and stored in the sign column in the edge table. The network can be further customized using this information: for example, a common approach is to label positive edges with green arrows, negative edges with red crosses, and unknown edges with grey diamonds to visualize interaction types (Fig. 3d, e). GENIST also includes a width value for each edge, which indicates the confidence of the algorithm in the inferred interaction. Commonly, this value is visualized with continuous mapping of the edge weight property.

3.7 Using Network Motifs to Select Candidate Genes for Further Research

Although node outdegree can be a strong indicator of biological importance, it is imperative to also consider involvement in sub-networks with key functionalities known as network motifs (*see* **Note 17**). To quantify the involvement of each gene in these network motifs, network topology and sub-networks are used to calculate a normalized motif score (NMS) for each gene [10]:

1. Install NetMatch* (*see* **Note 18**), an application within Cytoscape used to assess whether each network motif type is significantly overrepresented in the network by comparing its actual abundance to its abundance in randomly generated networks [25].
2. Use the built-in motif library and custom manually created motifs to construct query networks (*see* **Note 19**). These query networks are used to search for and count network motifs within the inferred network.
3. Evaluate the significance of the abundance of each motif with NetMatch* (Fig. 4) (*see* **Note 20**). Motifs can be searched for in a general manner (e.g., feedforward loop) or more specific manner (e.g., incoherent feedforward loop). A list of network motifs is found in **Note 21**.
4. Only include the significantly overrepresented motifs in the scoring process.
5. Copy and paste the abundance results of each motif into a spreadsheet software such as Excel (*see* **Note 20**).
6. Calculate the NMS for a gene as follows: count and normalize (from 0 to 1) the involvement of each gene in each of the selected motifs against the rest of the genes in the network. Sum these normalized counts to produce a final motif score for each gene, with higher values indicating higher importance within the network.

Applying NMS enables identification of important genes, which may have been overlooked otherwise due to low outdegree. Top scoring genes form strong candidates for future research (Box 2).

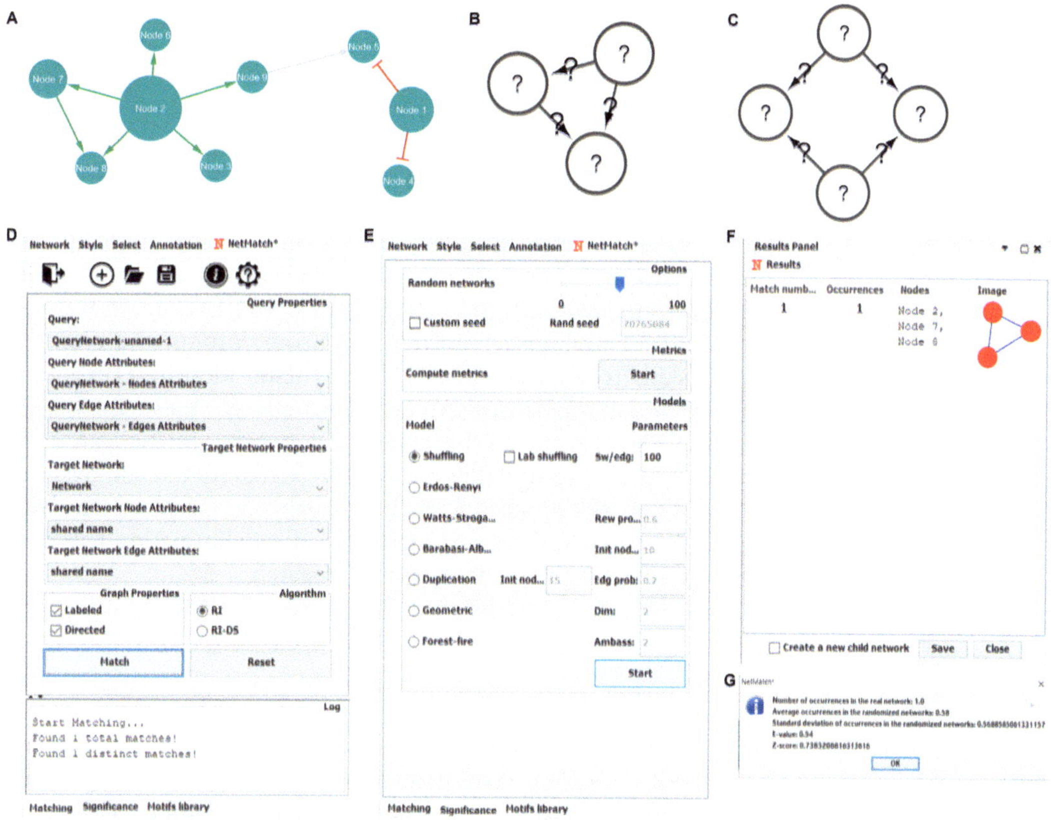

Fig. 4 Motif finding within a network with NetMatch* (*see* **Note 13**). (**a**) The inferred network visualized in Cytoscape. (**b**) Query network of a feedforward loop. (**c**) Query network of a bifan. (**d**) NetMatch* Matching tab to identify a query network within the inferred network. (**e**) NetMatch* Significance tab to determine whether the query network is significantly overrepresented within the inferred target network. (**f**) Results panel of the Matching tab identifying feedforward loops within the network presented in (**a**). (**g**) Popup screen with the results of the Significance tab

Box 2: In this chapter, we describe a workflow to computationally infer GRNs that includes TuxNet, a software specifically designed to combine RNA-seq processing and network prediction in a user-friendly interface. The workflow assists the biologist in setting up their experimental design, helps in the identification of the proper inference technique, and suggests visualization and downstream analysis techniques, taking into account prior knowledge, the biological question, and scale. As RNA-seq becomes more accessible and affordable, more sequencing data is produced that needs to be processed, analyzed, and interpreted. With this chapter, we offer a method for biologists with minimal programming and bioinformatics skills to analyze their RNA-seq data, identify DEGs, and predict causal regulatory interactions between these DEGs. By applying a system-wide approach with network inference, an unbiased selection of candidate genes can be achieved and the dynamic behavior of the system can be visualized. This system-wide approach assists in directing further studies, eliminating the need for large screenings and accelerating research.

4 Notes

1. RNA-seq has become the preferred method for investigating global transcriptional changes because it provides single base resolution across an entire reference genome with high throughput and low noise [28]. RNA-seq has the advantage of detecting the entire transcriptome with high sensitivity, while for example, proteomics and/or metabolomics detect a proportion of the whole proteome and/or metabolome due to the complex and diverse biochemical properties and inaccessibility of proteins and small molecules.
2. Steady-state data obtained from the comparison of a mutant or overexpression line to a wild type can address how the down-regulation or overexpression of a gene affects a specific biological function, respectively. However, rather than using steady overexpression lines, it could be advantageous to use an inducible system so that gene expression is readily controlled and operated when appropriate, such as during a specific developmental stage, to evaluate downstream transcriptional cascades (Fig. 1).
3. When interested in three or more developmental stages, this spatial data is considered temporal data and will have implications for downstream analyses, such as GRN inference (Fig. 1).

4. Processing steps of the TUX module:
   (a) Fastq-mcf (from the ea-utils package) is used to remove short and low quality reads as well as the Illumina adapters (specified in IlluminaAdapters.fasta), returning cleaned .fastq or .fq files [29, 30].
   (b) Hisat2 [31] is used to map the cleaned reads to the reference genome, returning aligned and sorted .sam files.
   (c) Cufflinks [32] is used to assemble the transcripts and estimate abundances, returning folders of output data.
   (d) Cuffmerge [32] is used to combine the assembled transcripts, returning a single folder.
   (e) Cuffdiff [32] is used to perform differential expression analysis of the mapped reads and combined transcripts, returning gene expression values, pairwise comparisons, and splicing information.
5. Two files are needed for TUX to correctly align the reads to the reference genome: a FASTA file (https://www.arabidopsis.org/download/index.jsp) and gtf/gff file (https://useast.ensembl.org/info/website/upload/gff.html).
6. The Threads option of the TUX tab of TuxNet should be used when multiple cores/processors are available on the user's computer. Each added thread beyond the first will find alignments in parallel on a different core/processor and will increase the speed of hisat2 linearly.
7. When a sample is run on different lanes, the .fastq or .fq files obtained from the different lanes need to be merged before processing. This can be done in a Unix command line using the `cat` command. For example, if merging sampleA_lane1.fastq and sampleA_lane2.fastq:

```
cat sampleA_lane1.fastq sampleA_lane2.fastq > sampleA.fastq
```

8. Several online Venn diagram programs are available, such as http://www.interactivenn.net/, http://bioinformatics.psb.ugent.be/webtools/Venn/, and https://bioinfogp.cnb.csic.es/tools/venny/.
9. TuxNet includes an Arabidopsis TF file and Gene Name file. For other species, the PlantTFDB is a good resource for acquiring a TF file (http://planttfdb.cbi.pku.edu.cn/).
10. The goal of clustering is to group objects (in this case genes) in a dataset in such a way that each object is more similar to other members of its cluster than to members in any other cluster. For GRN inference, clustering genes based on expression

improves accuracy and speed and is necessary for inference of large-scale networks [8]. The clusters themselves can also reveal groups of genes that are, for example, functionally similar or coexpressed either spatially or temporally.

11. DTW measures similarity between expression profiles through non-linear transformation and the clustering process aims to group genes in such a way that similarity is maximized for genes in the same group and minimized for genes in different groups. Linkage clustering begins with each object in its own cluster and then sequentially combines the closest clusters until all objects belong to the same cluster. The process can be visualized as a dendrogram, showing the order and distances of cluster combinations.
12. An inferred DBN represents a set of variables (genes) and their conditional dependencies (activation, repression, or undetermined). As GENIST relies on dynamic Bayesian probabilities, GENIST performs better in predicting linear pathways rather than cyclic network patterns such as feedforward loops.
13. RTP-STAR generates a decision tree of genes and interactions to quantify relationships between observations (gene expression values) and conclusions (gene interactions). This decision tree is, more specifically, a regression tree, which can be susceptible to overfitting, a form of error that results when an analysis too closely follows a given set of data and therefore cannot correctly provide predictions. To avoid overfitting, RTP-STAR utilizes a random forest approach, where regression trees are iteratively generated, updated, and averaged together to produce a final, more robust result.
14. The sparsity of a time course dataset refers to the regularity of the intervals between time points. A time course dataset with samples taken 10 min, 60 min, 12 h, 24 h, and 72 h is considered sparse due to its uneven intervals, while a dataset with samples taken every 30 min is considered not sparse due to its even intervals.
15. Precision = TP/(TP + FP), where TP = true positive edges and FP = false positive edges.
16. Outdegree and indegree refer to the number of outgoing and incoming interactions a given node has, respectively. Connectivity refers to the total number of neighbors of a given node and neighborhood connectivity refers to the average connectivity of all neighbors of a given node. More information can be found at: https://med.bioinf.mpi-inf.mpg.de/netanalyzer/help/2.7/index.html.

17. Common network motifs such as feedback and feedforward loops have been shown to be prevalent in GRNs in a variety of evolutionarily divergent species, indicating that these sub-networks are both critically important and optimal for performing specific functions [33]. For example, feedforward loops are shown to provide a rapid response to stimuli while eliminating the effect of background noise [34]. As such, it is necessary to quantify the involvement of each gene in these network motifs to capture a more complete view of the biological process a network represents.
18. Apps App Manager Search: NetMatchStar Install.
19. Apps NetMatch* Motifs library. Whenever the user selects a motif within the motif library, an example network is automatically made within Cytoscape. Small networks representing motifs can also be manually made: File New Network Empty, followed by adding nodes and edges: Right mouse click on the empty network Add Node Repeat this to increase the number of nodes Select a node Right mouse click on the node Add Edge Drag the edge to the downstream node.
20. Open NetMatch* and select your query network (the motif to find) and target network (the inferred network) at the *Matching* tab and press Match (Fig. 4d). A result panel will open automatically indicating the found motifs within the inferred network (Fig. 4f). Next go to the *Significance* tab and press start (Fig. 4e). The E-value of the query network abundance within the inferred network is calculated and given on a popup screen (Fig. 4g). When this value is significant, export the matching results as a .txt file onto your computer to calculate the motif score.
21. Feedback loops (three-cycle or two-cycle, positive or negative), feedforward loops (incoherent or coherent), bifans, m to n fans, cascades (3, 4, 5, etc. nodes), multi-input nodes [35] (Fig. 4b, c).

Acknowledgement

Support for this work was provided to R.S. by the National Science Foundation (NSF) (CAREER MCB-1453130).

References

1. Joshi R, Wani SH, Singh B et al (2016) Transcription factors and plants response to drought stress: current understanding and future directions. Front Plant Sci 7:1029
2. Vermeirssen V, De Clercq I, Van Parys T et al (2014) Arabidopsis ensemble reverse-engineered gene regulatory network discloses interconnected transcription factors in oxidative stress. Plant Cell 26:4656–4679
3. Miao Z, Xu W, Li D et al (2015) De novo transcriptome analysis of *Medicago falcata* reveals novel insights about the mechanisms

underlying abiotic stress-responsive pathway. BMC Genomics 16:818

4. Luo H, Zhao W, Wang Y et al (2016) SorGSD: a sorghum genome SNP database. Biotechnol Biofuels 9:6
5. Schlitt T, Brazma A (2007) Current approaches to gene regulatory network modelling. BMC Bioinformatics 8(Suppl 6):S9
6. Robinson MD, McCarthy DJ, Smyth GK (2010) edgeR: a Bioconductor package for differential expression analysis of digital gene expression data. Bioinformatics 26:139–140
7. Anders S, Huber W (2010) Differential expression analysis for sequence count data. Genome Biol 11:R106
8. de Luis Balaguer MA, Fisher AP, Clark NM et al (2017) Predicting gene regulatory networks by combining spatial and temporal gene expression data in Arabidopsis root stem cells. Proc Natl Acad Sci U S A 114: E7632–E7640
9. Spurney RJ, Van den Broeck L, Clark NM et al (2020) Tuxnet: a simple interface to process RNA sequencing data and infer gene regulatory networks. Plant J 101:716–730
10. Clark NM, Buckner E, Fisher AP et al (2019) Stem-cell-ubiquitous genes spatiotemporally coordinate division through regulation of stem-cell-specific gene networks. Nat Commun 10:5574
11. Ó'Maoiléidigh DS, Thomson B, Raganelli A et al (2015) Gene network analysis of *Arabidopsis thaliana* flower development through dynamic gene perturbations. Plant J 83:344–358
12. Nelissen H, Gonzalez N, Inzé D (2016) Leaf growth in dicots and monocots: so different yet so alike. Curr Opin Plant Biol 33:72–76
13. Huynh-Thu VA, Irrthum A, Wehenkel L, Geurts P (2010) Inferring regulatory networks from expression data using tree-based methods. PLoS One 5:e12776. https://doi.org/10.1371/journal.pone.0012776
14. Levesque MP, Vernoux T, Busch W et al (2006) Whole-genome analysis of the SHORT-ROOT developmental pathway in Arabidopsis. PLoS Biol 4:e143
15. Moreno-Risueno MA, Sozzani R, Yardımcı GG et al (2015) Transcriptional control of tissue formation throughout root development. Science 350:426–430
16. Sabatini S, Heidstra R, Wildwater M, Scheres B (2003) SCARECROW is involved in positioning the stem cell niche in the Arabidopsis root meristem. Genes Dev 17:354–358
17. Welch D, Hassan H, Blilou I et al (2007) Arabidopsis JACKDAW and MAGPIE zinc finger proteins delimit asymmetric cell division and stabilize tissue boundaries by restricting SHORT-ROOT action. Genes Dev 21:2196–2204
18. Ogasawara H, Kaimi R, Colasanti J, Kozaki A (2011) Activity of transcription factor JACKDAW is essential for SHR/SCR-dependent activation of SCARECROW and MAGPIE and is modulated by reciprocal interactions with MAGPIE, SCARECROW and SHORT ROOT. Plant Mol Biol 77:489–499
19. Long Y, Smet W, Cruz-Ramírez A et al (2015) Arabidopsis BIRD zinc finger proteins jointly stabilize tissue boundaries by confining the cell fate regulator SHORT-ROOT and contributing to fate specification. Plant Cell 27:1185–1199
20. Lopes CT, Franz M, Kazi F et al (2010) Cytoscape web: an interactive web-based network browser. Bioinformatics 26:2347–2348
21. Smoot ME, Ono K, Ruscheinski J et al (2011) Cytoscape 2.8: new features for data integration and network visualization. Bioinformatics 27:431–432
22. Su G, Morris JH, Demchak B, Bader GD (2014) Biological network exploration with Cytoscape 3. Curr Protoc Bioinformatics 47:8.13.1–8.1324
23. Shannon P, Markiel A, Ozier O et al (2003) Cytoscape: a software environment for integrated models of biomolecular interaction networks. Genome Res 13:2498–2504
24. Hu Z, Mellor J, Wu J et al (2005) VisANT: data-integrating visual framework for biological networks and modules. Nucleic Acids Res 33:W352–W357
25. Rinnone F, Micale G, Bonnici V et al (2015) NetMatchStar: an enhanced Cytoscape network querying app. F1000Res 4:479
26. Benjamini Y, Hochberg Y (1995) Controlling the false discovery rate: a practical and powerful approach to multiple testing. J R Stat Soc Series B Stat Methodol 57:289–300
27. Krouk G, Lingeman J, Colon AM et al (2013) Gene regulatory networks in plants: learning causality from time and perturbation. Genome Biol 14:123
28. Wang Z, Gerstein M, Snyder M (2009) RNA-Seq: a revolutionary tool for transcriptomics. Nat Rev Genet 10:57–63
29. Aronesty E (2013) Comparison of sequencing utility programs. Open Bioinforma J 7:1–8

30. ea-utils by ExpressionAnalysis. https://expressionanalysis.github.io/ea-utils/. Accessed 30 Apr 2020
31. Kim D, Langmead B, Salzberg SL (2015) HISAT: a fast spliced aligner with low memory requirements. Nat Methods 12:357–360
32. Trapnell C, Roberts A, Goff L et al (2012) Differential gene and transcript expression analysis of RNA-seq experiments with TopHat and cufflinks. Nat Protoc 7:562–578
33. Conant GC, Wagner A (2003) Convergent evolution of gene circuits. Nat Genet 34:264–266
34. Boyle AP, Araya CL, Brdlik C et al (2014) Comparative analysis of regulatory information and circuits across distant species. Nature 512:453–456
35. Kitchen JL, Allaby RG (2013) Systems modeling at multiple levels of regulation: linking systems and genetic networks to spatially explicit plant populations. Plan Theory 2:16–49

Chapter 5

Dynamic Modeling of Transcriptional Gene Regulatory Networks

Joanna E. Handzlik, Yen Lee Loh, and Manu

Abstract

Diverse cellular phenotypes are determined by groups of transcription factors (TFs) and other regulators that influence each others' gene expression, forming transcriptional gene regulatory networks (GRNs). In many biological contexts, especially in development and associated diseases, the expression of the genes in GRNs is not static but evolves in time. Modeling the dynamics of GRN state is an important approach for understanding diverse cellular phenomena such as cell-fate specification, pluripotency and cell-fate reprogramming, oncogenesis, and tissue regeneration. In this protocol, we describe how to model GRNs using a data-driven dynamic modeling methodology, gene circuits. Gene circuits do not require knowledge of the GRN topology and connectivity but instead learn them from training data, making them very general and applicable to diverse biological contexts. We utilize the MATLAB-based gene circuit modeling software Fast Inference of Gene Regulation (FIGR) for training the model on quantitative gene expression data and simulating the GRN. We describe all the steps in the modeling life cycle, from formulating the model, training the model using FIGR, simulating the GRN, to analyzing and interpreting the model output. This protocol highlights these steps with the example of a dynamical model of the gap gene GRN involved in *Drosophila* segmentation and includes example MATLAB statements for each step.

Key words Dynamical modeling, Gene regulatory networks, Differential equations, Transcriptional networks, Pattern formation, Binary classification, Parameter inference, Development, Cell fate, Differentiation

1 Introduction

Transcriptional gene regulatory networks (GRNs) underlie diverse biological phenomena such as pattern formation [5, 12, 31, 49], cell-fate specification [15, 33], pluripotency and cell-fate reprogramming [10, 30], oncogenesis [46], and regeneration [36]. GRNs are comprised of genes that encode proteins, most often transcription factors (TFs), that can directly or indirectly regulate the expression of other genes in the network. GRNs can be quite large, comprising tens to hundreds of genes [11], are highly interconnected, and are wired recursively since the genes encoding transcription factors (TFs) are themselves regulated by

Shahid Mukhtar (ed.), *Modeling Transcriptional Regulation: Methods and Protocols*, Methods in Molecular Biology, vol. 2328,
https://doi.org/10.1007/978-1-0716-1534-8_5,

other TFs or indirectly by non-TF gene products [27, 34]. The expression of the genes in such networks changes over time and, particularly in developmental contexts, the biological outcome is dictated by the time evolution of gene expression. Furthermore, the rate of change of the expression of any given gene depends on the expression of the other genes in the network—that is, the GRN *state*—since the former is regulated by the latter. In other words, GRNs function as *dynamical systems*.

Mathematical modeling of transcriptional GRNs is essential for understanding and predicting dynamical biological processes [7, 10, 15, 32, 47, 52]. Modeling is necessary because it is difficult, if not impossible, to reason qualitatively about the function of large and complex gene networks [11, 29]. Furthermore, among the different types of GRN models that are possible, *dynamical* models, which simulate the evolution of gene expression in time [32, 33], are the most appropriate for modeling biological processes.

There are several different frameworks available for modeling GRN dynamics. The choice of a modeling framework suited to a specific biological problem depends, among other considerations, on the questions being asked, the availability and type of quantitative measurements, and the availability of computational resources. Perhaps the first decision to be made is whether cell-to-cell or temporal variability in gene expression is pertinent to the biological questions being asked. If this is the case, then a *stochastic* modeling framework, where the state variables are random and the dynamical equations are stochastic, must be adopted [43, 50, 51]. If on the other hand variability is not central to the problem and we are interested in simulating the mean behavior of the GRN, then a *deterministic* framework, where the state variables are not random and the equations are deterministic, would suffice.

Second, one may decide whether the state variables, usually the concentrations of the products—mRNA or protein—of the GRN constituents, take discrete values, such as "ON" and "OFF," or continuous values, and also whether time takes discrete values or not. If the choice is made to represent the state and time with discrete variables, the resulting models are *Boolean networks* [6, 10, 35, 39, 44], which are fairly inexpensive computationally. Conversely, one may choose to represent time and GRN state with continuous variables, since mRNA and protein concentrations do in fact vary continuously [45], resulting in *ordinary differential equation (ODE)* or *partial differential equation (PDE)* models of GRNs [5, 21, 31, 46].

The most important choice to be made is that of the dynamical equations that determine the time evolution of the GRN state. The dynamical equations represent the key biological hypotheses about the GRN mathematically, such as the biochemistry of gene regulation, the identities of the regulators of specific genes, whether the regulators activate or repress gene expression, and whether the activity of a regulator is context-dependent, that is, whether it depends on the concentration of other regulators. In Boolean networks with discrete time, the equations update the GRN state at the next time step based on Boolean functions of the current gene states. In differential equation modeling, the time evolution of GRN state is determined by equations that describe how the rates of change of gene product concentrations depend on the current GRN state. In stochastic modeling, the dynamical equations represent stochastic processes, and different approaches such as the master equation, Fokker-Planck equation, Langevin equation, and Markov chains are available. Although the modeler has complete freedom to choose the equations based on the specific biological questions they wish to probe via the model, we mention here two types of choices that are contrasting in approach. First, if detailed knowledge about the direct *cis* regulation of the genes is available from prior empirical studies or the modeler wishes to test specific hypotheses about the regulatory connectivity of the GRN, then the modeler can formulate equations customized to their problem [6, 19, 35, 41]. In most GRNs however, detailed prior knowledge about the regulation of the genes is not available. In such situations, one may utilize a generalized modeling framework called *gene circuits* [12, 21, 32, 37] that represents gene product synthesis as a switch-like function of regulator concentrations and allows all pairwise interactions between the genes in a network. The connectivity of the GRN is inferred by training the model on quantitative gene expression time series data. In learning the GRN connectivity from data, gene circuit modeling is closely allied with modern machine learning techniques.

Another choice the modeler must make is how the free parameters of the model will be determined. In what we call *qualitative* modeling [18, 19, 28] here, the parameter space is uniformly sampled to identify broad regions in which the model's behavior qualitatively matches empirical observations. In the *data-driven* approach, in contrast, the model's free parameters are inferred by fitting the model to quantitative gene expression data [21, 32, 33, 37].

Given the multiplicity of choices at each step of the modeling pipeline, it is not possible to discuss all the combinations in one chapter. Instead, in this protocol, we focus on the gene circuit approach, that is easily generalizable to diverse GRNs and does not require deep prior empirical knowledge of the biological system in question. Gene circuits are a deterministic modeling framework that utilize coupled ODEs or PDEs to represent the time evolution of gene product concentrations. Furthermore, instead of hardwiring the regulatory connectivity between the genes, gene circuits learn it from gene expression time series data, which are relatively easy to acquire from most biological systems with functional genomics techniques such as RNA-seq. Until recently, the main weakness of the gene circuit approach was that training the model required high performance computing resources and specialized skills [8]. This obstacle was recently overcome by an optimization method, Fast Inference of Gene Regulation (FIGR), which allows gene circuit inference in minutes on desktop-class hardware [12]. In this protocol, we describe how FIGR can be used to formulate, infer, and simulate a dynamical model of GRN function. To illustrate these steps, we have used the gap gene network [20], involved in the anteroposterior patterning of the *Drosophila* embryo, as an example. In Sect. 2, we briefly describe the gene circuit approach and FIGR. The protocol is presented in the methods section (Sect. 4).

2 Gene Circuits

2.1 Gene Circuit Models of GRNs

In this section, we provide a brief description of the gene circuit method and FIGR (see Note 1). The reader is referred to Fehr et al. [12] for a more detailed description.

We consider a GRN of G genes whose state at time t is defined by the concentrations of the gene products, either mRNA or proteins, $x_g(t)$, $g = 1, 2, \cdots, G$. Gene circuits [37] describe the time evolution of $x_g(t)$ according to G coupled ordinary differential equations,

$$\frac{dx_g}{dt} = R_g\, S\left(\sum_{f=1}^{G} T_{gf} x_f + h_g\right) - \lambda_g x_g. \tag{1}$$

where R_g is the maximum synthesis rate of product g. T_{gf} are genetic interconnectivity coefficients describing the regulation of gene g by the product of gene f. Positive and negative values of T_{gf} signify activation and repression of gene g by gene f, respectively. The threshold h_g determines the basal synthesis rate, and λ_g is the degradation rate of product g. Nominally, all genes in the model also function as regulators, so that both g and f run over the range $1, 2, 3, \ldots, G$. Sometimes such gene networks include upstream regulators that are not themselves influenced by other gene products represented in the model. For example, in the *Drosophila* segmentation gene network, maternal proteins such as Bicoid activate the zygotically expressed genes, but are not regulated by their targets [2]. An upstream regulator g can be represented by setting $T_{gf} = 0$ for all f.

$S(u)$ is the regulation-expression function, which determines the fraction of the maximum synthesis rate attained by the gene given the total regulatory input $u = \sum_{f=1}^{G} T_{gf} x_f + h_g$. $S(u)$ is required to have a switch-like dependence on u and to take values between 0 and 1. A commonly utilized [21, 37] form of the regulation-expression function is the sigmoid function

$$S(u) = \sigma(u) \quad = \frac{1}{2}\left(\frac{u}{\sqrt{1+u^2}} + 1\right). \tag{2}$$

Another form of the regulation-expression that may be utilized is the Heaviside function,

$$S(u) = \Theta(u) = \begin{cases} 0 & \text{if } u < 0 \\ 1 & \text{if } u \geq 0, \end{cases} \tag{3}$$

so that the gene switches discretely between ON and OFF states. Although the sigmoid function is on the whole more realistic biologically, the Heaviside function allows for gene circuit inference that is orders of magnitude faster than traditional methods [12].

In spatially extended biological systems such as embryos, the gene circuit equations have to be reformulated as PDEs or ODEs that represent protein transport with Fickian diffusion [17]. For simplicity, we only show the simplest case here, that of protein transport between cells lying in a one-dimensional row. The gene circuit equations (Eq. 1) are modified so that

$$\begin{aligned} \frac{dx_{ng}}{dt} =& R_g\,\sigma\left(\sum\nolimits_{f=1}^{G} T_{gf}x_{nf} + h_g\right) \\ &+ D_g(x_{n-1_g} + x_{n+1_g} - 2x_{n_g}) - \lambda_g x_{ng}. \end{aligned} \tag{4}$$

Here $x_{ng}(t)$ is the expression level of gene product g in nucleus n at time t, D_g is the diffusion constant for protein g, and $\sigma(u)$ is the sigmoid regulation-expression function (Eq. 2). Zero-flux boundary conditions are used at the ends of the modeled region.

2.2 The Gap Gene GRN of *Drosophila*

Throughout this protocol, we will use the example of a dynamical model of the *Drosophila* gap gene GRN to illustrate the key steps of the protocol. We provide a very brief description of the system and refer the reader to the extensive gap gene literature for further details [20]. The segmentation proteins pattern the anteroposterior axis of the *Drosophila* embryo during the first three hours of embryogenesis. During this period, the embryo is a syncitium, so that nuclei lack cell membranes and undergo 13 mitotic divisions, termed cleavages. After the tenth cleavage cycle, the majority of the nuclei migrate to the periphery of the embryo and are arranged in a monolayer, forming a syncytial blastoderm.

Near the end of cleavage cycle 14, the segmentation genes are expressed in spatially resolved patterns that specify the position of each cell to an accuracy of one cell diameter. Segmentation gene expression is initiated by shallow protein gradients formed by the translation of localized mRNAs, such as *bicoid* (*bcd*) and *caudal* (*cad*), deposited in the oocyte by the mother. These maternal protein gradients regulate the gap genes, which commence mRNA expression in cleavage cycle 10–12 and are expressed in broad domains ~20 nuclei wide during cycle 14. The expression of the segmentation genes is largely symmetrical around the anteroposterior axis, and therefore can be modeled in a one-dimensional row of nuclei along the axis. All the maternal and gap proteins are known to act as transcription factors that regulate each others' expression in a complex GRN that has been modeled extensively [4, 13, 14, 21–23, 31, 32, 38, 48, 49, 53].

As an example, we illustrate the inference of a gene circuit for the gap genes *hunchback* (*hb*), *Krüppel* (*Kr*), *giant* (*gt*), and *knirps* (*kni*). The model includes the upstream regulators Bicoid (Bcd), Caudal (Cad), and Tailless (Tll), so that $G=7$. The gene circuit models the protein expression of these genes between 35% and 92% egg length during cleavage cycle 14 (Sect. 2.3.1).

2.3 Inference of GRN Connectivity and Parameters from Quantitative Gene Expression Data

GRN models have a large number of parameters, usually of biophysical or biochemical provenance, such as dissociation constants, rates of synthesis, and rates of degradation. For the vast majority of GRN models, the values of such parameters are difficult, if not impossible, to measure and must be inferred by training the model on quantitative gene expression data.

More specifically, gene circuits have several unknown parameters, the genetic interconnectivity coefficients T_{gf}, the thresholds h_g, the synthesis rates R_g, and the degradation rates λ_g, that must be inferred from data. One benefit of gene circuits is that the topology and connectivity of the GRN do not need to be specified beforehand but can be learned from fitting the model to the data. If a positive or negative value is inferred for T_{gf}, it implies that gene f activates or represses gene g, respectively. Furthermore, if $|T_{gf}|$ is small one may infer that gene f does not regulate gene g.

2.3.1 Training Data

Gene expression *time series* data are a requirement for training any dynamical GRN model. Gene circuits are flexible in their data requirements and may be trained on mRNA or protein concentrations measured by many methods such as RT-qPCR, in situ hybridization, microarrays, RNA-seq, western blots, immunofluorescence, immunohistochemistry, quantitative mass spectrometry, and fluorescent protein reporters.

Suppose that an experiment has produced a dataset of gene expression levels $\tilde{x}_{ng}(t_e)$ where $g=1, \ldots, G$ are gene indices and t_e are timestamps at N_t time points $e=1, \ldots, N_t$, which may or may not be equally spaced. In a spatially extended one-dimensional model, such as the *Drosophila* gap gene GRN model, $n=1, \ldots, N$ are cell indices. In cell-autonomous models, n can also be used to signify different experimental conditions or treatments. For example, for inferring the gap gene circuit in Sect. 4, we utilize a dataset of segmentation protein immunofluorescence measurements at the resolution of individual nuclei at eight time points during cleavage cycle 14 (Fig. 6a; [22]).

2.4 Fast Inference of Gene Regulation (FIGR)

Given a gene expression dataset $\{\tilde{x}_{ng}(t_e)\}$, FIGR infers values of gene circuit parameters $\tilde{T}_{gf}$, $\tilde{h}_g$, $\tilde{R}_g$, and $\tilde{\lambda}_g$. FIGR exploits the observation that the inference of the connectivity of a given gene can be rephrased as a supervised learning problem: to find a hyperplane in state space that classifies observations into two groups, one where the gene is ON and the other where the gene is OFF. The FIGR algorithm determines whether a gene is ON or OFF at a given observation point by computing the time derivative of concentrations in a numerically robust manner. It then performs classification using logistic regression to determine the equation of the switching hyperplane. The genetic interconnectivity can then be computed from the coefficients of the hyperplane equation in a straightforward manner. Until this point FIGR works under the assumptions that the regulation-expression function is Heaviside (Eq. 3), that is, genes switch ON and OFF discretely and that gene products do not diffuse between cells (Eq. 1). If a model with the sigmoid regulation-expression function (Eq. 2) or diffusing gene products or both is desired, then FIGR can perform an optional refinement of the parameters by fitting the solutions of the full differential equations (Eq. 8) to data using local search. We briefly describe the algorithm below and refer the reader to Fehr et al. [12] for details.

2.4.1 Determining the ON/OFF State of the Genes

The first step in FIGR is to determine whether a gene is ON or OFF at each observation of the GRN state. This is accomplished by differentiating a spline fit to the time series data and inspecting its sign.

Spline Fits

FIGR uses the MATLAB function `csaps`, which takes a set of data points (t_j, x_j) and constructs a cubic smoothing spline function $f(t)$ by minimizing a cost function $p\sum_j w_j |x_j - f(t_j)|^2 + (1-p) \times \int dt\, \lambda(t) \left|\frac{d^2 f}{dt^2}\right|^2$. Here, p is a parameter in [0, 1] such that $p = 1$ (no smoothing) gives an ordinary cubic spline passing through all the data points, whereas $p = 0$ (extreme smoothing) gives a least-squares straight-line fit (Fig. 1). In the present context, given the time series data $x_{ng}(t_e)$ and smoothing parameters p_g, FIGR constructs smooth functions $x_{ng}(t)$ on the interval $t \in [t_1, t_{N_t}]$.

Velocities

The next step is to differentiate the cubic spline function with respect to time. From a mathematical point of view, gene expression levels play the role of coordinates in a G-dimensional state space. By analogy, we define the "velocity" of gene g as

$$v_g(t) = \frac{d}{dt} x_{ng}(t). \tag{5}$$

For concreteness, FIGR evaluates the velocities at the original time points, that is, $v_{ng}(t_e)$.

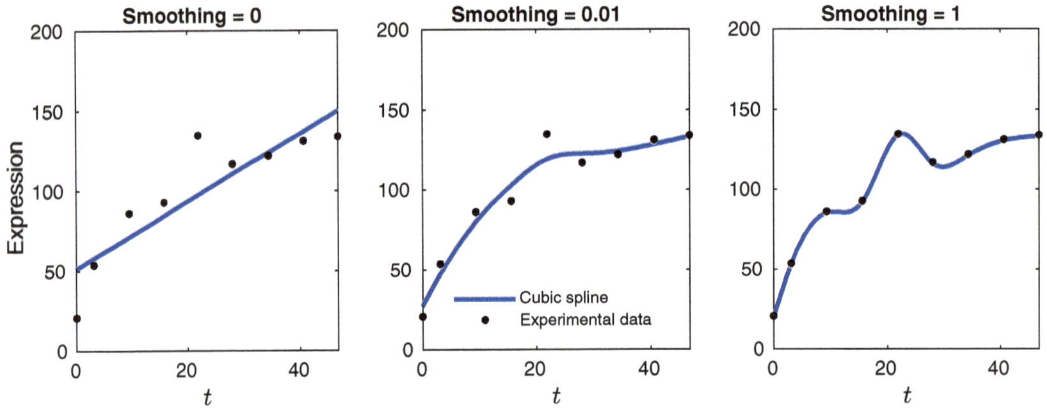

Fig. 1 The effect of the smoothing parameter p_g. Data are from example 3 of the FIGR package. The black dots show the expression data of gene $g=2$ (*Krüppel*) in nucleus $n=10$ for all nine timepoints t_e ($e=1, 2, \ldots, 9$). The blue curves show the trajectory functions $x_{n=10,g=2}(t)$ determined using cubic spline interpolation with smoothing parameter values $p_2=0$, 0.01, and 1

Assigning ON/OFF Gene States

Having computed the velocity, FIGR ascertains whether a gene g is ON or OFF at a particular time point as follows. If the absolute value of the velocity v_g is greater than *velocity threshold* v_g^c, then the ON/OFF state of the gene $y_g = \pm 1$ is determined from the sign of the velocity. That is, a steep upward slope is interpreted as ON, and a steep plunge is interpreted as OFF.

If a gene started in a steady state, or it converged to a steady state, then its *velocity* will be close to zero, and the previous criterion is not useful. In that case, FIGR examines the *expression* itself. If the expression x_g is higher than the *expression threshold* x_g^c, FIGR assumes that gene g is ON, otherwise it is OFF. In summary, the ON/OFF state y_g is computed using

$$y_g = \begin{cases} \operatorname{sgn}\dfrac{dx_g}{dt}, & \text{if } \left|\dfrac{dx_g}{dt}\right| \geq v_g^c \\ \operatorname{sgn}(x_g - x_g^c), & \text{if } \left|\dfrac{dx_g}{dt}\right| < v_g^c. \end{cases} \tag{6}$$

Figures 2 and 3 illustrate the classification of data points along a trajectory as ON/OFF based on the velocity or expression threshold, respectively.

2.4.2 Inference of Gene Circuit Parameters

We distinguish between the *regulatory parameters* T_{gf} and h_g, which determine how the genes are regulated, and the *kinetic parameters* R_g and λ_g, which determine the maximum expression and half-life of the genes.

Binary Classification to Infer the Regulatory Parameters

At this stage of the algorithm we have a set of "points" in gene expression space, one point for each observation, with coordinates x_{nge}. Each point is associated with an ON/OFF state y_{nge}. From this

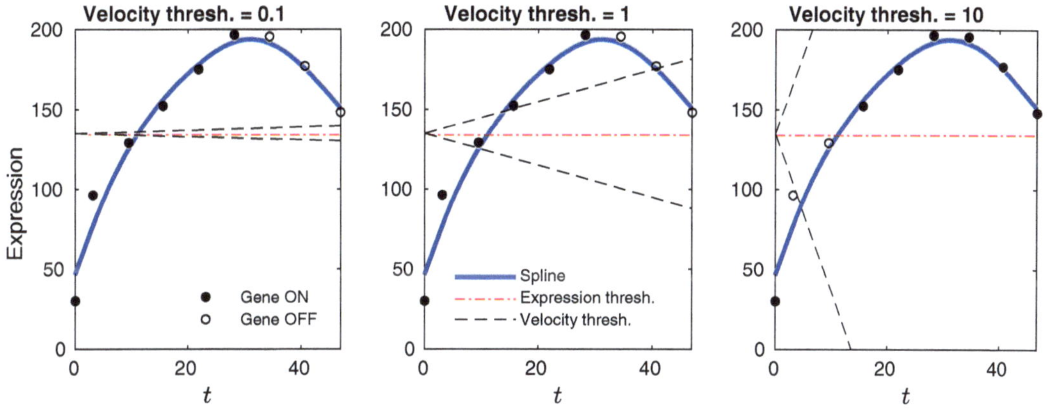

Fig. 2 The effect of velocity threshold parameter v_g^c. The circles and blue curves show data and splines respectively for gene $g = 2$ (Krüppel) in nucleus $n = 20$. **Good choice:** In the first two panels, the velocity threshold v_g^c is set to a low level, or high sensitivity, indicated by shallow black dashed lines. FIGR applies the velocity criterion to correctly determine that the gene is ON (filled circles) during the first six timepoints ($t < 30$), and has switched OFF (empty circles) during the last three timepoints. **Bad choice:** In the third panel, the slope threshold is set too high, so that the expression criterion is applied by FIGR for all the timepoints except the first one. With the expression threshold at $x_g^c = 135$ (red dash-dotted line), FIGR marks the gene—incorrectly—as OFF whenever $x_g < 135$ and ON when $x_g > 135$

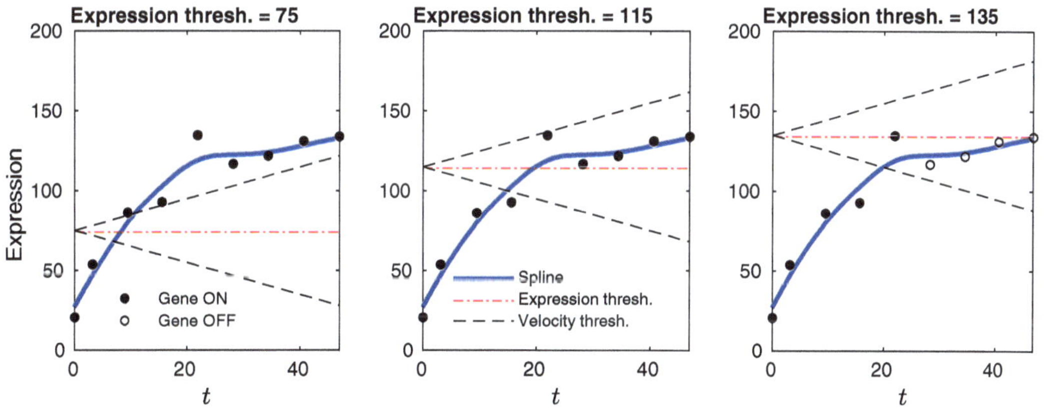

Fig. 3 The effect of the expression threshold parameter x_g^c. Here $v_g^c = 1$ (black dashed line) and $x_g^c = 75$, 115, or 135 (red dash-dotted lines). Filled or open circles indicate data points where gene $g = 2$ (Krüppel) is classified as ON or OFF in nucleus $n = 10$, respectively. Gene expression is roughly constant at the last four timepoints so that $|v_g| < v_g^c$ and FIGR utilizes the expression criterion to classify those points. The first two panels show valid settings for the expression threshold. In the third panel the expression threshold is set too high and the last four timepoints are incorrectly classified as OFF

point on we lump the nucleus index n and time point index e together to form a composite datapoint index $p = 1, \cdots, P$, where $P = N \times N_t$ is the total number of observations per gene. Thus the data may be described as x_{pg} and y_{pg}. Furthermore, we view x_{pg} as a set of vectors $\mathbf{x}_p$.

For each gene g, FIGR attempts to find a hyperplane $\mathbf{T}_g \cdot \mathbf{x} + h_g = 0$ in state space that separates the ON/OFF points as cleanly as possible. The parameters of this hyperplane $\mathbf{T}_g$ and h_g are inferred using either logistic regression or support vector machine (SVM) classification methods. See Fehr et al. for details [12].

Inference of the Kinetic Parameters

In gene circuit models without diffusion (Eq. 1), the velocities $v_{pg} = \frac{dx_{ng}}{dt}(t_e)$ satisfy the system of equations

$$v_{pg} = \begin{cases} R_g - \lambda_g x_{pg} & \text{if } y_{pg} = +1, \\ -\lambda_g x_{pg} & \text{if } y_{pg} = -1. \end{cases} \tag{7}$$

The velocities are known, having been computed by differentiating the spline fits (Sect. 2.4.1), while the concentrations of the gene products, x_{pg}, are also known. Therefore, the above linear system is overdetermined, having P equations and only two unknowns, R_g and λ_g. The best estimates of R_g and λ_g can be determined by linear least-squares regression. In practice, the error in the spline, and hence in v_{pg}, is the largest when a gene is switching states. In order to avoid these high-error points, FIGR excludes a user-configurable number of time points nearest to switching events (`Rld_tsafety`; Table 1). This method is implemented as the "`slope`" method of FIGR. Alternatively, R and λ can also be determined by fitting the solutions of Equation 1 to the concentration data ([12], "`conc`" method).

In gene circuit models with diffusion (Eq. 4), the diffusion constants D_g have to be estimated in addition to R_g and λ_g. FIGR exploits so-called kink solutions to the gene circuit equations [48] to estimate the kinetic parameters. Let gene g be ON in a domain $[l, r]$, where l and r are the positions of cells in a one-dimensional row, so that there is net diffusion out of the domain into surrounding OFF nuclei. Then the balance of synthesis, degradation, and diffusion will establish a stable gradient,

$$x_{ng}(t) = \begin{cases} \dfrac{R_g}{2\lambda_g} e^{-\gamma_g (n-r)}, & \text{if } n > r, \\ \dfrac{R_g}{2\lambda_g} e^{-\gamma_g (l-n)}, & \text{if } n < l, \end{cases} \tag{8}$$

outside the domain. Here, $\gamma_g = \sqrt{\frac{\lambda_g}{D_g}}$ is the characteristic length scale of the gradient at steady state. FIGR fits the observed one-dimensional spatial pattern to the kink equations (Eq. 8) for $n > r$ and $n < l$ using MATLAB's `lsqnonlin` function. This is implemented as the "`kink`" method of FIGR [12].

2.4.3 Refinement

Classification-based inference, as described above, optimizes GRN model parameters T_{gf}, h_g, R_g, and λ_g to fit gene product velocities $v_{ng}(t_e)$ as functions of gene expression $x_{ng}(t_e)$. Therefore, it fits the

Table 1
User-defined options and parameters utilized in FIGR code. Parameters that modify FIGR's behavior. See Sect. 2 for their meaning. The rightmost column lists option values used in example 3 of the FIGR package, which was used to infer the *Drosophila* gap gene network

Option	Description	Acceptable values	Value in Example 3
Determining regulatory parameters			
`splinesmoothing` (p_c)	Spline smoothing parameter for determining velocities	[0, 1]	0.01
`slopethresh` (v_g^c)	Velocity threshold for determining on/off state	≥ 0	1
`exprthresh` (x_g^c)	Expression threshold for determining on/off state	>0	100
Determining kinetic parameters			
`Rld_method`	Method for determining the kinetic parameters	`slope` or `kink` or `conc`	`kink`
Determining kinetic parameters by "slope" method			
`Rld_tsafety`	Number of unreliable velocity estimates (for points around maxima and minima of time series) to exclude	≥ 0	NA
Determining kinetic parameters by "kink" method			
`spatialsmoothing`	Spline smoothing parameter for identifying spatial expression domains and border positions	[0, 1]	0.5
`minborder_expr_ratio`	Expression threshold above which points are included in fitting the kink equations, expressed as fraction of maximum domain expression	(0, 1)	0.01

(continued)

Table 1
(continued)

Option	Description	Acceptable values	Value in Example 3
Recomputing gene trajectories for validation			
`synthesisfunction`	Switch-like function for synthesis	`synthesis_heaviside` or `synthesis_sigmoid_sqrt`	`synthesis_sigmoid_sqrt`
`ODEsolver`	MATLAB ODE solver	Supported MATLAB solvers	`ode45`
`ODEAbsTol`	Tolerance for MATLAB ODE solver	Arbitrary	10^{-3}

differential equations directly to the data. This is *fast* and *reproducible*, because it corresponds to a convex optimization problem that has a unique solution and also does not require repeatedly solving large systems of coupled ODEs as is done in all other approaches for inferring gene circuits [1, 9, 24, 25]. There are however two situations in which further optimization is warranted. First, if the goal is to infer a gene circuit model with a sigmoid regulation-expression function (Eq. 2), so that genes switch smoothly between ON and OFF, then fine tuning the parameters will yield better fits. Second, if a gene circuit model with diffusion is desired, further refinement is beneficial since the correspondence between the velocity and the ON/OFF state is not exact in the presence of diffusion.

During refinement, the parameter values estimated using classification serve as a starting point for an unconstrained local search using the Nelder-Mead algorithm implemented in MATLAB's `fminsearch` function. In contrast to classification-based inference, which fit the differential equations, refinement fits the solutions of the DEs, $x_{ng}(t)$, as functions of initial conditions $\tilde{x}_{ng}(t = 0)$. The cost function is

$$\chi^2 = \sum_{ngt} (\tilde{x}_{ng}(t) - x_{ng}(t))^2. \tag{9}$$

Here, $\tilde{x}_{ng}(t)$ are data and $x_{ng}(t)$ are solutions of Equation (4) or Equation (1) depending on whether the gene circuit incorporates gene product diffusion or not respectively. Since refinement requires repeatedly solving the ODEs given a set of initial conditions $x_{ng}(t=0)$, it is important to choose an appropriate ODE solver. If the gene circuit is based on the Heaviside regulation-expression function (Eq. 3), which is discontinuous, we recommend using a lower-order ODE solver such as a third-order Runge-Kutta algorithm implemented in MATLAB as `ode23`. If, on the other hand, the gene circuit is based on the continuous sigmoid regulation-expression function (Eq. 2), we recommend using a higher-order ODE solver such as MATLAB's `ode45`.

3 Materials

FIGR is a supervised classification-based method for inferring dynamical models of gene regulatory networks implemented in MATLAB and publicly available at https://github.com/mlekkha/FIGR. FIGR requires MATLAB R2018 or newer. The example of inference of the gap gene network acting during *Drosophila* segmentation described in Methods is available at https://github.com/mlekkha/FIGR/blob/master/example3_fly.m.

4 Methods

This protocol describes the work flow of FIGR and navigates the user through the basic functions used during the inference of dynamical GRNs. It guides the user in tuning FIGR-specific parameters for optimal results and proposes ways of visualizing and interpreting the output. Since the inference process requires some amount of repetition to identify the optimal values for the thresholds (Sect. 2.4.1), writing a MATLAB script that can be run repeatedly can help streamline the procedure. The FIGR distribution contains an example script, `example3_fly.m`, that can serve as a starting point for the user's script. The FIGR distribution is under active development and receives new features and bug fixes regularly. We refer the user to the `README.md` file in the FIGR package for these updates.

4.1 Choosing the GRN and Experimental Design

We assume that the user has chosen a set of G genes to model based on the goals of the project and prior genetic and biochemical evidence. We also assume that the user has either conducted a time series experiment themselves or has downloaded a dataset where the RNA or protein expression of the G genes has been measured at N_t time points and N conditions or, in a spatially extended system, in N cells. Although designing experiments is outside the scope of this protocol, one important consideration is worth mentioning. It is important to ensure that the inference problem is not *underdetermined*, that is the number of parameters is not greater than the number of data points. There is a risk of *overfitting*—fitting to the peculiarities rather than the general features of a dataset—in an underdetermined problem, which results in models with poor predictive ability [16]. The number of parameters is either $(G-G_e)(G+3)$ or $(G-G_e)(G+4)$ for cell-autonomous or spatially extended systems respectively, where G_e is the number of upstream regulators. The number of datapoints is $(G-G_e)\times(N_t-1)\times N$. The gap gene problem (Sect. 2.2), with $G=7$, $G_e=3$, $N_t=9$, and $N=58$, has 44 free parameters and 1,856 data points and is, therefore, a very well-constrained inference problem. The number of genes that can be modeled is limited by the amount of data available and one must increase the number of time points or conditions/cells or both in order to increase the size of the GRN.

4.2 Obtaining FIGR

FIGR may be downloaded from https://github.com/mlekkha/FIGR by clicking the `Clone or download` button and choosing "`Download ZIP`". Decompressing the ZIP file will yield a directory, `FIGR-master`, which can be placed in a location of the user's choosing. In order to access the FIGR code, the user must navigate to the directory in MATLAB by executing `cd` at the MATLAB prompt.

```
cd '/path/to/FIGR-master'
```

Alternatively, the user may add the directory to MATLAB path using the function `addpath( )`.

```
addpath('/path/to/FIGR-master')
```

In the commands above `/path/to/FIGR-master` should be replaced by the location of the FIGR directory. If the user does not add the directory to the MATLAB path, then all data files must be places in the FIGR directory.

4.3 Defining FIGR Options

FIGR's behavior can be modified by several parameters (Table 1), which should be assigned values at the beginning of the inference script. The first seven parameters influence the behavior of the main FIGR function, `infer( )`. In the present version of FIGR, `splinesmoothing`, `slopethresh`, and `exprthresh` are assumed to be the same for all genes; future releases will allow velocity and expression thresholds to be set on a per-gene basis. The other options influence the behavior of the utility function `computeTrajs( )` that solves the ODEs to obtain gene trajectories. `computeTrajs( )` is called by `initChiSquare( )` and by the refinement routine `refineFIGRParams( )`.

4.4 Supplying Input Data

As described in Sect. 2, FIGR accepts any time-series data derived from experiments where mRNA or protein concentrations have been measured over time. FIGR's main MATLAB routine, `infer( )`, requires as arguments an N_t-element input vector `tt` of time points and an $N \times N_t \times G$ array `xntg` of expression levels $x_{ng}(t_e)$. These data are supplied to FIGR in two files, one for the time points and the other for the expression data.

4.4.1 Time Point Data File Format

The time point data file contains header information describing a 2D array of dimensions $N_t \times 1$, which is equivalent to an N_t-element column vector in MATLAB, followed by a list of timepoints delimited by white space in one row.

Time point data file format

```
Line 1:2 dims
Line 2:N_t elems
Line 3:1 elems
```

$$\text{Line 4: } \overbrace{t_1 \quad t_2 \quad \cdots \quad t_{N_t}}^{N_t \text{ timepoints}}$$

For example, in the gap gene problem $N_t = 9$ and the timepoints file, available as `fly_tt.txt` in the FIGR directory, takes the following form:

Gap gene problem time point data file

```
2 dims
9 elems
1 elems
0.00 3.12 9.38 15.62 21.88 28.12 34.38 40.62 46.88
```

4.4.2 Gene Expression Data File Format

The expression data are provided to FIGR in a file in which three-dimensional data are flattened into two dimensions. The number of dimensions are specified on the first line. Lines 2–4 specify, in order, the number of conditions/cells, the number of time points, and the number of genes. The expression data are provided from line 5 onward, each line containing expression in N conditions or cells separated by spaces. The t_e index varies next and the g index varies the slowest.

Gene expression data file format

```
Line    1:3 dims
Line    2:N elems
Line    3:N_t elems
Line    4:G elems
Line 5-end:
```

$$
\overbrace{\hphantom{x_{11}(t_1)\quad x_{21}(t_1)\quad \cdots \quad x_{N1}(t_1)}}^{N \text{ conditions/cells}}
$$

$$
\begin{array}{cc}
\begin{array}{c} g=1 \\ N_t \text{ timepoints} \end{array} & \left\{\begin{array}{cccc}
x_{11}(t_1) & x_{21}(t_1) & \cdots & x_{N1}(t_1) \\
x_{11}(t_2) & x_{21}(t_2) & \cdots & x_{N1}(t_2) \\
\vdots & \vdots & \ddots & \vdots \\
x_{11}(t_{N_t}) & x_{21}(t_{N_t}) & \cdots & x_{N1}(t_{N_t})
\end{array}\right. \\
\begin{array}{c} g=2 \\ N_t \text{ timepoints} \end{array} & \left\{\begin{array}{cccc}
x_{12}(t_1) & x_{22}(t_1) & \cdots & x_{N2}(t_1) \\
x_{12}(t_2) & x_{22}(t_2) & \cdots & x_{N2}(t_2) \\
\vdots & \vdots & \ddots & \vdots \\
x_{12}(t_{N_t}) & x_{22}(t_{N_t}) & \cdots & x_{N2}(t_{N_t})
\end{array}\right. \\
 & \vdots \\
\begin{array}{c} g=G \\ N_t \text{ timepoints} \end{array} & \left\{\begin{array}{cccc}
x_{1G}(t_1) & x_{2G}(t_1) & \cdots & x_{NG}(t_1) \\
x_{1G}(t_2) & x_{2G}(t_2) & \cdots & x_{NG}(t_2) \\
\vdots & \vdots & \ddots & \vdots \\
x_{1G}(t_{N_t}) & x_{2G}(t_{N_t}) & \cdots & x_{NG}(t_{N_t})
\end{array}\right.
\end{array}
$$

In the gap gene problem, the number of cells (nuclei) $N=58$, the number of time points $N_t=9$, and the number of genes $G=7$.

$n = 1$ and $n = 58$ correspond to the nuclei at 35% and 92% egg length along the anteroposterior axis, respectively. The expression data file, which is available as `fly_xntg.txt` in the FIGR directory, takes the following form:

Gap gene problem expression data file

```
3 dims

58 elems

9 elems

7 elems

89.95 88.49 86.03 ... 1.76

.

.

.

144.47 140.80 138.00 ... 10.90

.

.

.

0.00 0.00 0.00 ... 49.76

.

.

.

0.66 0.00 0.55 ... 107.69
```

4.4.3 Reading the Files into MATLAB Workspace

The data files may be read in using built-in MATLAB routines, or using the `readArray` utility function provided in `example3_-fly.m`. In that example, the gene concentrations and timepoints are stored in the `xntgEXPT` and `tt` variables using

```
xntgEXPT = readArray ('fly_xntg.txt');
tt       = readArray ('fly_tt.txt');
```

4.5 Inferring the GRN Using `Infer( )`

Once the data have been imported, one may proceed straight to the inference of the parameters. The FIGR methodology described in Sect. 2.4 is implemented in the function `infer`, which is called with the FIGR options structure `opts`, the expression array `xntg`, time point array `tt`, and the number of genes in the GRN (G) `numGenes`, as arguments.

```
[grn, diagnostics] = infer (opts, xntg, tt, numGenes);
```

`infer` returns a structure, `grn`, with five fields, `Tgg`, `hg`, `Rg`, `lambdag`, and `Dg`, corresponding to the GRN parameters inferred by FIGR. Optionally, `infer` also returns a structure, `diagnostics`, which contains debugging information. The diagnostic structure currently includes intermediate results such as the ON/OFF states $y_{ng}(t_e)$. In future releases of FIGR, the diagnostics structure will include warning and error codes.

4.6 Optional Refinement of the GRN Using `refineFIGRParams( )`

If the regulation-expression function is sigmoid or if the model includes diffusion, the GRN parameters returned by `infer( )` may be optionally refined (Sect. 2.4.3). `refineFIGRParams( )` takes the inferred GRN structure `grn`, a flattened two-dimensional matrix containing the expression data `xntgFLAT`, and the time point array `tt` as arguments and returns the refined parameter structure `grnREF`.

```
grnREF = refineFIGRParams(grn, xntgFLAT, tt)
```

The flattened expression data matrix has dimensions $NN_t \times G$ and may be generated using MATLAB's inbuilt `reshape( )` function. For example, in the gap gene problem, in which $NN_t = 522$ and $G = 7$, the following command would generate the flattened matrix.

```
xntgFLAT = reshape(xntgEXPT, 522, 7);
```

The options for the Nelder-Mead optimization performed by `refineFIGRParams( )` may be set using MATLAB's `optimset( )` function. It is important to set the stopping criteria `MaxFunEvals` and `MaxIter` which determine the maximum number of cost function evaluations and iterations, respectively. The default choice is to set `MaxFunEvals` and `MaxIter` to 200 times the number of parameters, which may be determined by flattening the parameter structure `grn` into a vector using `packParams()` and computing its length.

```
packed_paramvec = packParams(grn, numGenes);
optimopts = optimset( 'Display', 'Iter', ...
                      'MaxFunEvals', 200*length(packed_paramvec),
    ...
                      'MaxIter', 200*length(packed_paramvec));
```

Finally, it is worth mentioning that `refineFIGRParams()` is the most computationally intensive part of FIGR and a tenfold speedup may be achieved by compiling the function. The instructions for doing so are provided in `README.md` of the FIGR distribution. The compiled function should be called (Sect. 4.8) instead of `refineFIGRParams()` in order to avail oneself of the speedup.

4.7 Simulating and Analyzing the GRN

4.7.1 Evaluating How Well the Model Fits the Data

The goodness of fit—how well the obtained model fits the data—is crucial for determining whether the model is suitable for further analysis or not. One important way of evaluating the goodness of fit is by computing the root mean square (RMS) discrepancy

$$\mathrm{RMS} = \sqrt{\frac{\chi^2(\tilde{x}_{ng}(t), x_{ng}(t))}{N(N_t - 1)(G - G_e)}}, \tag{10}$$

where $\chi^2(\tilde{x}_{ng}(t), x_{ng}(t))$ is the squared difference between data $\tilde{x}_{ng}(t)$ and model output $x_{ng}(t)$ (Eq. 9). The RMS normalizes the squared difference to the number of observations $N(N_t - 1)(G - G_e)$, allowing for the comparison of models derived from different datasets. For example, the gap gene circuit obtained by FIGR has an RMS of 13.29 [12], which is about ~5% of peak expression. The discrepancy is therefore well below the level of experimental error ~10% [42], suggesting that the gene circuit model fits the data well. The RMS may be calculated by calling the `initChiSquare()` function as follows:

```
[xntgREF] = computeTrajs (opts, grn, xntgEXPT, tt);
```

Above, `opts` is the FIGR options structure, `grn` is the parameter structure containing the inferred values, `xntgEXPT` is the 3D array of expression data, and `tt` is the time point array. `initChiSquare()` returns the χ^2 score and RMS in `init_chisq` and `init_rms`, respectively.

Although the RMS score is an important criterion for evaluating the goodness of fit, it averages the discrepancies over all time-points and conditions or cells. It is possible for a model to have a low RMS but have a large discrepancy in a subset of cells or time-points that are biologically significant. In order to avoid selecting a model with potentially fatal flaws, it is important to visualize and compare model output with data in time and/or space and not rely exclusively on the RMS. The trajectories are calculated by solving the differential equations starting from the initial conditions. This is implemented in the `computeTrajs()` function, which takes the FIGR options structure, `opts`, the inferred parameters structure `grn`, the 3D array of expression data `xntgEXPT`, and the time point array `tt` and returns a 3D array, `xntgREF`, containing the model output.

```
[init_chisq, init_rms] = initChiSquare(opts, grn, xntgEXPT, tt);
```

The computed trajectories may be compared with the experimental data by using MATLAB's extensive plotting functionality to visualize expression with respect to time (Fig. 4), to spatial/condition index (Fig. 5), or to both (Fig. 6)

4.7.2 Inferring Genetic Interactions in the GRN

The genetic interconnectivity matrix **T** is particularly important for understanding the architecture of the GRN and the causal relationships between the genes [21, 32, 37]. The GRN parameters returned by `infer()` or `refineFIGRParams()` may be inspected from the MATLAB command line or in the variable inspector.

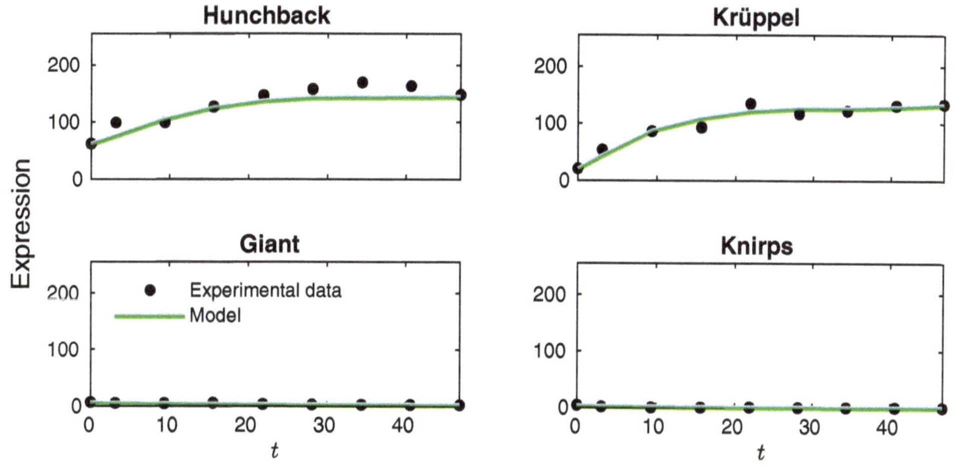

Fig. 4 Comparison of gap gene data and model output in time. Data and model output are from nucleus $n = 10$. The black dots show experimental gene expression data $\tilde{x}_{n=10,g}(t)$. The green curves show gene expression computed by solving the ODEs ($x_{n=10,g}(t)$) using the parameter values inferred by FIGR

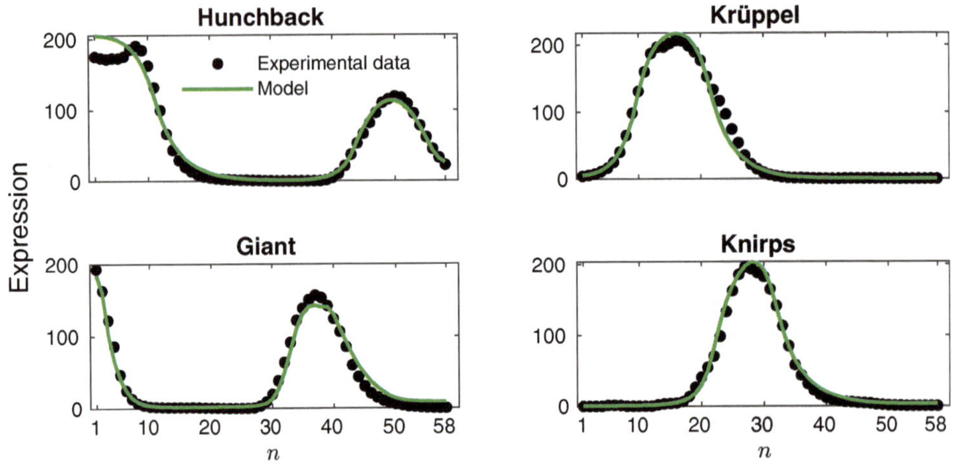

Fig. 5 Comparison of gap gene data and model output with nuclear position at time point $t_8 = 40.62$ **min.** The black dots show experimental gene expression data $\tilde{x}_{ng}(t = t_8)$. The green curves show gene expression patterns computed by solving the ODEs ($x_{ng}(t = t_8)$) using the parameter values inferred by FIGR

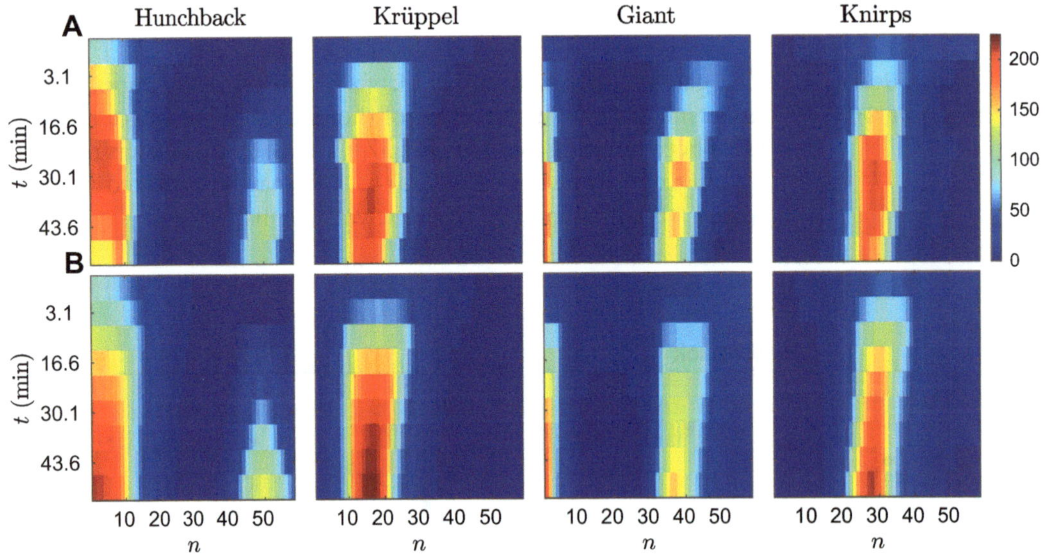

Fig. 6 Comparison of gap gene data and model output in space and time. (**a**) Experimental expression data $\tilde{x}_{ng}(t)$ for all four gap genes are visualized as heatmaps with position $n = 1, 2, \ldots, 58$ on the horizontal axis and time $t = 0, 3.12, \ldots, 46.88$ on the vertical axis. (**b**) Model output $x_{ng}(t)$ computed by solving the ODEs using the parameter values inferred by FIGR. Hot (cold) colors represent high (low) expression

```
grnREF =
          Tgg               hg          Rg        lambdag         Dg

      ------------     -------     ------     --------     --------

      [1x7 double]      -3.752     10.498     0.044051      0.44232

      [1x7 double]     -4.7434     14.892     0.056893      0.35558

      [1x7 double]     -11.069     7.8378       0.0255     0.076815

      [1x7 double]     -7.2588     11.854     0.038369      0.29139


grnREF.Tgg =
                 Hunchback       Kruppel        Giant         Knirps        Bicoid      Caudal        Tailless
                 ----------     ---------     ---------     ----------     --------    --------     ----------

Hunchback         0.036058     -0.011437     0.0029486       -0.18655     0.011469     0.03825     -0.0054774
Kruppel         -0.0038011      0.014416      -0.46074      -0.030918      0.14385    0.065907       -0.17654
Giant             -0.04855      -0.56805      0.013625       0.031512      0.56387    0.089211      -0.056328
Knirps            -0.82928      -0.02585     -0.025507     -0.0025754        1.258    0.028393       -0.29289
```

It is more intuitive to represent the parameters visually as network graphs (Fig. 7), which often yield biological insight into GRN function. Software such as Cytoscape [40] may be used to plot the network graph based on the parameter values. In Fig. 7, circles represent the upstream regulators, squares represent the gap genes, while blue and red arrows represent activation and repression, respectively. The color intensity of the edges varies between

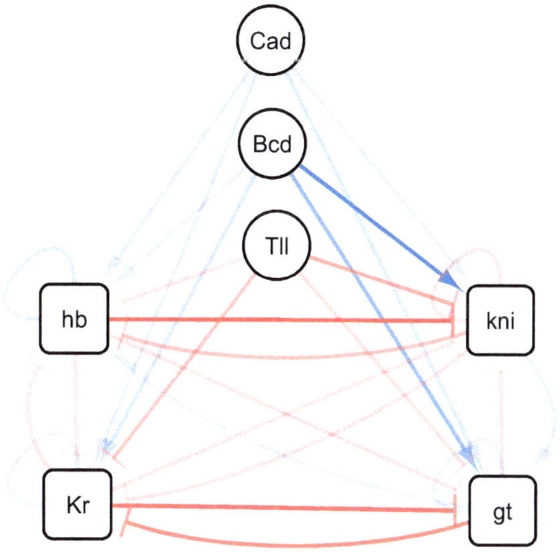

Fig. 7 *Cytoscape* visualization of the interconnectivity parameters T_{gf} of the gap gene network

60 and 255 and is proportional to the interaction strength. Visualizing the gap gene GRN in this manner (Fig. 7) shows that there is strong cross-repression between pairs of gap genes expressed in mutually exclusive domains, *Kr* and *gt*, and, *hb* and *kni*, a network motif referred to as "alternating cushions" [26]. A key dynamical feature reproduced by FIGR output is the movement of posterior gap gene domains to the anterior during cycle 14 (Fig. 6; [22]). These shifts have been understood to occur in a cell-autonomous manner [22] due to asymmetric weak repression between gap genes expressed in adjacent domains, Kr ⊢ Kni ⊢ Gt [31], another motif apparent in the gene circuit graph (Fig. 7).

Even though the interconnectivity matrix and its representation as a graph are static in time and space, the regulation of each gene, in fact, varies with time and space since the concentrations of the regulators vary in time and space. Although it is outside the scope of this protocol, we mention that it is possible to visualize the regulatory connections as they vary in time or space to arrive at a more fine-grained understanding about how each gene is regulated in the GRN [22, 32]. A final aspect of regulatory analysis is that the value of each inferred parameter has some associated uncertainty arising from the uncertainty in the observations as well as the sensitivity of the model output to the parameter. Estimating this uncertainty can help with interpreting the GRN structure so that strong conclusions need not be drawn about highly uncertain parameters. *Parameter identifiability analysis* [3] is a technique for estimating the uncertainty in the parameter values and may be performed optionally during the analysis of the model.

4.7.3 Perturbations and Predictions

One of the most important applications of dynamical models is that they can be used to predict the behavior of GRNs under environmental or genetic perturbations. There is not a standard recipe for simulating perturbations since the precise nature of the perturbation and how to simulate it depends on the biological question being asked. For example, a knockout of gene $g = k$ may be simulated by setting its synthesis rate to 0, $R_k = 0$, and setting the initial concentration to 0 as well, $\tilde{x}_{nk}(t = 0) = 0, \quad \forall n$. In the gap gene network, gene circuits have been used to simulate embryo-to-embryo variation in upstream regulator spatial distribution [32], the effect of different embryo lengths [32, 54], and the effect of upstream regulator dose [54].

4.8 Example Script for Inferring and Simulating the Gap Gene GRN

Here we demonstrate a full-blown example of GRN inference with the variable definitions and sequence of function calls necessary for FIGR execution. The steps below can be incorporated into a single script or executed separately on the MATLAB command line, although the former streamlines the work flow. Users may also modify, according to their needs, the example scripts that are provided in the FIGR package. This example is based on `example3_fly.m`.

1. Declare global structures and variables for FIGR-specific options, ODE solver options, and optimization options used in the refinement process.

```
global opts;        % Structure containing options determining FIGR's behavior
global ODEopts;     % Structure containing options for ODE solvers
global optimopts; % Structure containing options for optimization
```

2. Set global FIGR-specific options and parameters.

```
opts = struct( 'debug', 0, ...
               'slopethresh', 1.0, ...
               'exprthresh', 100.0, ...
               'splinesmoothing', 0.01, ...
               'spatialsmoothing', 0.5, ...
               'minborder_expr_ratio', 0.01, ...
               'Rld_method', 'kink', ...
               'Rld_tsafety', 3, ...
               'synthesisfunction', 'synthesis_sigmoid_sqrt', ...
               'ODEAbsTol', 1e-3, ...
               'ODEsolver', 'ode45');
```

The `debug` option can take values between 0 (no debugging messages) and 2 (verbose debugging mes sages), and the optional parameter `geneNames` stores the names of the genes, while the rest of the options are described in Sect. 4.5 and Table 1. The optimal values of some of these parameters, such as `splinesmoothing`, `slopethresh`, and `exprthresh`, depend on the problem and are chosen by inspecting intermediate steps of the algorithm (Sect. 2.4.1).

3. Set ODE options.

   Since MATLAB's ODE solvers are adaptive, the tolerance of the solutions must be specified. In this example, the absolute tolerance is set in the `ODEopts` structure. However, the user may use relative tolerance (`RelTol`) if so desired.

```
ODEopts = odeset('AbsTol', opts.ODEAbsTol);
```

4. Specify the number of genes that are not upstream regulators.

```
numGenes = 4;
```

5. Read input data.

```
xntgEXPT = readArray (file);
tt       = readArray (file);
```

   The format of input files is described in Sect. 4.4.

6. Infer GRN parameters.

```
[grnFIGR, diagnostics] = infer (opts, xntgEXPT, tt, numGenes);
```

7. Refine GRN. This optional step may be performed when inferring models that use the sigmoid regulation-expression function and/or include diffusion (Sect. 2.4.3). The gap gene model (Eq. 4) meets both criteria and is refined here. The first step in refinement is to set the stopping criteria, `MaxFunEvals` and `MaxIter`, in the optimization options as 200 times the number of parameters (see Sect. 4.6 for details).

```
packed_paramvec = packParams(grnFIGR, numGenes);
optimopts = optimset( 'Display', 'Iter', ...
                      'MaxFunEvals', 200*length(packed_paramvec
    ), ...
                      'MaxIter', 200*length(packed_paramvec));
```

   The `refineFIGRParams( )` function is called after setting the optimization options. Running a compiled version (`MEX`) of this function leads to a significant speed up. Here, we run the `MEX` if it exists, otherwise we run the interpreted MATLAB code.

```
if (exist('refineFIGRParams_mex') == 3)
    disp('MEX file for refinement found. Running compiled code
   ...');
    grnREF = refineFIGRParams_mex(grnFIGR, xntgFLAT, tt);
else
    fprintf(1, ['MEX file for refinement ' ...
    '(refineFIGRParams_mex.{mexa64/mexmaci64/mexw64})' ...
    ' not found.\n See README.md for instructions for compiling
    the ' ...
    ' MEX file.\n\n Running interpreted (but slower) .m code...
   ']);
    grnREF = refineFIGRParams(grnFIGR, xntgFLAT, tt);
end
```

8. Calculate model output based on the inferred parameters.

```
[xntgREF] = computeTrajs (opts, grnREF, xntgEXPT, tt);
```

9. Evaluate the goodness of fit of the obtained model (Sect. 4.7). One may compute the RMS score of the inferred model using the `initChiSquare( )` function.

```
[init_chisq, init_rms] = initChiSquare(opts, grn, xntgEXPT, tt);
```

Another way of evaluating model accuracy is by comparing plots of experimental and recomputed trajectories in time (Fig. 4) or space (Fig. 5). As an example, one may plot the spatial expression pattern at time point $t_7 = 34.375$ min for $g = 4$ as follows:

```
plot([1:58], xntgEXPT(:,7,4), 'or'); % plot data, there are 58
   nuclei
hold on;                              % next plot will be
   overlaid
plot([1:58], xntgREF(:,7,4), 'b');   % plot model output
```

Similarly, the trajectory of gene $g=2$ in the 10th cell may be plotted as follows:

```
plot(tt, xntgEXPT(10,:,2), 'or'); % plot data, tt is timepoints
hold on;                          % next plot will be overlaid
plot(tt, xntgREF(10,:,2), 'b');   % plot model output
```

The third example shows how to plot the model output for $g=2$ as a 2D space-time heatmap.

```
imagesc(flipud(xntgREF(:,:,2)'), [0 225]);
```

Note 1: *FIGR implements a novel approach for GRN inference using binary classification that results in a considerable speedup over global nonlinear optimization techniques such as simulated annealing or genetic algorithms. Whereas global nonlinear optimization techniques require parallel computing for all but the smallest GRNs, FIGR runs in minutes on consumer desktop hardware. The increased computational efficiency is necessary for inferring and modeling larger (>10 genes) GRNs. Finding optimal models that recapitulate the data is usually an iterative process and therefore the near real-time feedback provided by FIGR further speeds up the inference procedure. Another advantage resulting from FIGR's efficiency is that technical knowledge of parallel computing or access to high-performance computing facilities are not requirements, making it relatively easy to adopt.*

Bibliography

1. Abdol AM, Cicin-Sain D, Kaandorp JA, Crombach A (2017) Scatter search applied to the inference of a development gene network. Computation 5(2). https://doi.org/10.3390/computation5020022. https://www.mdpi.com/2079-3197/5/2/22
2. Akam M (1987) The molecular basis for metameric pattern in the *Drosophila* embryo. Development 101:1–22
3. Ashyraliyev M, Jaeger J, Blom JG (2008) Parameter estimation and determinability analysis applied to drosophila gap gene circuits. BMC Syst Biol 2:83. https://doi.org/10.1186/1752-0509-2-83
4. Ashyraliyev M, Siggens K, Janssens H, Blom J, Akam M, Jaeger J (2009) Gene circuit analysis of the terminal gap gene huckebein. PLoS Comput Biol 5(10):e1000548. https://doi.org/10.1371/journal.pcbi.1000548
5. Balaskas N, Ribeiro A, Panovska J, Dessaud E, Sasai N, Page KM, Briscoe J, Ribes V (2012) Gene regulatory logic for reading the sonic hedgehog signaling gradient in the vertebrate neural tube. Cell 148(1–2):273–284. https://doi.org/10.1016/j.cell.2011.10.047
6. Bonzanni N, Garg A, Feenstra KA, Schütte J, Kinston S, Miranda-Saavedra D, Heringa J, Xenarios I, Göttgens B (2013) Hard-wired heterogeneity in blood stem cells revealed using a dynamic regulatory network model. Bioinformatics 29(13):i80–i88. https://doi.org/10.1093/bioinformatics/btt243
7. Chickarmane V, Enver T, Peterson C (2009) Computational modeling of the hematopoietic erythroid-myeloid switch reveals insights into cooperativity, priming, and irreversibility. PLoS Comput Biol 5(1):e1000268. https://doi.org/10.1371/journal.pcbi.1000268

8. Chu KW (2001) Optimal parallelization of simulated annealing by state mixing. PhD Thesis, Department of Applied Mathematics and Statistics. Stony Brook University
9. Chu KW, Deng Y, Reinitz J (1999) Parallel simulated annealing by mixing of states. J Comput Phys 148:646–662
10. Collombet S, van Oevelen C, Sardina Ortega JL, Abou-Jaoudé W, Di Stefano B, Thomas-Chollier M, Graf T, Thieffry D (2017) Logical modeling of lymphoid and myeloid cell specification and transdifferentiation. Proc Natl Acad Sci U S A 114(23):5792–5799. https://doi.org/10.1073/pnas.1610622114
11. Davidson EH, Rast JP, Oliveri P, Ransick A, Calestani C, Yuh CH, Minokawa T, Amore G, Hinman V, Arenas-Mena C, Otim O, Brown CT, Livi CB, Lee PY, Revilla R, Rust AG, Pan ZJ, Schilstra MJ, Clarke PJC, Arnone MI, Rowen L, Cameron RA, McClay DR, Hood L, Bolouri H (2002) A genomic regulatory network for development. Science 295 (5560):1669–1678. https://doi.org/10.1126/science.1069883
12. Fehr DA, Handzlik JE, Manu, Loh YL (2019) Classification-based inference of dynamical models of gene regulatory networks. G3 (Bethesda) 9(12):4183–4195. https://doi.org/10.1534/g3.119.400603
13. Gursky VV, Kozlov KN, Samsonov AM, Reinitz J (2008) A model with asymptotically stable dynamics for the network of *Drosophila* gap genes. Biophysics (Biofizika) 53:164–176
14. Gursky VV, Panok L, Myasnikova EM, Manu, Samsonova MG, Reinitz J, Samsonov AM (2011) Mechanisms of gap gene expression canalization in the drosophila blastoderm. BMC Syst Biol 5(1):118. https://doi.org/10.1186/1752-0509-5-118
15. Hamey FK, Nestorowa S, Kinston SJ, Kent DG, Wilson NK, Göttgens B (2017) Reconstructing blood stem cell regulatory network models from single-cell molecular profiles. Proc Natl Acad Sci U S A 114 (23):5822–5829. https://doi.org/10.1073/pnas.1610609114
16. Hastie TJ, Tibshirani RJ, Friedman JH (2009) The elements of statistical learning: data mining, inference, and prediction. Springer, New York
17. Hengenius JB, Gribskov M, Rundell AE, Fowlkes CC, Umulis DM (2011) Analysis of gap gene regulation in a 3d organism-scale model of the drosophila melanogaster embryo. PLoS One 6(11):e26797. https://doi.org/10.1371/journal.pone.0026797
18. Hong T, Xing J, Li L, Tyson JJ (2012) A simple theoretical framework for understanding heterogeneous differentiation of cd4+ t cells. BMC Syst Biol 6:66. https://doi.org/10.1186/1752-0509-6-66
19. Huang S, Guo Y, May G, Enver T (2007) Bifurcation dynamics in lineage-commitment in bipotent progenitor cells. Dev Biol 305:695–713
20. Jaeger J (2011) The gap gene network. Cell Mol Life Sci 68(2):243–274. https://doi.org/10.1007/s00018-010-0536-y
21. Jaeger J, Blagov M, Kosman D, Kozlov KN, Manu, Myasnikova E, Surkova S, Vanario-Alonso CE, Samsonova M, Sharp DH, Reinitz J (2004) Dynamical analysis of regulatory interactions in the gap gene system of *Drosophila melanogaster*. Genetics 167:1721–1737
22. Jaeger J, Surkova S, Blagov M, Janssens H, Kosman D, Kozlov KN, Manu, Myasnikova E, Vanario-Alonso CE, Samsonova M, Sharp DH, Reinitz J (2004) Dynamic control of positional information in the early *Drosophila* embryo. Nature 430:368–371
23. Jaeger J, Sharp DH, Reinitz J (2007) Known maternal gradients are not sufficient for the establishment of gap domains in *Drosophila melanogaster*. Mech Dev 124:108–128
24. Kozlov K, Samsonov A (2009) Deep—differential evolution entirely parallel method for gene regulatory networks. In: Malyshkin V (ed) Parallel computing technologies. Springer, Berlin/Heidelberg, pp 126–132
25. Kozlov K, Surkova S, Myasnikova E, Reinitz J, Samsonova M (2012) Modeling of gap gene expression in drosophila kruppel mutants. PLoS Comput Biol 8(8):e1002635. https://doi.org/10.1371/journal.pcbi.1002635
26. Kraut R, Levine M (1991) Mutually repressive interactions between the gap genes *giant* and *Krüppel* define middle body regions of the *Drosophila* embryo. Development 111:611–621
27. Kueh HY, Champhekhar A, Nutt SL, Elowitz MB, Rothenberg EV (2013) Positive feedback between pu.1 and the cell cycle controls myeloid differentiation. Science. https://doi.org/10.1126/science.1240831
28. Laslo P, Spooner CJ, Warmflash A, Lancki DW, Lee HJ, Sciammas R, Gantner BN, Dinner AR, Singh H (2006) Multilineage transcriptional priming and determination of alternate hematopoietic cell fates. Cell 126(4):755–766. https://doi.org/10.1016/j.cell.2006.06.052
29. Levine M, Davidson EH (2005) Gene regulatory networks for development. Proc Natl Acad Sci U S A 102(14):4936–4942. https://doi.org/10.1073/pnas.0408031102

30. Li C, Wang J (2013) Quantifying cell fate decisions for differentiation and reprogramming of a human stem cell network: landscape and biological paths. PLoS Comput Biol 9(8): e1003165 EP –. https://doi.org/10.1371%2Fjournal.pcbi.1003165
31. Manu, Surkova S, Spirov AV, Gursky V, Janssens H, Kim A, Radulescu O, Vanario-Alonso CE, Sharp DH, Samsonova M, Reinitz J (2009) Canalization of gene expression and domain shifts in the *Drosophila* blastoderm by dynamical attractors. PLoS Comput Biol 5: e1000303. https://doi.org/10.1371/journal.pcbi.1000303
32. Manu, Surkova S, Spirov AV, Gursky V, Janssens H, Kim A, Radulescu O, Vanario-Alonso CE, Sharp DH, Samsonova M, Reinitz J (2009) Canalization of gene expression in the *Drosophila* blastoderm by gap gene cross regulation. PLoS Biol 7:e1000049. https://doi.org/10.371/journal.pbio.1000049
33. May G, Soneji S, Tipping AJ, Teles J, McGowan SJ, Wu M, Guo Y, Fugazza C, Brown J, Karlsson G, Pina C, Olariu V, Taylor S, Tenen DG, Peterson C, Enver T (2013) Dynamic analysis of gene expression and genome-wide transcription factor binding during lineage specification of multipotent progenitors. Cell Stem Cell 13(6):754–768. https://doi.org/10.1016/j.stem.2013.09.003
34. Palani S, Sarkar CA (2008) Positive receptor feedback during lineage commitment can generate ultrasensitivity to ligand and confer robustness to a bistable switch. Biophys J 95 (4):1575–1589. https://doi.org/10.1529/biophysj.107.120600
35. Peter IS, Faure E, Davidson EH (2012) Predictive computation of genomic logic processing functions in embryonic development. Proc Natl Acad Sci U S A 109 (41):16434–16442. https://doi.org/10.1073/pnas.1207852109
36. Pietak A, Bischof J, LaPalme J, Morokuma J, Levin M (2019) Neural control of body-plan axis in regenerating planaria. PLoS Comput Biol 15(4):e1006904. https://doi.org/10.1371/journal.pcbi.1006904
37. Reinitz J, Sharp DH (1995) Mechanism of *eve* stripe formation. Mech Dev 49:133–158
38. Reinitz J, Mjolsness E, Sharp DH (1995) Cooperative control of positional information in *Drosophila* by *bicoid* and maternal *hunchback*. J Exp Zool 271:47–56
39. Sánchez L, Thieffry D (2003) Segmenting the fly embryo: a logical analysis of the *pair-rule* cross-regulatory module. J Theor Biol 224:517–537
40. Shannon P, Markiel A, Ozier O, Baliga NS, Wang JT, Ramage D, Amin N, Schwikowski B, Ideker T (2003) Cytoscape: a software environment for integrated models of biomolecular interaction networks. Genome Res 13(11):2498–2504. https://doi.org/10.1101/gr.1239303. https://pubmed.ncbi.nlm.nih.gov/14597658. 14597658[pmid]
41. Shea MA, Ackers GK (1985) The OR control system of bacteriophage lambda. A physical-chemical model for gene regulation. J Mol Biol 181:211–230
42. Surkova S, Kosman D, Kozlov K, Manu, Myasnikova E, Samsonova A, Spirov A, Vanario-Alonso CE, Samsonova M, Reinitz J (2008) Characterization of the *Drosophila* segment determination morphome. Dev Biol 313 (2):844–862
43. Thattai M, van Oudenaarden A (2001) Intrinsic noise in gene regulatory networks. Proc Natl Acad Sci U S A 98:8614–8619
44. Theiffry D, Colet M, Thomas R (1993) Formalization of regulatory networks: a logical method and its automatization. Math Model Sci Comput 2:144–151
45. Tusi BK, Wolock SL, Weinreb C, Hwang Y, Hidalgo D, Zilionis R, Waisman A, Huh JR, Klein AM, Socolovsky M (2018) Population snapshots predict early haematopoietic and erythroid hierarchies. Nature 555(7694):54–60. https://doi.org/10.1038/nature25741
46. Tyson JJ, Baumann WT, Chen C, Verdugo A, Tavassoly I, Wang Y, Weiner LM, Clarke R (2011) Dynamic modelling of oestrogen signalling and cell fate in breast cancer cells. Nat Rev Cancer 11(7):523–532. https://doi.org/10.1038/nrc3081
47. Umulis DM, Serpe M, O'Connor MB, Othmer HG (2006) Robust, bistable patterning of the dorsal surface of the *Drosophila* embryo. Proc Natl Acad Sci U S A 103(31):11613–11618
48. Vakulenko S, Manu, Reinitz J, Radulescu O (2009) Size regulation in the segmentation of *Drosophila*: interacting interfaces between localized domains of gene expression ensure robust spatial patterning. Phys Rev Lett 103 (16):168102
49. Verd B, Clark E, Wotton KR, Janssens H, Jiménez-Guri E, Crombach A, Jaeger J (2018) A damped oscillator imposes temporal order on posterior gap gene expression in drosophila. PLoS Biol 16(2):e2003174. https://doi.org/10.1371/journal.pbio.2003174
50. Volfson D, Marciniak J, Blake WJ, Ostroff N, Tsimring LS, Hasty J (2006) Origins of extrinsic variability in eukaryotic gene expression.

Nature 439(7078):861–864. https://doi.org/10.1038/nature04281

51. Wang J, Xu L, Wang E (2008) Potential landscape and flux framework of nonequilibrium networks: robustness, dissipation, and coherence of biochemical oscillations. Proc Natl Acad Sci U S A 105(34):12271–12276. https://doi.org/10.1073/pnas.0800579105

52. Weston BR, Li L, Tyson JJ (2018) Mathematical analysis of cytokine-induced differentiation of granulocyte-monocyte progenitor cells. Front Immunol 9:2048. https://doi.org/10.3389/fimmu.2018.02048

53. Wotton KR, Jiménez-Guri E, Crombach A, Janssens H, Alcaine-Colet A, Lemke S, Schmidt-Ott U, Jaeger J (2015) Quantitative system drift compensates for altered maternal inputs to the gap gene network of the scuttle fly megaselia abdita. Elife 4. https://doi.org/10.7554/eLife.04785

54. Wu H, Manu, Jiao R, Ma J (2015) Temporal and spatial dynamics of scaling-specific features of a gene regulatory network in drosophila. Nat Commun 6:10031. https://doi.org/10.1038/ncomms10031

Chapter 6

Mathematical Programming for Modeling Expression of a Gene Using Gurobi Optimizer to Identify Its Transcriptional Regulators

Vijaykumar Yogesh Muley

Abstract

The cell expresses various genes in specific contexts with respect to internal and external perturbations to invoke appropriate responses. Transcription factors (TFs) orchestrate and define the expression level of genes by binding to their regulatory regions. Dysregulated expression of TFs often leads to aberrant expression changes of their target genes and is responsible for several diseases including cancers. In the last two decades, several studies experimentally identified target genes of several TFs. However, these studies are limited to a small fraction of the total TFs encoded by an organism, and only for those amenable to experimental settings. Experimental limitations lead to many computational techniques having been proposed to predict target genes of TFs. Linear modeling of gene expression is one of the most promising computational approaches, readily applicable to the thousands of expression datasets available in the public domain across diverse phenotypes. Linear models assume that the expression of a gene is the sum of expression of TFs regulating it. In this chapter, I introduce mathematical programming for the linear modeling of gene expression, which has certain advantages over the conventional statistical modeling approaches. It is fast, scalable to genome level and most importantly, allows mixed integer programming to tune the model outcome with prior knowledge on gene regulation.

Key words Gene expression, Gene regulation, Gene regulatory networks, Gurobi, Linear programming, Transcription factors, Transcriptional regulation, Transcriptional regulatory networks

1 Introduction

Genes are transcribed into RNA molecules, and a portion of them is then translated into proteins. This vital process is known as *gene expression* (GE) [1, 2]. It maintains the required amount of RNA molecules in the cell and hence controls the rate of protein synthesis. Several experimental methods exist for measuring GE in a tissue or a single cell and are referred to as *GE profiling* methods [3]. Expression profiling data across several samples provides information on spatio-temporal functional activities performed by genes in an organism. The number of RNA molecules originating from a

Shahid Mukhtar (ed.), *Modeling Transcriptional Regulation: Methods and Protocols*, Methods in Molecular Biology, vol. 2328, https://doi.org/10.1007/978-1-0716-1534-8_6, © Springer Science+Business Media, LLC, part of Springer Nature 2021

gene depends in large parts on the binding affinity of the transcription factors (TFs) to its regulatory region and TF availability in the cell [2]. Therefore, expression changes of TFs reflect in the expression changes of their target genes. This relationship can be statistically or mathematically retraced by linear modeling. This is a simplified assumption because a linear relationship may not exist between the expression of genes and their transcriptional regulators. GE is also influenced by several other cellular factors, and also by post-transcriptional and post-translational modifications [2]. However, the linear relationship assumption provides a first step towards understanding the highly dynamic and complex process of gene regulation. Furthermore, linear models are among the most used statistical and mathematical methods and have been successfully applied on GE profiling data [4–9]. The model system simplifies complex real systems and allows manipulations that are otherwise limited in the real system.

1.1 Linear Model Function

Mathematical modeling originally began with the invention of linear programming (LP) in 1947 by George B. Dantzig. It has three major components: The formulation of a real-world problem in detailed mathematical terms as equation is called the *model* (models). The model needs to be solved in a reasonable amount of time and in an efficient way, which needs algorithms for optimization of the model, and use of suitable software and hardware to run the algorithms. Hence, mathematical modeling is also called *mathematical optimization.*

Suppose expression profiling of genes and TFs, carried out in several physiological samples or time-points. The expression profiling data can be organized into a matrix as shown in Table 1. In the table, columns represent variables (i.e., genes and TFs) and rows represent systems (or samples) in which the variables were measured. The numbers in the table are expression levels of a gene of interest and five TFs in eight samples. This data can be used to build a linear model.

The simplest linear model finds the relationship between one independent variable, which is also called the predictor, and the dependent (or response) variable as a straight line according to the values of two constant parameters. This model can be written in the familiar mathematical form:

$$\tilde{y}_j = \beta_0 + \beta_t \times x_{t,j} \tag{1}$$

where y is a response variable (predicted expression of a gene), whose output depends on the predictor variable x ($x_{t,j}$ is the expression of TF t in j^{th} sample), and β parameters. β_0 is the value of y when $x = 0$ (y-intercept), and β_t is the degree to which y changes per unit of change in x_t (gradient of the line). The goal of the linear model is to use the j independent measurements to determine a

Table 1
An example of a gene expression data matrix for linear modeling

	Target gene (response variable)	TFs (predictor variables)				
Samples	**Gata6**	**Atrx**	**Adnp**	**Foxf2**	**Pax9**	**Cdx1**
M9.5a	0.2932	6.3713	1.7527	0.2276	0.0817	1.0849
M9.5b	0.6881	6.4895	2.1796	0.4272	0.0734	0.9254
M9.5c	0.4341	6.5375	2.3263	0.1032	0.1032	0.881
M9.5d	0.2926	6.5326	2.1637	0.1073	0.1073	0.5186
M10.5a	0	6.536	1.9664	1.2989	0.1381	0.3149
M10.5b	0.0737	6.5425	1.8371	0.8658	0	0.4288
M10.5c	0.3466	6.2628	2.0849	1.1875	0	0.2706
M10.5d	0	6.245	1.8874	0.7816	0	0

Columns represent variables (a gene of interest and TFs)
Rows represent systems (or samples) in which variables were measured
The numbers in the table represent expression levels of the variables

mathematical function that describes the relationship between the response (i.e., expression of the gene) and the predictor variable (expression of the TF). Since there are j measurements to estimate the parameter coefficients (β values) which can take up any positive or negative number, the model needs to explore gigantic combinations of numbers and find out the five coefficients which will provide the best solution of Eq. (1). Hence, this becomes a mathematical optimization problem and needs an objective function which will keep track of the best solution over a combination of numbers used as coefficients in Eq. (1). In this case, the objective is to minimize the difference between the measured (real) and the predicted GE. It can be mathematically formulated as shown below:

$$\min \sum_{j=1}^{S} \left| y_j - \widetilde{y}_j \right| = \sum_{j=1}^{S} e_j \tag{2}$$

where S is the total number of samples or measurements, y_j and $\widetilde{y}_j$ is the measured and the predicted expression of a gene in the jth sample, respectively. e_j is the difference between the measured and the predicted GE in the jth sample, conventionally referred to as an error.

A linear optimizer is required to solve the objective function. The Gurobi is a widely used optimizer, which uses highly efficient mathematical programming to solve objective function [10].

1.2 Linear Model for Multiple Independent Variables

Before going into details on how to solve Eq. (2) (i.e., the objective function) using Gurobi, let's generalize the simple linear model for multiple predictor variables (i.e., as many TFs as possible). It is just a simple addition of a new predictor variable to Eq. (1) as follows:

$$\widetilde{y}_j = \beta_0 + \beta_1 \times x_{1,j} + \beta_2 \times x_{2,j} + \ldots\ldots. \beta_n \times x_{n,j} \quad (3)$$

where n is the total number of TFs used as predictor variables, and all β are optimization parameters. The equation can be summarized in mathematical form as follows:

$$\widetilde{y}_j = \beta_0 + \sum_{t=1}^{n} \beta_t \times x_{t,j} \quad (4)$$

The objective function to optimize this linear model is the same as has been shown in Eq. (2).

1.3 Gurobi Optimizer for Linear Modeling

Gurobi has been used in several industries for mathematical programming to solve complex problems [10]. Gurobi is a highly configured tool for mathematical optimization problems. It captures the key features of an optimization problem effectively, and efficiently solves them in a reasonable amount of time. Gurobi uses leading-edge mathematical and computer science technology in solving optimization problems and has perhaps the best performance. Users do not need to worry about the mathematical background of how to solve the optimization problems. This is in-built in the Gurobi optimizer. The mathematics and computer technology behind Gurobi optimization is rather complex and details are beyond the scope of this chapter. However, users are encouraged to explore documentation and tutorials available at Gurobi website. The users only need to efficiently formulate the mathematical model that captures the main characteristics of the optimization problem and the required data for the model. Gurobi optimizes the model automatically behind the scenes. Gurobi cannot handle absolute values or terms as shown in Eq. (2) [11]. Therefore, it is essential to transform Eq. (2) into two inequalities for each sample j as shown below:

$$y_j - \widetilde{y}_j - e_j \leq 0 \quad (5)$$

$$-y_j + \widetilde{y}_j - e_j \leq 0 \quad (6)$$

With this, the optimization problem of modeling GE is formulated in the mathematical form which Gurobi can access. The subsequent sections provide a detailed workflow to solve this optimization problem practically, and to tune the model with prior information on the gene regulation.

2 Material

1. The workflow can be performed on a standard UNIX or MacOS laptop with 4–8 Gb of RAM. Gurobi automatically uses available computational processing and users do not need to worry about it. It should not be difficult to adopt this workflow on Windows OS.
2. It is expected that users are familiar with at least one programming language to convert gene and TF expression profiles into equations in the specific format which Gurobi can handle.
3. Users lacking basic programming skills are encouraged to collaborate with good computer programmers.
4. Expression profiling data for (a) gene(s) of interest and TFs as shown in Table 1 should originate from the same source.

3 Methods

3.1 Gurobi Optimizer Installation

1. Go to https://www.gurobi.com/
   (a) Click on *Downloads and Licenses.*
   (b) Click on *Gurobi Optimizer-Download Software*, accept the terms and conditions, then download the version appropriate for your operating system.
   (c) Install the software by following instructions given on the Gurobi webpage for the choice of your operating system.
   (d) Click on *Academic license*, accept the terms and conditions, and generate an academic license.
2. Open command prompt or terminal and type command *grbgetkey* followed by license key code and hit enter to activate the license.
   (a) Gurobi will ask where to save the license file. It is recommended to save the file to its default location by hitting enter key.
   (b) The license key code can be obtained from the *Academic license* menu of the Gurobi webpage after its creation.
   (c) The command looks like "*grbgetkey XXXXXXXX-XXXX-XXXX-XXXX-XXXXXXXXXXXX,*" where Xs represent license key code.
3. To test if Gurobi works fine, type *gurobi_cl* command in the terminal and hit the enter key. *gurobi_cl* is a Gurobi executable file, which seeks for the input model file, solves optimization problems therein and writes the output solution file. Some typical errors or warnings can be expected (*see* **Note 1**) but usually Gurobi installation is straightforward.

3.2 Gene Expression Data

1. The minimal requirement for linear modeling is a matrix containing the expression levels of a gene of interest and for the set of TFs across several samples (Table 1 in this case).
   (a) Many databases provide ready-to-use expression matrices in diverse physiological conditions if users do not have such data readily available (*see* **Note 2**).
2. Likewise, a list of TFs encoded by an organism of interest can be obtained from many resources (*see* **Note 3**).
3. It totally depends on prospective users to either pre-select important TFs that are known to regulate the gene or choose all TFs encoded by an organism (*see* **Note 4**). Both strategies have their advantages and disadvantages as described below, and users can choose the best strategy appropriate for their work.
   (a) When using only TFs known to regulate the gene of interest, the model will identify which TFs regulate the gene under the physiological condition from which expression data was derived. Basically, new findings from such models are condition-specificity of TF-gene regulation. However, only a small fraction of TFs is well studied for their target genes, even in the highly studied organisms [12]. Hence, the model will be essentially biased towards well-studied TFs.
   (b) When used with all TFs, the linear model will identify all potential TFs (known and unknown regulators) that can impact the gene of interest. However, many TFs could have a linear expression relationship with a gene even though they do not regulate it. This can lead to a high number of false predictions.
   (c) I prefer the latter strategy because the modeling will be data-dependent, and not biased towards limited knowledge on gene regulation. Furthermore, it can identify novel TFs regulating a gene.
4. For brevity, conceptualization, and demonstration purposes, I chose expression measurements of the Gata6 gene and a set of five TFs from murine telencephalon tissue at embryonic days E9.5 and E10.5, each with four replicates (Table 1), which was obtained from Ref. 13. The selected TFs play crucial roles during these developmental time-points. The idea is to model Gata6 GE to identify which of the five TFs governs its expression and are more likely to be its dominating regulator.

3.3 Preparing the Gurobi Input Model File

1. This workflow uses the Gurobi command-line version for scalability. In addition, the input model file format captures an optimization model in a way that is easier for the user to understand and can often be more natural to produce.

A) Gurobi model file format

```
Minimize

e1 + e2 + e3 + e4 + e5 + e6 + e7 + e8

Subject To

e1 + b0 + 6.3713 bTF_Atrx + 1.7527 bTF_Adnp + 0.2276 bTF_Foxf2 + 0.0817 bTF_Pax9 + 1.0849 bTF_Cdx1 >=  0.2932
e2 + b0 + 6.4895 bTF_Atrx + 2.1796 bTF_Adnp + 0.4272 bTF_Foxf2 + 0.0734 bTF_Pax9 + 0.9254 bTF_Cdx1 >=  0.6881
e3 + b0 + 6.5375 bTF_Atrx + 2.3263 bTF_Adnp + 0.1032 bTF_Foxf2 + 0.1032 bTF_Pax9 + 0.8810 bTF_Cdx1 >=  0.4341
e4 + b0 + 6.5326 bTF_Atrx + 2.1637 bTF_Adnp + 0.1073 bTF_Foxf2 + 0.1073 bTF_Pax9 + 0.5186 bTF_Cdx1 >=  0.2926
e5 + b0 + 6.5360 bTF_Atrx + 1.9664 bTF_Adnp + 1.2989 bTF_Foxf2 + 0.1381 bTF_Pax9 + 0.3149 bTF_Cdx1 >=  0.0000
e6 + b0 + 6.5425 bTF_Atrx + 1.8371 bTF_Adnp + 0.8658 bTF_Foxf2 + 0.0000 bTF_Pax9 + 0.4288 bTF_Cdx1 >=  0.0737
e7 + b0 + 6.2628 bTF_Atrx + 2.0849 bTF_Adnp + 1.1875 bTF_Foxf2 + 0.0000 bTF_Pax9 + 0.2706 bTF_Cdx1 >=  0.3466
e8 + b0 + 6.2450 bTF_Atrx + 1.8874 bTF_Adnp + 0.7816 bTF_Foxf2 + 0.0000 bTF_Pax9 + 0.0000 bTF_Cdx1 >=  0.0000

e1 - b0 - 6.3713 bTF_Atrx - 1.7527 bTF_Adnp - 0.2276 bTF_Foxf2 - 0.0817 bTF_Pax9 - 1.0849 bTF_Cdx1 >=  -0.2932
e2 - b0 - 6.4895 bTF_Atrx - 2.1796 bTF_Adnp - 0.4272 bTF_Foxf2 - 0.0734 bTF_Pax9 - 0.9254 bTF_Cdx1 >=  -0.6881
e3 - b0 - 6.5375 bTF_Atrx - 2.3263 bTF_Adnp - 0.1032 bTF_Foxf2 - 0.1032 bTF_Pax9 - 0.8810 bTF_Cdx1 >=  -0.4341
e4 - b0 - 6.5326 bTF_Atrx - 2.1637 bTF_Adnp - 0.1073 bTF_Foxf2 - 0.1073 bTF_Pax9 - 0.5186 bTF_Cdx1 >=  -0.2926
e5 - b0 - 6.5360 bTF_Atrx - 1.9664 bTF_Adnp - 1.2989 bTF_Foxf2 - 0.1381 bTF_Pax9 - 0.3149 bTF_Cdx1 >=  0.0000
e6 - b0 - 6.5425 bTF_Atrx - 1.8371 bTF_Adnp - 0.8658 bTF_Foxf2 + 0.0000 bTF_Pax9 - 0.4288 bTF_Cdx1 >=  -0.0737
e7 - b0 - 6.2628 bTF_Atrx - 2.0849 bTF_Adnp - 1.1875 bTF_Foxf2 + 0.0000 bTF_Pax9 - 0.2706 bTF_Cdx1 >=  -0.3466
e8 - b0 - 6.2450 bTF_Atrx - 1.8874 bTF_Adnp - 0.7816 bTF_Foxf2 + 0.0000 bTF_Pax9 + 0.0000 bTF_Cdx1 >=  0.0000

Bounds

bTF_Atrx free
bTF_Adnp free
bTF_Foxf2 free
bTF_Pax9 free
bTF_Cdx1 free

End
```

B) Gurobi output file format

```
# Objective value = 8.3454237785580559e-02
e1 0
e2 8.1412372410549522e-02
e3 2.0418653750310443e-03
e4 0
e5 0
e6 0
e7 0
e8 0
b 7.2908425343035912e+00
bTF_Atrx -3.4897721513445690e-01
bTF_Adnp -1.3643476178457858e-01
bTF_Foxf2 -8.7933167818503369e-03
bTF_Pax9 2.2273346475829282e-02
bTF_Cdx1 1.5974452019068985e-01
```

Fig. 1 An example of the Gurobi linear model input file format and its output for gene expression modeling. (**a**) Gurobi input file format containing the objective function in the **Minimize** section, mandatory constraints in the **Subject To** section, and optional constraints in the **Bounds** section. Gurobi estimates coefficients for *b* and *bTF* prefixed variables. Variables are set free in the bound section, which means they can have positive or negative coefficients. (**b**) Gurobi solution file format containing the value of the objective function (i.e., overall modeling error), followed by samples contributing to the objective function value (sample-wise errors), and then estimated β coefficients for β_0 and TFs, prefixed with *bTF_* followed by their names

Experienced programmers may explore the implementation of Gurobi optimization problems in Python or R.

2. Gurobi reads a model from a file, optimizes it, and writes the solution to a file. The model input file can be written in various formats, but the LP format is simple and easy to implement.
3. An example LP input file format file is shown in Fig. 1a, which contains a structured list of sections, where each section captures a logical piece of the whole optimization model. Sections begin with particular keywords and must generally be in the fixed order.
4. The first section in an example LP file is the objective section.
   (a) The goal is to minimize the errors between observed and predicted GE, hence it begins with the term **Minimize** on its own line.
   (b) Then the next line specifies the equation containing the sum of eight variables, which represent the sum of errors that need to be minimized.
5. The second section is the mandatory constraint section and begins with the term **Subject To**, followed by a linear combination of variables and parameters that need to be estimated.
   (a) Each sample is represented by two equations written in the LP format syntax, which are equivalent to Eq. (5) (first block of eight equations) and Eq. (6) (second block of eight equations). The aim here is to restrict the predicted expression of the gene from too great deviations in either direction (i.e., positive or negative) from the measured expression. That's the reason there are two equations for each sample.
   (b) Briefly, e1 to e8 represent errors for eight samples, then the parameter specification begins represented by β coefficients which will be estimated as part of the optimization solution. Except for β_0, all β are represented with the prefix "bTF_", and their suffix being the name of the corresponding TF, preceded by the expression value. For example, "1.7527 bTF_Adnp," equivalent to $x_{t,\ j} \times \beta_t$, represents the expression value (1.7527) of TF Adnp in the sample j (which is the first sample in Table 1 and Fig. 1), multiplied (denoted with a space) by its β coefficient (represented bTF_Adnp). Please *see* Eqs. (1) and (4) for understanding.
   (c) Every equation in this section ends with the number (preceded by $\geq$operator) representing the measured expression of the gene whose expression needs to be modeled; in this case, Gata6.
6. The optional bounds section follows the mandatory constraint section. It begins with the word **Bounds**, on its own line, and is followed by a list of variable bounds, each on its own line.

(a) The idea here is to put restrictions on a lower and upper value (ranges) on the parameters (coefficients) being estimated. This is particularly important for GE modeling since TFs can activate or repress the expression of the gene [12].

(b) For example, activator TFs can be set within positive bounds and repressors within negative bounds. In this example, however, each variable is declared free, meaning that it is unbounded in either direction, i.e., can assume positive or negative values. This is often a good choice and the model will be more dependent on the data than prior assumption. Enforcing bounds on coefficients may not work for many reasons but the most important are:
   - Biological systems are not rigid, in the sense that a gene can be a target of a TF in one particular physiological condition but may not be in another.
   - Restricting coefficients to have desired values could prevent the model from optimizing.

(c) Therefore, I would not recommend playing with bounds unless there is thorough knowledge of the biological context and expected output of modeling.

7. The last line in an LP format file should be a word **End**, to conclude the end of model formulation.
8. The file should be saved with the *lp* extension, so that Gurobi understands it; for example, *Gata6.lp*.
9. It is worth mentioning that the whitespace between two variables or constants is treated as multiplication sign (resulting in their product being calculated), and the backslash symbol starts a comment and the remainder of that line is ignored by Gurobi. Understanding the LP file format will take users way ahead in formulating various optimization problems in Gurobi (please see user manual on the Gurobi website).

3.4 Running Gurobi with Input Model File

1. The optimizer Gurobi can be executed by the simple command given below with default settings (Recommended, *see* **Note 5**). It is necessary to provide name of the Gurobi output file with extension *sol* (for example *output.sol*), which stands for solution information format, which can be set with *ResultFile* command, followed by the input file name, as shown below:

```
gurobi_cl ResultFile=output_filename.sol input_
filename.lp
```

2. When the above command is executed, Gurobi should write an optimization solution to the output file. If you have an infeasible model, it writes an Irreducible Inconsistent Subsystem instead (*see* **Note 6**).

3.5 Interpreting the Gurobi Output

1. The Gurobi output file contains three components (Fig. 1b).
   (a) First, the objective function value, which is the sum of the errors in all samples. Since it is a minimization problem, a smaller value (close to zero) indicates a good optimal solution.
   (b) The second component enlists how much each measurement or sample (e1 to e8) contributed to the objective function value on its own line. When the objective function value is too high for the minimization problem, the samples with high error rates may be removed.
   (c) The third and very important component represents the estimated β coefficients for each of the TFs, and the additive offset value of β_0.
2. The β coefficients of TFs in a positive direction can generally be interpreted as the activating effect on the GE, whereas the coefficients with negative values indicate inhibitory effect. This can be peculiar because it can be in the opposite direction too, as we shall see.
3. For example, Atrx has negative estimated coefficient ($\beta = -0.3769$), which is consistent with its role in global silencing of GE [14], while Cdx1 has been shown to induce Gata6 expression [15], which is also consistent with its positive estimated coefficient ($\beta = 0.5397$).
4. On the other hand, Adnp knockdown is known to induce Gata6 expression [16], which suggests that Adnp could be a repressor of Gata6. However, its coefficient ($\beta = 0.9717$) indicates that it has an activating effect on Gata6 expression. This reveals a very important aspect of linear modeling, especially when interpreting β coefficients: It is not considered wise to interpret coefficient values estimated by linear models. The modeling works reasonably well but it is not infallible. From a biological perspective, both results could be true, if in different biological contexts, unless confirmed otherwise. However, experimental evidence is always favored over predictions when contradictory results are obtained. From a mathematical perspective, the result could arise due to multicollinearity. The essence of this example is to make users aware that linear models work reasonably well do not necessarily provide absolute accuracy in all cases (*see* **Notes** 7 and **8** for details).

3.6 Predicting Gene Expression

1. GE can be predicted easily by plugging the estimated β coefficients in Eq. (4) and solving it for each sample. For example, the expression of Gata6 gene in M9.5a sample can be calculated by summing up the products of the expression values of TFs and their estimated β coefficients plus β_0 as follows:

$$\widetilde{y}_{\text{M9.5a}} = \beta_0 + (\beta_{\text{Atrx}} \times 6.3713) + \left(\beta_{\text{Adnp}} \times 1.7527\right) + (\beta_{\text{Foxf2}} \times 0.2276) + (\beta_{\text{Pax9}} \times 0.0817) + (\beta_{\text{Cdx1}} \times 1.0849)$$

2. When solving the above equation by substituting the estimated β coefficients corresponding to each TF and β_0 estimated by Gurobi, the predicted expression of Gata6 gene in M9.5a sample ($\widetilde{y}_{\text{M9.5a}}$) is 0.2933, which is almost exactly its measured expression level of 0.2932. Table 2 shows a comparison of predicted and measured expression of Gata6 for all samples.
3. The estimated β coefficients can also be used with other expression datasets to predict Gata6 expression in them and actually check how well the model performs on unseen data. It is a direct way to validate the model.

3.7 Manipulating Linear Model with Prior Information

1. In the above example, all predictor variables were unbounded, that is they were declared free in the bound section of the model file. It means, their estimated β coefficients can have infinite positive or negative values. In this case, the optimization solution should be closely similar to results obtained by statistical linear modeling, such as linear modeling performed by the *lm* function in R.
2. However, Gurobi has a certain advantage over statistical modeling. Among other reasons, the model can be constraint with prior information and according to the desired output.
3. For instance, this linear model can be bounded by the information available prior for Adnp and Cdx1 in repressing and activating Gata6 expression, respectively. This can be done by:
   (a) replacing "bTF_Adnp free" statement in the Bounds section of the input model file by "bTF_Adnp <= 0" because Adnp is assumed to be a repressor, and hence it should have negative influence on the Gata6 expression.
   (b) replacing "bTF_Cdx1 free" statement by "bTF_Cdx1 >= 0" because Cdx1 is assumed to be an activator, and should have a positive influence on Gata6 expression.
4. Upon execution of the above new model in Gurobi, the objective function value becomes 0.850 (with bounds) compared to 0.308 of the unbounded model. The difference between measured and predicted expression is greater in the bounded model, meaning an unbounded model has a better solution (*see* **Note 8**).

Table 2
A comparison between measured and predicted expression of Gata6 gene using linear programming

Samples	Measured expression	Predicted expression
M9.5a	0.2932	0.2933
M9.5b	0.6881	0.5945
M9.5c	0.4341	0.6488
M9.5d	0.2926	0.2925
M10.5a	0	0
M10.5b	0.0737	0.0737
M10.5c	0.3466	0.3467
M10.5d	0	0

5. Gurobi allows special-ordered set (SOS) constraints in the model, where variables can be assigned with weights based on their importance in the model, or even a switch to choose a specific variable with better coefficient over another. These special constraints can be added in the SOS section in the model file, and worth exploring in detail to truly harness the power of mix integer programming in Gurobi.
6. Manipulating a model is easy; however, it should be done carefully with detailed knowledge of context and expectations from the model output.

4 Notes

1. After installing Gurobi, it may return an error that the *gurobi_cl* Gurobi "executable file not found." In this case, you need to set a global path for the Gurobi library. It can be done easily by following the installation guide provided by the Gurobi developers on their webpage.
2. Users lacking in-house expression datasets are encouraged to explore the Gene Expression Omnibus (https://www.ncbi.nlm.nih.gov/geo/) and Sequence Read Archive (https://www.ncbi.nlm.nih.gov/sra) databases available at NCBI. The GEO database provides thousands of publicly available raw as well as processed GE datasets, while the SRA database provides next-generation sequencing datasets. There are many more databases available and worth exploring.

3. Model organism-specific TF databases are too many to list here. However, the DBD database (www.transcriptionfactor.org) is a good place to look for predicted TFs in several completely sequenced genomes.
4. Linear modeling tools assume that the predictor variables are independent, and hence fail to handle variables that are co-linear or strongly correlated at expression levels, a circumstance called multicollinearity. However, it is often not obvious how each of the variables depends on others, especially in biological systems. Gurobi handles multicollinearity issues quite efficiently by its cutting plane-based algorithm and will find a solution anyhow. However, Gurobi underscores variables having multicollinearity, especially on GE datasets with small numbers of samples (personal experience). The variance inflation factor is a widely used diagnostics tool for multicollinearity. Another approach to handle such a situation is the clustering of TFs based on their expression similarity and keeping one representative TF from each cluster for modeling.
5. Gurobi has deterministic and non-deterministic algorithms to solve models, which can be set by the *Method* parameter. The former gives the exact same result each time you run, while the latter can produce different optimal bases when running multiple times. However, it is recommended not to change this parameter since there is no big gain over the default algorithm. One useful parameter is *TimeLimit*, when the samples and TFs are in large numbers, and the model would take a lot of time for optimization depending on the computational infrastructure. Gurobi will terminate optimization when the time expended exceeds the value specified in the *TimeLimit* parameter.
6. When the model shows the Irreducible Inconsistent Subsystem (IIS) message in the solution file, it is possible to identify the cause of the infeasibility. Users can run Gurobi again by setting one more output file, with *ilp* extension, in the *ResultFile* option. Gurobi attempts to solve the model and will automatically compute an IIS and write it to the requested file name (.ilp). This file contains the details on the subset of the constraints and variable bounds causing the infeasibility, which when removed may render the model feasible. It is possible that the model has multiple IISs. The one returned by Gurobi is not necessarily the one with minimum cardinality. Users need to play with the model and identify the problems.
7. In general, linear models work quite well. However, GE data is always subject to technical and biological variations. Genes and TFs can have co-linear expression patterns if they are required at same time-point, even though these TFs have no role in regulating the genes. Furthermore, GE is controlled by a

concerted action of a combination of TFs. The addition or removal of one factor can inverse the original control over GE. Some recommendations can be proposed to achieve better solutions.

(a) Selection of expression samples derived from diverse physiological conditions. It is favorable using as many samples as possible while excluding highly similar ones identified by clustering approaches.

(b) For modeling of one or few genes, it is possible to exclude one TF or a combination at a time, and see whether the objective function value improves [11].

(c) Multicollinearity can hamper results and cause models to display entirely different behavior than expected. In such cases, removing multicollinear independent variables may improve the solution (*see* **Note 4**).

8. The data used in this workflow is from telencephalon tissues obtained in specific developmental stages. Biologically, it is possible that Adnp can be a repressor at a particular physiological condition or time-point and an activator in others. This situation exists more often in biological systems than can be expected by chance [12, 17]. However, this could be an artifact of linear modeling due to multicollinearity due to small number of samples in GE data. Therefore, considering the mathematical and biological aspects helps to interpret results in a better way.

Acknowledgments

This work was supported by DGAPA-UNAM grant IA203920 to V.Y.M. The author would like to sincerely thank Anne Hahn (Queensland Brain Institute, Australia) for critical reading of the manuscript.

References

1. CRICK FH (1958) On protein synthesis. Symp Soc Exp Biol 12:138–163. http://www.ncbi.nlm.nih.gov/pubmed/13580867
2. Muley VY, Pathania A (2017) Gene Expression. In: Vonk J, Shackelford T (eds) Encycl Anim Cogn Behav. Springer, Cham
3. Pathania A, Muley VY (2017) Gene expression profiling. In: Vonk J, Shackelford T (eds) Encycl Anim Cogn Behav. Springer, Cham
4. Law CW, Alhamdoosh M, Su S et al (2018) RNA-seq analysis is easy as 1-2-3 with limma, Glimma and edgeR. F1000 Res 5:1408. https://doi.org/10.12688/f1000research.9005.3
5. Cheng C, Alexander R, Min R, Leng J, Yip KY, Rozowsky J et al (2012) Understanding transcriptional regulation by integrative analysis of TF binding data. Genome Res 22 (9):1658–1667
6. Taylor RC, Acquaah-Mensah G, Singhal M, Malhotra D, Biswal S (2008) Network inference algorithms elucidate Nrf2 regulation of mouse lung oxidative stress. PLoS Comput Biol 4(8):e1000166
7. Setty M, Helmy K, Khan AA, Silber J, Arvey A, Neezen F et al (2012) Inferring transcriptional and microRNA-mediated regulatory programs in glioblastoma. Mol Syst Biol 8:605

8. Poos AM, Maicher A, Dieckmann AK, Oswald M, Eils R, Kupiec M et al (2016) Mixed Integer Linear Programming based machine learning approach identifies regulators of telomerase in yeast. Nucleic Acids Res 44: e93. https://doi.org/10.1093/nar/gkw111
9. Marbach D, Costello JC, Küffner R, Vega NM, Prill RJ, Camacho DM et al (2012) Wisdom of crowds for robust gene network inference. Nat Methods 9:796–804
10. Gurobi Optimization LLC. Gurobi optimizer reference manual. 2020. http://www.gurobi.com
11. Schacht T, Oswald M, Eils R, Eichmüller SB, König R (2014) Estimating the activity of TFs by the effect on their target genes. Bioinformatics 30:i401–i407. https://doi.org/10.1093/bioinformatics/btu446
12. Vaquerizas JM, Kummerfeld SK, Teichmann SA, Luscombe NM (2009) A census of human TFs: Function, expression and evolution. Nat Rev Genet 10:252–263. http://www.ncbi.nlm.nih.gov/pubmed/19274049
13. Muley VY, López-Victorio CJ, Ayala-Sumuano JT, González-Gallardo A, González-Santos L, Lozano-Flores C et al (2020) Conserved and divergent expression dynamics during early patterning of the telencephalon in mouse and chick embryos. Prog Neurobiol 186:101735
14. Kernohan KD, Jiang Y, Tremblay DC, Bonvissuto AC, Eubanks JH, Mann MRW et al (2010) ATRX Partners with Cohesin and MeCP2 and Contributes to Developmental Silencing of Imprinted Genes in the Brain. Dev Cell 18:191–202. https://linkinghub.elsevier.com/retrieve/pii/S153458071000016X
15. Fujii Y, Yoshihashi K, Suzuki H, Tsutsumi S, Mutoh H, Maeda S et al (2012) CDX1 confers intestinal phenotype on gastric epithelial cells via induction of stemness-associated reprogramming factors SALL4 and KLF5. Proc Natl Acad Sci U S A 109:20584–20589. http://www.ncbi.nlm.nih.gov/pubmed/23112162
16. Ostapcuk V, Mohn F, Carl SH, Basters A, Hess D, Iesmantavicius V et al (2018) Activity-dependent neuroprotective protein recruits HP1 and CHD4 to control lineage-specifying genes. Nature 557:739–743. http://www.nature.com/articles/s41586-018-0153-8
17. Gerstein MB, Kundaje A, Hariharan M, Landt SG, Yan KK, Cheng C et al (2012) Architecture of the human regulatory network derived from ENCODE data. Nature 489:91–100

Chapter 7

Multiscale Modeling of Cross-Regulatory Transcript and Protein Influences

Megan L. Matthews and Cranos M. Williams

Abstract

With the popularity of high-throughput transcriptomic techniques like RNAseq, models of gene regulatory networks have been important tools for understanding how genes are regulated. These transcriptomic datasets are usually assumed to reflect their associated proteins. This assumption, however, ignores post-transcriptional, translational, and post-translational regulatory mechanisms that regulate protein abundance but not transcript abundance. Here we describe a method to model cross-regulatory influences between the transcripts and proteins of a set of genes using abundance data collected from a series of transgenic experiments. The developed model can capture the effects of regulation that impacts transcription as well as regulatory mechanisms occurring after transcription. This approach uses a sparse maximum likelihood algorithm to determine relationships that influence transcript and protein abundance. An example of how to explore the network topology of this type of model is also presented. This model can be used to predict how the transcript and protein abundances will change in novel transgenic modification strategies.

Key words Multiscale modeling, Cross-regulation, Transcript regulation, Protein regulation

1 Introduction

Due to the abundance of transcriptomics data that can now be readily obtained through methods like RNAseq, models of gene regulatory networks (GRNs) have been an important and popular tool for understanding how biological systems are regulated under abiotic stress, biotic stress, and gene manipulations. The transcriptomic data sets, and the gene regulatory networks that are estimated from them, typically assume that the transcript abundances serve as good proxies for protein abundances. However, this ignores any regulatory mechanisms occurring after transcription such as post-transcriptional, translational, and post-translational regulatory mechanisms. These mechanisms have been shown to play a substantial role in controlling steady-state protein abundances [1]. Multiscale models that capture transcriptional

Shahid Mukhtar (ed.), *Modeling Transcriptional Regulation: Methods and Protocols*, Methods in Molecular Biology, vol. 2328, https://doi.org/10.1007/978-1-0716-1534-8_7,

regulation and regulation after transcription that impact protein abundances are needed to understand how biological systems respond to stressors and gene modifications. Further, these models can be used to help identify strategies for improving these responses. Recently we developed a model capturing the cross-regulatory influences between the transcripts and proteins associated with the monolignol genes in *Populus trichocarpa* [2]. In this chapter we describe the steps for inferring the relationships that make up this model, using this model to predict the impact of new transgenic knockdowns, and identifying the highly influential relationships between the components in the model. We then provide an example implementation using the monolignol transcript and protein abundance data used in [2].

2 Materials

1. All of the code and data needed to implement the example used here can be found at https://github.com/leighmatth/Cross-regulatory-transcript-protein-modeling.git.
2. Transcript and protein abundance data for a set of genes that are from a series of wild-type and transgenic experiments that target these genes individually or combinatorially to modify their expression. We recommend that each gene is targeted in at least one of the experiments. For the example used here, we included transcript and protein abundance data of the 20 monolignol genes that were measured in 18 wild-type and 207 transgenics, which systematically knocked down the monolignol genes in *Populus trichocarpa* [2, 3].
3. MATLAB®. The results presented in this chapter were run using MATLAB version R2019a and a personal computer with 2.8 GHz Intel Core i7 processor and 16 GB RAM.

3 Methods

3.1 Model Development

1. Clone or download the code located at https://github.com/leighmatth/Cross-regulatory-transcript-protein-modeling.git.
2. Open MATLAB and navigate to the location of the downloaded code and/or add it (and the sub-folders) to your MATLAB path.
3. Load your set of transcript and protein abundances and save them as an $M \times N_d$ matrix, `Y`, with each row representing a transcript or protein and each column representing a wild-type or transgenic experiment (Table 1; see Notes 1-3).

Table 1
Example `Y` matrix containing transcript and protein abundances. Columns correspond with the experiments, and rows correspond with the transcripts and proteins. WT, kdA, kdB, and kdC in this example correspond to wild-type and knockdown experiments of genes A, B, and C respectively

	WT	kdA	kdB	kdC
Transcript A	100	10	100	100
Transcript B	200	245	40	200
Transcript C	80	125	80	15
Protein A	1000	100	1000	675
Protein B	2000	2675	400	1675
Protein C	800	1250	800	150

Table 2
Example mask matrix, `Xmask`, corresponding with Table 1

	WT	kdA	kdB	kdC
Transcript A	0	1	0	0
Transcript B	0	0	1	0
Transcript C	0	0	0	1
Protein A	0	0	0	0
Protein B	0	0	0	0
Protein C	0	0	0	0

4. (Optional) If you have multiple replicates of the same experiment, you may want to average the replicates before implementing the SML algorithm (see Note 4).
5. Create a binary "mask" matrix, `Xmask`, that is the same $M \times N_d$ dimensions of your abundance matrix, `Y` (see Note 5). For each column, set a value of 1 to indicate that the corresponding transcript was targeted in that experiment. All other, un-targeted, transcript and proteins in that column should be indicated with a 0 (Table 2).
6. Specify the parameters used for the cross-validation (see Note 6) in the SML algorithm, described in the *Cross-validation parameters* box, as fields within a struct `parameters`. For example, store the number of cross-validation folds, Kcv, in the variable `parameters.Kcv`, and the vector of rho_factors in the variable `parameters.rho_factors`.

Cross-validation parameters	
Kcv	Number of cross-validation folds, typically 10 or 5.
rho_factors	Vector containing a range of potential regularization parameter values for the ridge regression (see Note 7). Cross validation is used to identify which of these values will be used.
lambda_factors	Vector containing a range of potential regularization parameter values for the SML algorithm. Cross validation is used to identify which of these values will be used.
maxiter	Maximum number of iterations of coordinate ascent algorithm used in SML approach
cv_its	Number of times to repeat the Kcv-fold cross-validations. The data is sorted into new folds each iteration.

7. Run the SML algorithm (see Note 8) using the `SML_wrapper` function provided in the github repository. This wrapper function calls sub-functions in the *Sub-functions of SML algorithm* box as described in Fig. 1. The outputs of this function are described in the *SML_wrapper parameter* box below.

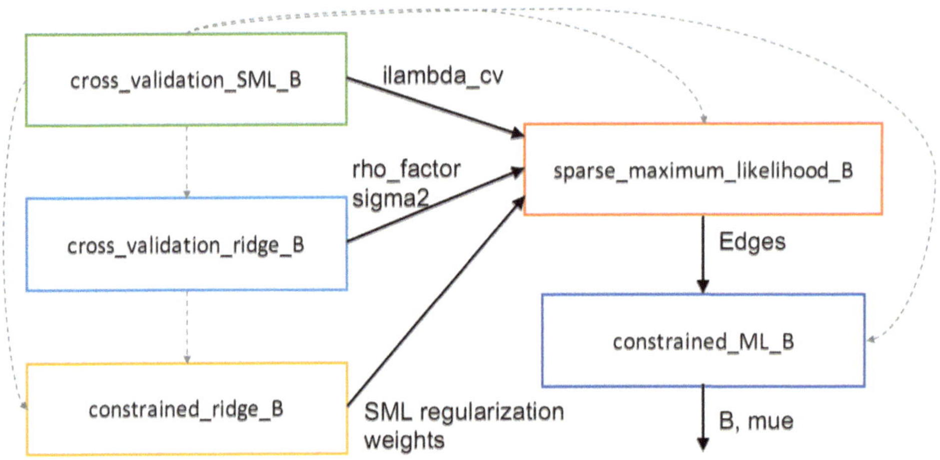

Fig. 1 Flow chart of SML algorithm functions. The black arrows show the important parameters calculated from each function and where these parameters use these parameters. The cross-validation functions call several of these functions within the cross-validation, we indicate the functions called within these cross-validation schemes with the dashed gray arrows

Sub-functions of SML algorithm	
`constrained_ML_B`	This function performs a maximum likelihood regression to solve for the weights of only the specified elements of **B** where the unspecified elements of **B** are 0.
`constrained_ridge_B`	This function performs a ridge regression to solve for the weights of **B**. These weights are used as part of the regularization term of the SML algorithm.
`cross_validation_ridge_B`	This function randomly sorts the data into Kcv folds and performs ridge regression over a cross validation scheme.
`cross_validation_SML_B`	This function performs the SML algorithm over a k-fold cross validation scheme.
`sparse_maximum_likelihood_B`	This function performs a maximum likelihood regression with an ℓ_1-norm penalizing term to solve for the weights of **B**.

SML_wrapper parameters

Inputs

`Y`	$M \times N_d$ matrix containing the measured transcript and protein abundances.
`Xmask`	$M \times N_d$ binary matrix, where a 1 indicates that transcripts and/or proteins targeted in a knockdown experiment, and a 0 for all other, un-targeted, transcripts and proteins.
`params`	Struct containing the cross-validation parameters (see Step 6 above).

Outputs

`B`	$M \times M$ matrix containing the inferred cross-regulatory influences. Corresponds to **B** in Eq 1.
`mue`	$M \times 1$ vector containing the constant terms for the model. Corresponds to $\boldsymbol{\mu}$ in Eq 1.
`ilambda_cv`	The index of the SML regularization parameter from the vector of defined lambda_factors that produced the minimum cross-validation error.
`rho_factor`	The ridge regression regularization parameter from the vector of the defined rho_factors that produced the minimum cross-validation error.
`sigma2`	$M \times 1$ vector of the estimated noise variance for each transcript and protein. Corresponds with σ_2 in Eq 1.

3.2 Model Prediction

This section describes how to use the **B** and $\boldsymbol{\mu}$ found in the previous section to predict how the transcript and protein abundances change under novel perturbations. The functions described in this section are located in the same repository as the previous section.

1. In MATLAB define the inputs to the single or combinatorial perturbation experiment(s) that you want to simulate as described in the box *Model_Prediction parameters.*

Model_Prediction parameters	
Inputs	
`B`	An $M \times M$ matrix that is output from the `SML_wrapper` function. Defines the relationships between the transcripts and proteins as determined by the SML algorithm.
`mue`	An $M \times 1$ vector that is output from the `SML_wrapper` function. Defines the constant portion of model.
`Xtarg`	An $M \times N_p$ matrix containing the desired abundances for the transcripts that are being targeted for each of the N_p experiments being simulated (see Note 9). While the untargeted transcript and the protein abundance values are ignored for ease of interpretation we recommend setting these values to 0 (Table 3).
`Xmask`	An $M \times N_p$ binary matrix, with a 1 indicating the targeted transcript(s), and a 0 for all other transcripts and all proteins. `Xmask` is needed in addition to `Xtarg` for when you want to simulate a complete knockout (Table 2).
`targ_prot_flag`	Binary flag; 1 for targeting the protein as well as the transcript, 0 for only targeting the transcript. If 1, the protein abundances of the targeted genes are set to the same percentage of their wildtype as the associated transcript abundances. For example, if we are decreasing Transcript A to 25% of its wildtype abundance, then Protein A will also be decreased to 25% of its wildtype abundance.
`Ywt`	(Optional) $M \times 1$ vector of the wildtype abundances (Table 4). Required when `targ_prot_flag` is set to 1. Used to determine the protein abundances of the targeted transcripts.
Outputs	
`Ypred`	An $M \times N_p$ matrix of the predicted abundances for each of the N_p experiments simulated.

Table 3
Example of `xtarg` matrix to predict the abundances in Table 1

	WT	kdA	kdB	kdC
Transcript A	0	10	0	0
Transcript B	0	0	40	0
Transcript C	0	0	0	15
Protein A	0	0	0	0
Protein B	0	0	0	0
Protein C	0	0	0	0

Table 4
Example of `Ywt` vector based on the example in Table 1

	WT
Transcript A	100
Transcript B	200
Transcript C	80
Protein A	1000
Protein B	2000
Protein C	800

2. Call the `Model_Prediction` function with the inputs you defined above.

3.3 Network Topology Analysis

It can be useful to look at different network topology features to gain insight into the specific relationships identified in our model. While the connections in the model could be explored as is, the number of inferred relationships and the inter-connectedness of the model can make it difficult to draw conclusions. Further, it is important to consider that the nodes of the network represent both the transcripts and proteins of the genes, which can further complicate analyzing the network topology characteristics. We recommend identifying the inferred relationships in your network that highly contribute to the net change in a transcript or protein's abundance from its wild-type levels. The topology characteristics of that sub-network can then be explored. Some network topology characteristics that may be of interest include the in and out degrees of the transcripts and proteins and different network motifs. These characteristics may provide insight into post-transcriptional or post-translational regulatory influences [2].

1. (Optional) For each transcript and protein, filter the experiments in `Y` (Sect. 3.1) to just the experiments where that transcript or protein is significantly differentially expressed (based on a differential expression analysis, see Note 10) and/or the experiments that are predicted "well" (see Note 11) when using the model to emulate the results from these experiments (Sects. 3.2 and 3.4.2).
2. Create a table of the inferred relationships that highly contribute to the changes of a transcript or protein from their wild-type levels. There are multiple ways one could define what is an influential relationship. One method is to define any relationship that contributes at least a certain percentage (e.g., ± 50%) of the net change from wild-type levels in at least one experiment as influential for that transcript or protein. An example of this method is shown in the example implementation in Sect. 3.4.3. We refer to the identified influential relationships as edges of the network.
3. Using this table of edges, calculate different network topology measures, such as in degrees (the number of other nodes influencing each node), out degrees (the number of nodes each node influences), and network motifs (repeated sub-networks). An example of calculating some of these topology measures are shown in the example implementation.

3.4 Example Implementation

3.4.1 Model Development

1. Clone or download the github repository at https://github.com/leighmatth/Cross-regulatory-transcript-protein-modeling.git
2. Open MATLAB and navigate to the location of the directory you saved to your local computer. The MATLAB code shown throughout Sects. 3.4.1, 3.4.2 and 3.4.3 can be found in the file `Main_Code.m`. The transcript and protein abundance data used in this example is stored in the file `LigninTranscriptsProteins.mat`.
3. Implement the code in *Box 1* to load the transcripts and proteins abundances and average the replicates for each experimental line.

Box 1

```
load LigninTranscriptsProteins.mat

GeneNames = Y_all.Properties.VariableNames;
Experiments = Y_all.Properties.RowNames;

% Group the experiments by their experimental lines
Exp_Lines = cellfun(@(x) x(6:end-2),Experiments,'UniformOutput',0);
[G_lines, ID_lines] = findgroups(Exp_Lines);

% Take the average of the replicates for each line
Ytable = splitapply(@(x) mean(x,1),Y_all{:,:},G_lines);
Ytable = array2table(Ytable,'VariableNames',GeneNames,'RowNames'
    ,ID_lines);

% Each replicate for a line is the same so taking the average of the
    replicates
% will give a mask for the averagehas the same Create a mask for each
    experimental line by taking the average
Xmask_table = splitapply(@(x) mean(x,1),X_all_mask{:,:},G_lines);
Xmask_table = array2table(Xmask_table,'VariableNames',GeneNames,'RowNames'
    ,ID_lines);

% Transpose Y and X so that each experiment represents a column and each
    row a
% transcript or protein
Y = Ytable{:,:}';
Xmask = Xmask_table{:,:}';
```

4. Implement the code in *Box 2* to perform the two cross-validation steps to identify the regularization parameters (see Note 14). The code in *Box 2* seeds the random number generator (see Note 15), sets the cross-validation parameters, and calls the functions for performing the cross-validation of the SML algorithm and the cross-validation of the ridge regression algorithm.

Box 2

```
% Set seed for random number generator
rng_seed = 123456;
rng(rng_seed);

% Set the cross-validation parameters
params.Kcv = 5; % # of cross-validation folds
params.rho_factors = 10.^(-6:0.1:1); %Ridge regularization parameters
params.lambda_factors = 10.^(0:-.1:-3); %SML regularization parameters
params.maxiter = 1000; % max iterations for SML algorithm
params.cv_its = 1; % # of iterations of the kcv-fold cross-validations
```

5. Implement the SML algorithm using the `SML_wrapper` function as shown in *Box 3* calling the variables `Y` and `Xmask` defined in *Box 1* and the algorithm parameters, `params`, defined in *Box 2*.

Box 3

```
% Call SML algorithm
[B,mue,ilambda_cv,rho_factor,sigma2] = SML_wrapper(Y,Xmask,params);
```

6. Run the code in *Box 4* to check that a minimum was found for the SML and ridge regularization parameters and that the algorithm did not run up against the bounds set in `params.lambda_factors` and `params.rho_factors` in *Box 2*.

Box 4

```
sml_regparam = [ilambda_cv > 1, ilambda_cv < length(params.lambda_factors)
    ]; % Check if SML regularization parameter is either the first or
    last input parameter, indicating a minimum was not found
ridge_regparam = [rho_factor > params.rho_factors(1), rho_factor < params.
    rho_factors(end)]; % Check if ridge regression regularization
    parameter is either the first or last input parameter, indicating a
    minimum was not found

display(sml_regparam) % display result to command window; 1 if TRUE, 0 if
    FALSE
display(ridge_regparam) % display result to command window; 1 if TRUE, 0
    if FALSE
```

7. In this example, both `sml_regparam` and `ridge_regparam` should be `[TRUE, TRUE]` (Fig. 2), and nothing further needs to be done. If, however, either of these values are `FALSE`, then the range of the `params.lambda_factors` and/or `params.rho_factors` set in *Box 2* needs to be increased and the `SML_wrapper` function needs to be re-run with these new parameter ranges (*Box 3*). For example, if `sml_regparam` is `[TRUE, FALSE]`, then change `params.lambda_factors` from `10.^(0:-1:3)` to `10.^(0:-1:-4)`, increasing the range from [1, 0.001] to [1, 0.0001], and re-run the `SML_wrapper` function in *Box 3*.
8. Check the number of relationships inferred by the algorithm and view a heatmap of the negative and positive relationships detected by implementing the code in *Box 5*. If the random number generator seed of 123456 was used as shown in *Box 2*, then your `num_relationships` should be 233 (Fig. 3) and your heatmap should be the same as shown in Fig. 4.

```
Command Window

  sml_regparam =

    1×2 logical array

     1   1


  ridge_regparam =

    1×2 logical array

     1   1

fx >> |
```

Fig. 2 Command window output when code in Box 4 is implemented

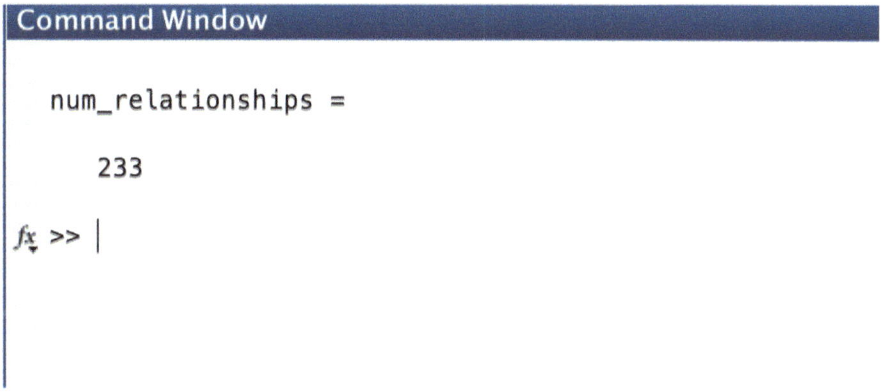

Fig. 3 Command window output when code in Box 5 is implemented

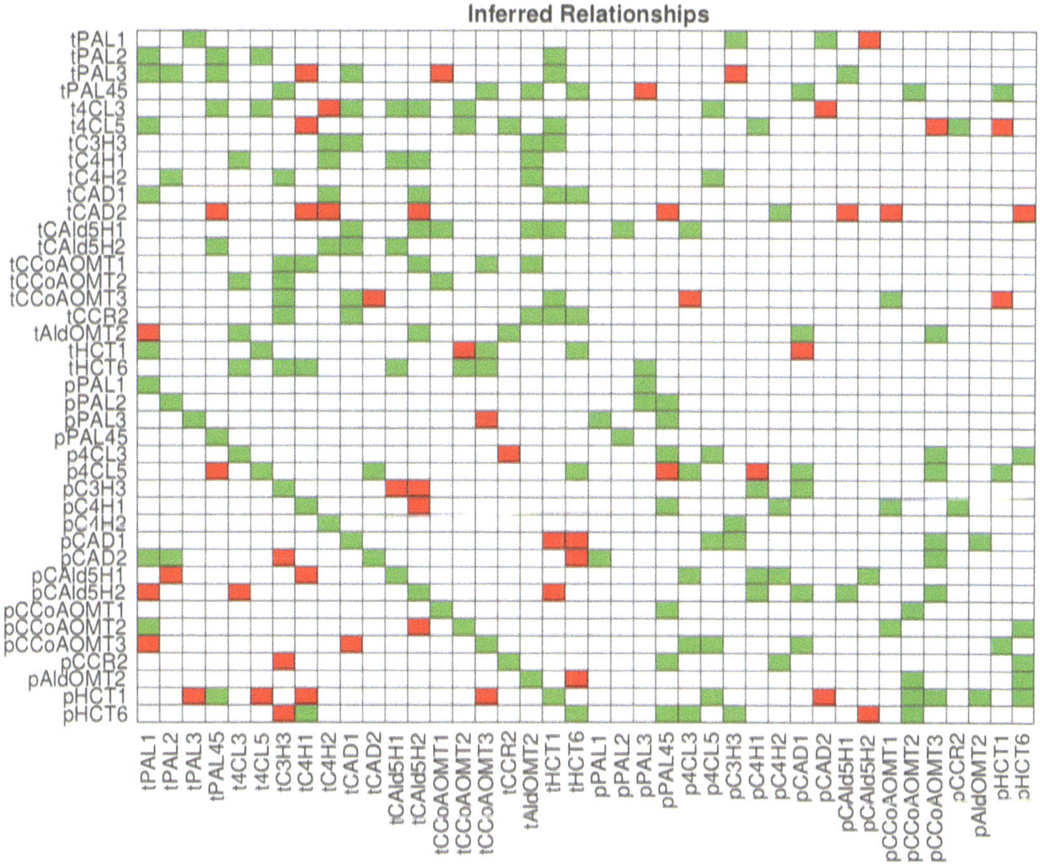

Fig. 4 Heatmap produced with code in Box 5 is implemented. The top left quadrant contains the transcript to transcript influences, the top right quadrant the protein to transcript influences, the bottom left quadrant the transcript to protein influences, and bottom right quadrant the protein to protein influences. For example the top row of this plot indicates positive influences from *PtrPAL3* transcript, *PtrC3H3* protein, and *PtrCAD2* protein on the *PtrPAL1* transcript, and a negative influence from the *PtrCAld5H2* protein on the *PtrPAL1* transcript

Box 5

```
Inferred_Relationships = abs(sign(B));
num_relationships = sum(sum(Inferred_Relationships)); % Number of
    relationships detected
display(num_relationships)

% Plot heatmap
colormap = [1 0 0; % red for negative relationships
   1 1 1; % white for no relationship detected
   0 1 0]; % green for positive relationships

figure;
heatmap(GeneNames,GeneNames,sign(B),'Colormap',colormap,'ColorbarVisible',
    'off');
title('Inferred Relationships')
```

3.4.2 Model Prediction

In this section we will use our model to emulate the experiments in the `LigninTranscriptsProteins.mat` file (see Note 12).

1. Define the inputs needed for the `Model_Prediction` function (Step 3.2.1). `B` and `mue` are outputs from the `SML_wrapper` function. Define the remaining parameters `Xmask`, `Xtarg`, and `Ywt` as shown in *Box 6*. In this example we will consider the proteins to also be targeted at this step.

Box 6

```
Xmask = X_all_mask{:,:}'; % Mask indicating the targeted transcripts for
    all of the experiments
Xtarg = Xmask.*Y_all{:,:}'; % Desired knockdown levels of targeted
    transcripts
Ywt = Ytable{"WT_CK",:}'; % Average wildtype abundances
targ_prot_flag = 1; % Consider proteins targeted as well as transcripts
```

2. Call the `Model_Prediction` function as shown in *Box 7* (see Note 13).

Box 7

```
% Call the function to predict the un-targeted transcript and protein
% abundances
Ypred = Model_Prediction(B,mue,Xtarg,Xmask,targ_prot_flag,Ywt);
Ypred_table = array2table(Ypred','VariableNames',GeneNames,'RowNames',
    Experiments); % Save in table format with Gene names as columns and
    Experiments as rows
```

3. The `Knockdown_Visualization` function has been provided to help visualize how well our model is able to emulate the experimental results. This function will plot the predicted and experimental transcript and protein abundances for a specified monolignol gene for a specified construct (i.e., targeted knockdown), as well as the transcript abundances for the gene(s) targeted in that experiment. *Box 8* shows the code needed to plot the *PtrC3H3* transcripts and proteins when *PtrCAld5H1* and *PtrCAld5H2* are knocked down (Construct i29). The output plots are shown in Figs. 5, 6 and 7.

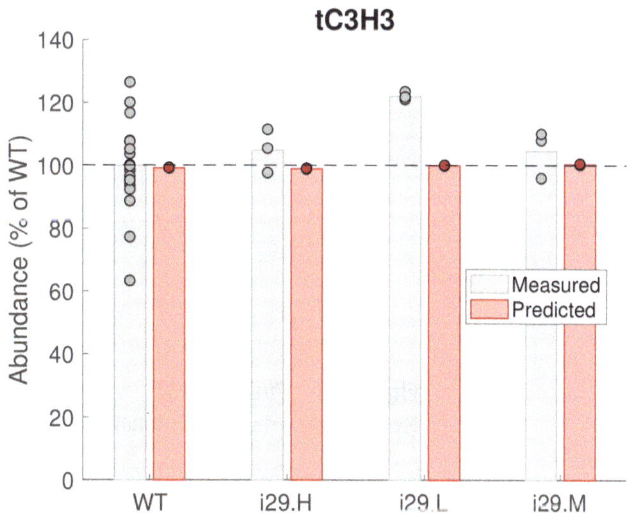

Fig. 5 Predicted and experimental *PtrC3H3* transcript abundances when *PtrCAld5H1&2* are knocked down (Construct i29)

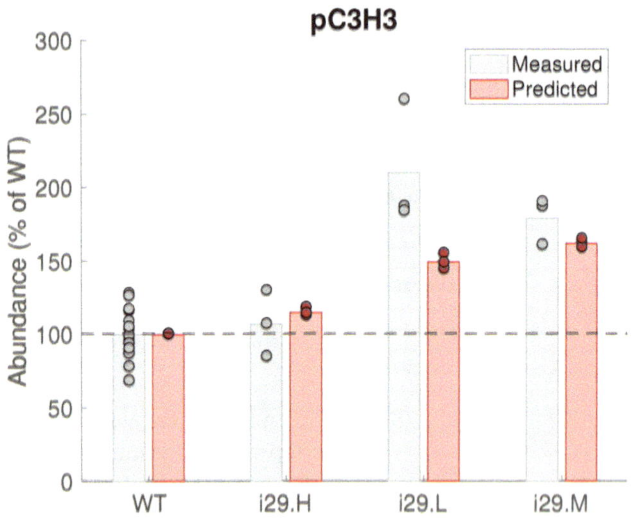

Fig. 6 Predicted and experimental *PtrC3H3* protein abundances when *PtrCAld5H1&2* are knocked down (Construct i29)

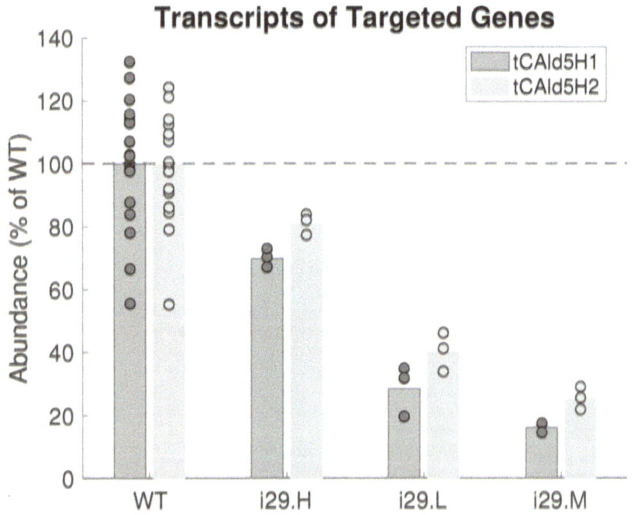

Fig. 7 *PtrCAld5H1&2* transcript abundances input to the `Model_Prediction` function for Construct i29

Box 8

```
gene_of_interest='C3H3'; % Gene you want to plot
knockdown_of_interest='i29'; % Construct/experiment you want to plot

Knockdown_Visualization(gene_of_interest,knockdown_of_interest,Y_all,
    X_all_mask,Ypred)
```

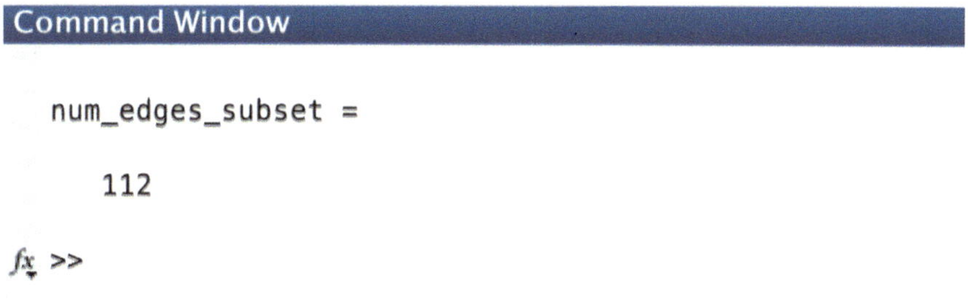

Fig. 8 Command window output when code in Box 10 is implemented

3.4.3 Network Topology Analysis

1. For each experiment determine which transcripts and proteins were predicted "well" using our model. What is considered predicted "well" is up to the user. Here, we consider a transcript or protein to be "well" predicted if the average predicted error over all of the experimental lines is within ±25% of its wild-type abundance. For example, if the wild-type abundance is 100, and the experimental observation is 150, then a predicted abundance between 125 and 175 would be considered "well" predicted. To obtain this information, implement the code in *Box 9*. We recommend using a fairly loose definition of "well" predicted. You can increase or decrease the number of "well" predicted experiments by adjusting the `err_thresh` term in *Box 9*. The `RMSE` table returned from the `PredictedWell` function can be used to help determine what a good `err_thresh` value is (Fig. 8).

Box 9

```
err_thresh = 25; %error threshfold for being considered a 'good' fit

[Gc, Const] = findgroups(cellfun(@(x) x(6:8),Experiments,'UniformOutput',
    false)); %Group experiments into their different constructs (targeted
     knockdowns) (e.g., a01)
[Gcl, ConstLines] = findgroups(cellfun(@(x) x(6:end-2),Experiments,'
    UniformOutput',false)); %Group replicates of the same experimental
    line (e.g., a01.01 for construct a01 line 1)

[fit_table, RMSE] = PredictedWell(err_thresh,Ypred,Y,Ywt,Const,ConstLines,
    Gcl,GeneNames);
```

2. Load the `SigDEtable` table provided in the `SigDEtable.mat` file (*Box 10*). This table indicates which transcripts and proteins were determined to be differentially expressed [2]. There are several method papers on differential expression analysis, so we will not go into that here, but instead refer you to [4–7]. The differentially expressed transcripts and proteins indicated in the provided `SigDEtable` can be seen in Fig 2, and S1–S4 Figs in [2].

3. Call the `CreateEdgeTable` function as shown in *Box 10* to identify which edges are highly influential on changing a given transcript or protein from its wild-type abundance in the experiments where that transcript or protein was both differentially expressed and predicted "well." Here, we define an edge as highly influential if it contributes at least ±50% of the net change in the transcript or protein abundance. The `EdgeTable` returned from the `CreateEdgeTable` function contains a list of all these edges as defined by the source of the edge, the target of the edge, and the sign of the edge (positive or negative). This table can be saved as a CSV file and uploaded into programs such as Cytoscape to view the resulting network.

Box 10

```
load SigDEtable.mat
cutoff=0.5; %percentage of change from wt to identify highly influencing
    relationships (e.g., cutoff=0.5 -> 50% of total change amount)
select_table=fit_table(2:end,:); % creates table with columns and row
    names

select_table{:,:}=fit_table{2:end,:}.*SigDEtable{:,:}; % Look at
    experiments that were BOTH predicted 'well' and differentially
    expressed for each transcript/protein element

% Create EdgeTable with highly influencing relationships (aka edges)
EdgeTable = CreateEdgeTable(B,cutoff,select_table,Y_all,X_all_mask,Ypred,
    Ywt,Gc,Const,Gcl,GeneNames);

num_edges_subset = size(EdgeTable,1); % number of edges considered highly
    influencing
display(num_edges_subset)
```

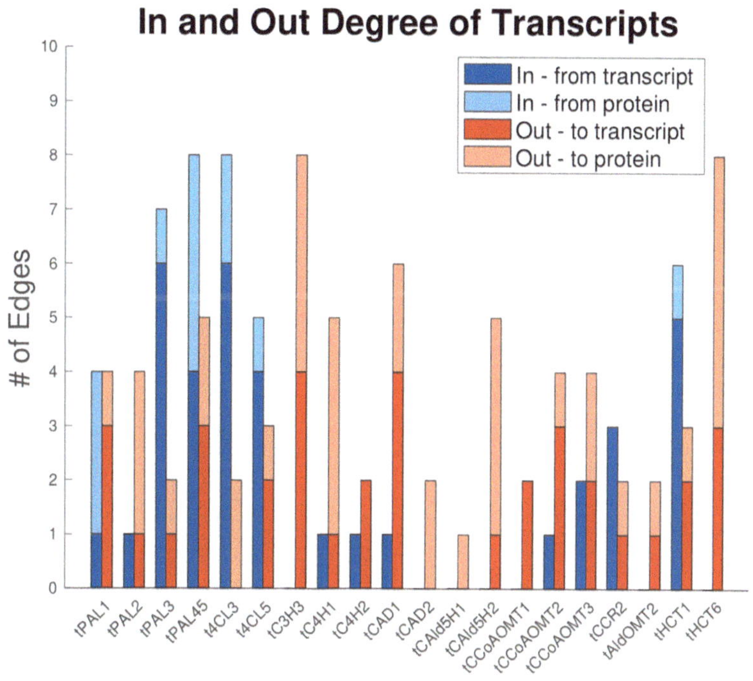

Fig. 9 In and out degrees of the transcript nodes when code in Box 11 is implemented

4. Implement the code in *Box 11* to plot the node degrees from the network of highly influential edges. The resulting plots are shown in Figs. 9 and 10. These plots use the *plotBarStackGroups* function [8]. This function has been provided with the code for this paper, but it can also be installed from MathWorks File Exchange: https://www.mathworks.com/matlabcentral/fileexchange/32884-plot-groups-of-stacked-bars.

Box 11

```
NodeDegrees(EdgeTable,GeneNames) % Plots In-degrees (# of edges
    influencing a node (i.e., transcript/protein)) and Out-degrees (# of
    edges leaving a node)
```

5. Two network motifs were identified in [2] that suggest potential post-transcriptional or post-translational regulatory mechanisms. These motifs feature either a transcript or protein having an opposite influence on the transcript and protein pair of another gene (`DblEdges_out`), or a transcript or protein that

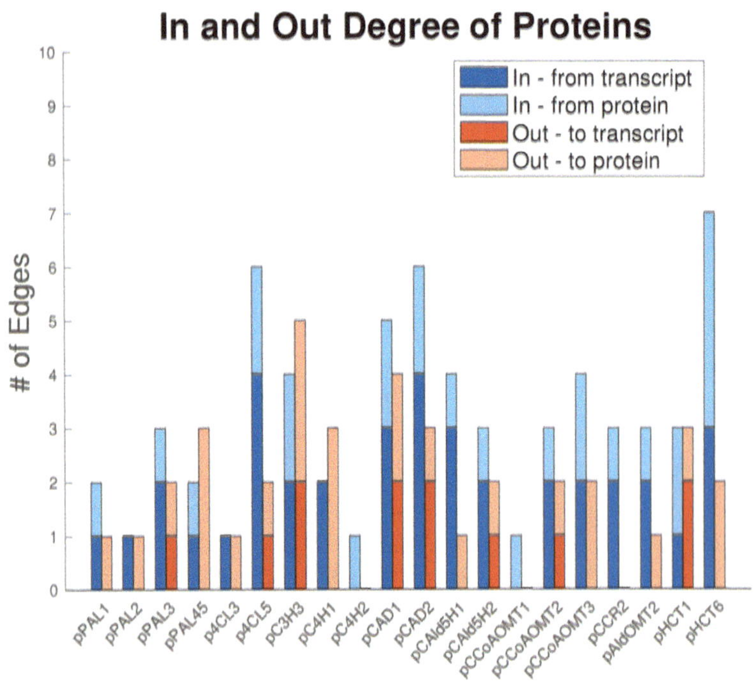

Fig. 10 In and out degrees of the protein nodes when code in Box 11 is implemented

is influenced in opposite ways by the transcript and protein pair of another gene (`DblEdges_in`). See [2] for more information on these motifs. Implement the code in *Box 12* to view these motifs in our network. The list of the edges from our `EdgeTable` that make up these motifs are shown in Figs. 11 and 12.

Box 12

```
% Find the double edge motifs
[DblEdges_in, DblEdges_out]=DblEdge_Motif(EdgeTable,GeneNames);


display(DblEdges_in)
display(DblEdges_out)
```

```
Command Window

  DblEdges_in =

    10×3 table

      Source      Target         Sign
      _______     ___________    ____

      'tHCT1'     't4CL5'          1
      'pHCT1'     't4CL5'         -1
      'tC4H1'     'pCAld5H1'      -1
      'pC4H1'     'pCAld5H1'       1
      'tCAD1'     'pCCoAOMT3'     -1
      'pCAD1'     'pCCoAOMT3'      1
      'tHCT6'     'pAldOMT2'      -1
      'pHCT6'     'pAldOMT2'       1
      'tC3H3'     'pHCT6'         -1
      'pC3H3'     'pHCT6'          1

fx >>
```

Fig. 11 Command window output of the `DblEdgess_in` table when code in Box 12 is implemented

4 Notes

1. *M*: Number of transcripts and proteins.
2. N_d: Number of experiments used for training and developing the model.
3. If your data was taken from multiple batches, then make sure to remove batch effects such as by standardizing to the batch wild-types.
4. SML: Sparsity aware maximum likelihood algorithm that maximizes the maximum likelihood function, but is penalized by the ℓ_1-norm of the parameter vector.
5. This binary matrix specifies which transcripts were targeted for knockdown in each experiment. Since the targeted abundances are experimentally set via gene perturbation strategies, this matrix tells the SML algorithm to ignore the measured abundances of the targeted transcripts when inferring the relationships that impact the abundance of those transcripts.
6. *k*-fold cross-validation: A model validation approach to estimate how well a model will perform on an independent data set. In *k*-fold cross-validation the data is split into *k* groups. A

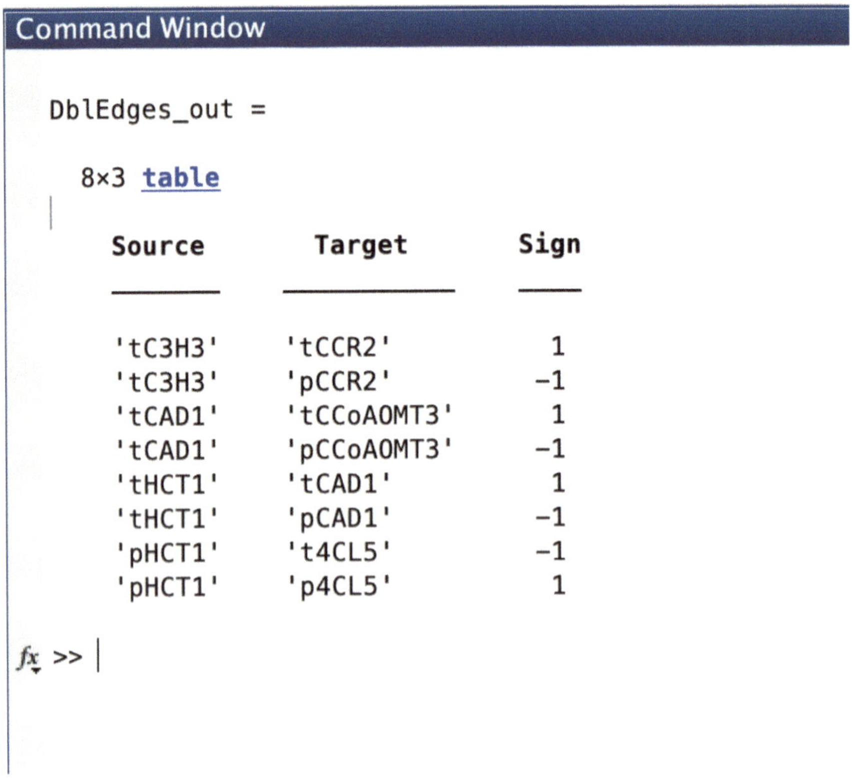

Fig. 12 Command window output of the `DblEdges_out` table when code in Box 12 is implemented

model is then trained on k-1 of the groups and then validated on the kth group of data. This is repeated such that each group of data is the validation set one time.

7. Ridge regression: A regression algorithm that minimizes the least squares problem with a penalty term defined by the ℓ_2-norm of the parameter vector. Ridge regression is useful when the number of parameters, p, are greater than the number of experiments N_d, which is common in biological datasets.

8. The SML code presented in this chapter and [2] is adapted from the code developed by Cai et al. in [9]. The differences between the original code from [9] and the code used here are due to differences in the model formulation. Cai et al. were looking at using eQTL data to help identify relationships between genes and used the model $\mathbf{Y} = \mathbf{BY} + \mathbf{FX} + \boldsymbol{\mu} + \mathbf{E}$. Where $\mathbf{X}$ represented the exogenous eQTL information, and $\mathbf{F}$ their influence on gene expression.

   With the transgenic knockdown experiments we do not have a measure of the exogenous influence (i.e., outside force) modifying the expression of the targeted gene. We only have the final abundance measurement $\mathbf{Y}$. Due to this, we remove the $\mathbf{FX}$ term from our model (Eq. 1).

$$\mathbf{Y} = \mathbf{B}\mathbf{Y} + \boldsymbol{\mu} + \mathbf{E} \tag{1}$$

However, we needed to remove the abundances of the targeted genes when fitting our model, as the exogenous influences are responsible for their change in abundance. For more information on the adjustments to the SML algorithm and the math behind them, see the Methods section of [2] and its supplemental material.

9. N_p: Number of experiments simulated in the model prediction steps.
10. There are a variety of existing methods and papers for doing differential expression analyses to identify differentially expressed genes, so we will not go into detail on these methods and instead refer you to the following references [4–7].
11. What you may consider to be "well" predicted is relative. We recommend being fairly loose on what is considered well predicted. One example for determining what is well predicted would be to consider the experiments that are predicted to be within a given tolerance to be well predicted. An example of this is shown in the example implementation section.
12. Even though our model was trained on the experiments that we are emulating, there is a difference between the data used to train the model (experimental abundances for all transcripts and proteins) and the data used to simulate the model (experimental abundances for only the transcripts of the targeted genes). By emulating these experiments we can compare how well our model is able to predict the changes in abundance of the un-targeted transcripts and proteins. This is not a validation of the model on independent data, which should be performed either using a separate validation set, or an outer cross-validation of the model development as shown in [2].
13. It is possible for this type of model to predict negative abundances which do not make biological sense. If only one or two abundance go negative, you can re-run the `Model_Prediction` algorithm with the negative elements also declared as targets in `Xmask` and set their abundance to 0 in `Xtarg`.
14. This algorithm involves nested cross-validation schemes which can be computationally time intensive. This portion of the code took approximately 20 min to run on the authors 2019 MacBook Pro with 2.8 GHz processor and 16 GB of RAM.
15. The cross-validation schemes randomly sort the data which can lead to slightly different results. To reproduce the exact results shown in this chapter, it is important to seed your random number generator as shown in *Box 2*.

References

1. Vogel C, Marcotte EM (2012) Insights into the regulation of protein abundance from proteomic and transcriptomic analyses. Nat Rev Genet 13 (4):227–232
2. Matthews ML, Wang JP, Sederoff R et al (2020) Modeling cross-regulatory influences on monolignol transcripts and proteins under single and combinatorial gene knockdowns in *Populus trichocarpa*. PLoS Comput Biol 16(4):e1007197
3. Wang JP, Matthews ML, Williams CM et al (2018) Improving wood properties for wood utilization through multi-omics integration in lignin biosynthesis. Nat Commun 9:1579
4. Anders S, McCarthy DJ, Chen Y et al (2013) Count-based differential expression analysis of RNA sequencing data using R and Bioconductor. Nat Protoc 8(9):1765–1786
5. Kammers K, Cole RN, Tiengwe C et al (2015) Detecting significant changes in protein abundance. EuPA Open Proteom 7:11–19
6. Van den Berge K, Hembach KM, Soneson C et al (2019) RNA sequencing data: Hitchhiker's guide to expression analysis. Ann Rev Biomed Data Sci 2(1):139–173
7. Love MI, Huber W, Anders S (2014) Moderated estimation of fold change and dispersion for RNA-seq data with DESeq2. Genome Biol 15:550
8. Bollig E (2011) Plot groups of stacked bars. https://www.mathworks.com/matlabcentral/fileexchange/32884-plot-groups-of-stacked-bars. Accessed 24 Sept 2020
9. Cai X, Bazerque JA, Giannakis GB (2013) Inference of gene regulatory networks with sparse structural equation models exploiting genetic perturbations. PLoS Comput Biol 9(5):e1003068

Chapter 8

Biological Network Mining

Zongliang Yue, Da Yan, Guimu Guo, and Jake Y. Chen

Abstract

In this book chapter, we introduce a pipeline to mine significant biomedical entities (or bioentities) in biological networks. Our focus is on prioritizing both bioentities themselves and the associations between bioentities in order to reveal their biological functions. We will introduce three tools BEERE, WIPER, and PAGER 2.0 that can be used together for network analysis and function interpretation: (1) BEERE is a network analysis tool for "Biomedical Entity Expansion, Ranking and Explorations," (2) WIPER is an entity-to-entity association ranking tool, and (3) PAGER 2.0 is a service for gene enrichment analysis.

Key words Network mining, Gene, Microarray, Protein, Protein–protein interaction (PPI)

1 Introduction

High-throughput biological studies using Genome-Wide Association Studies (GWAS) or RNA-sequencing yield an overwhelming amount of candidate genetic variants and candidate genes that are prohibitive for manual examinations, thus calling for the help of gene prioritization analysis [1]. Given the results of high-throughput biology experiments, systems biology analysis focuses on solving problems such as biomolecule prioritization and finding biological functions. Biomolecule prioritization usually provides a ranking list of genes or proteins, which biomedical researchers can refer to for further clinical validations. Biological function mining and translational biomedical research enable the discovery of critical bioentities such as genes, diseases, drugs, phenotypic features, and clinical attributes from a candidate bioentity list, which is an important step in the conventional hypothesis-driven analysis.

Various prioritization algorithms have been developed to prioritize significant biomolecules and associations. Bioinformaticians typically apply gene ranking, protein–protein interaction (PPI) ranking and/or gene set based enrichment analysis for subsequent experimental validations [2–5]. However, in the past decade, only a limited number of gene prioritization tools have been developed

Shahid Mukhtar (ed.), *Modeling Transcriptional Regulation: Methods and Protocols*, Methods in Molecular Biology, vol. 2328, https://doi.org/10.1007/978-1-0716-1534-8_8,

such as PINTA [6], ToppGene [7], SUSPECTS [8], PROSPECTR [9], and ENDEAVOUR [10]. These tools have been used for the statistical characterizations of genetic linkage patterns, sequence annotations, gene co-expression patterns, protein–protein network linkage patterns, or correlated pathways [11–16]. In particular, network-based gene prioritization methods have been shown to demonstrate an overall high accuracy and low system-level bias, thanks to the integration of scored gene-to-gene association relationships from a comprehensive set of sources into knowledge bases such as the STRING database [17] and the HAPPI 2.0 database [18]. Examples of network-based gene prioritization applications include discovering disease genes for complex human genetic disorders [19], finding drug targets [20, 21], and repositioning drugs [22].

A comprehensive biomedical entity-to-entity association knowledge base can significantly enhance the effectiveness of network-based gene prioritization methods in finding biological functions [23]. However, extracting semantic information from the biomedical literature or PubMed-scale clinical databases has been challenging for several reasons: (1) a lack of standard ontology and its application to annotate all PubMed sentences [23], (2) a lack of advanced natural language processing (NLP) techniques that have been tested at scale [24], (3) a lack of comprehensive application tools on top for further data analysis [25]. Furthermore, there is an urgent need for tools to help prioritize bioentities across the broad concept-categories of the Unified Medical Language System (UMLS). Before our work, only Phenolyzer covers two categories—"diseases" and "genes" [26].

In this chapter, we will introduce three tools BEERE, WIPER, and PAGER 2.0 that can be used together for network analysis and function interpretation: (1) BEERE is a network analysis tool for "Biomedical Entity Expansion, Ranking and Explorations," (2) WIPER is an entity-to-entity association ranking tool, and (3) PAGER 2.0 is a service for gene enrichment analysis.

Specifically, to perform general-purpose web-based "biomedical term prioritizations," **B**iomedical **E**ntity **E**xpansion, **R**anking, and **E**xplorations (**BEERE**) has been developed to integrate advanced entity disambiguation techniques [27–30], network-based prioritization techniques, and biomedical semantic relationship database repositories [23]. BEERE works in two input modes, i.e., a gene input mode and a term input mode. We will show how to use the tool BEERE to prioritize user-provided biomedical entities for detailed investigations of the related concepts, known associative relationships among them, supporting literature evidence, their relative significance to one another, and the relationship network context that they reside in.

Further, we will illustrate how to prioritize network edges using **W**eighted **I**n-**P**ath **E**dge **R**anking (**WIPER**) [31], which can effectively utilize weighted edges along with the graph topology to rank network edges. Therefore, BEERE can help solve biomedical problems such as finding a therapeutic strategy by "targeting the right interactions in the interactome" [32].

We will also apply **P**athway, **A**nnotated list and **G**ene signature **E**lectronic **R**epository (**PAGER**) [33] to find the gene functions in the network by performing enrichment analysis. PAGER is a novel and comprehensive database infrastructure by integrating PAGs—a new unified data structure to represent heterogeneous Pathways (P-type), Annotated-lists (A-type), and Gene-signatures (G-type) [34]. The PAGs were derived from 24 different data sources that cover, for example, human diseases, published gene expression signatures, known gene lists affected by shared drugs, pathways, shared miRNA-gene interaction targets, tissue-specifically co-expressed genes, and all genes sharing common protein functional annotations. PAGER also provides enrichment analysis for genes using the hypergeometric test. The input of PAGER is a gene list, and the output is a list of enriched PAGs ranked by p-values.

2 Materials

2.1 BEERE Webserver

BEERE provides a webserver (http://discovery.informatics.uab.edu/BEERE) to allow users to perform the network construction and node prioritization online. Users need to enter either a gene list or a biomedical term list as the input.

2.2 WIPER API Service

WIPER provides online documentation (http://discovery.informatics.uab.edu/wiper/) to explain the usage of the API services. Specifically, the webpage uses an example to illustrate how to format an API request using the Python and R languages.

2.3 PAGER Webserver

PAGER provides a web portal (http://discovery.informatics.uab.edu/PAGER/) to allow users to perform gene set enrichment analysis online.

3 Method

3.1 Ranking Biomedical Entities and Visualizing Networks

1. **Enter a list of terms or genes.** Go to the website "http://discovery.informatics.uab.edu/BEERE/" and click on the "try the tool" button to start a search. Select the type of entries and enter a list of terms or genes. When users select "Gene," BEERE will retrieve the matched entities and relationships from the databases HAPPI-2.0 [18]. On the other hand, when users select "Term," BEERE will retrieve the matched entities and relationships from the databases SemMedDB [35]. Click on either of the two examples on the searching

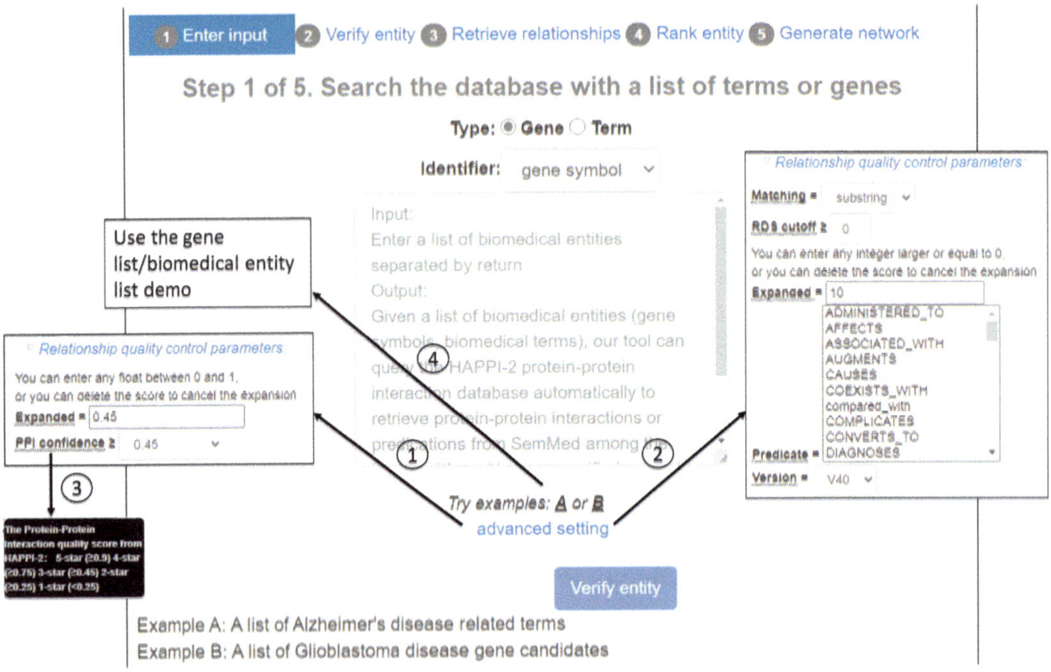

Fig. 1 The page for entering a list of genes or terms. (1–2) When users select "Gene"/"Term" as the identifier and click on the "advanced setting," the window with relationship quality-control parameters will appear. (3) All the parameters with dashed lines will show the details on mouseover events. (4) Users can click on "A" or "B" (Example A: a list of Alzheimer's disease-related terms, Example B: a list of glioblastoma disease gene candidates) to use the examples provided by BEERE

page to use the demo (*see* **Note 1** and Fig. 1). Click on the advanced parameter setting to control relationship quality. For the two types to choose from for an input list, two different parameter sets are provided. In PPI retrieval, the unique parameter "PPI confidence" provides three PPI cutoffs "0.45," "0.75," and "0.9" and a "customized cutoff" option. Those three PPI cutoffs are equivalent to the 3-star, 4-star, and 5-star PPI quality in the HAPPI-2.0 database. In term-to-term relationship retrieval, there are three unique parameters "relation density score (RDS)," "predicate," and "matching." In the "relation density score (RDS)" parameter, enter a number varying from 0 to the maximum value of the RDS to set an RDS cutoff. In the parameter "predicate," do a single predicate selection or multiple selections. In the parameter "matching," select one of the options "fuzzy matching," "substring matching," and "exact matching" to perform the term matching. Enter a score in the parameter "expanded" to enable the option for one-layer network expansion in both PPI and term-to-term relationship retrieval. This "expanded" option will potentially increase the index of aggregation (IOA) (*see* **Note 2**) by introducing "bridge" nodes to the network. If users do not want to

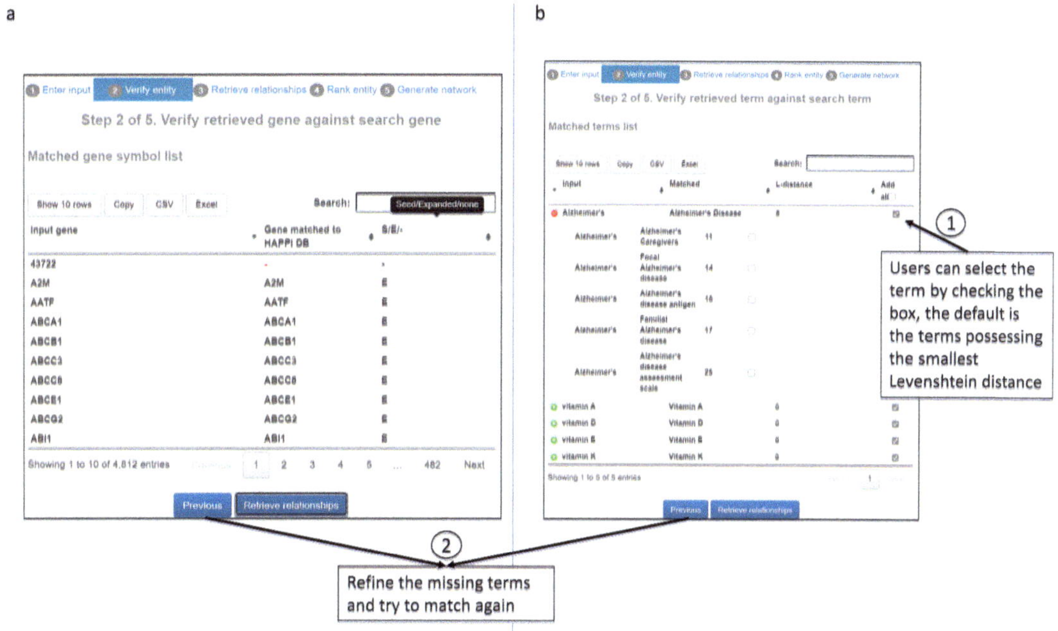

Fig. 2 The page for verifying genes/terms against search genes/terms. (1) Panel (**a**) displays the verified gene symbol list in the table and provides the expansion tags ("S": seed, "E": expanded and "-": none). Panel (**b**) displays the verified terms with Levenshtein distances. Users can click on the expansion button in green color to check or uncheck the matching terms. (2) Users can also click on the "previous" button to refine the missing terms and try to match again

do an expansion, simply delete the score in the "expanded" box and click on any other place to cancel the expansion.

2. **Verify the retrieved terms against search terms**. Check and verify the retrieved biomedical entities in the matched-entity table (*see* **Note 2** and Fig. 2). In the case of a gene list, the gene-matching table shows the queried gene symbols, matched genes, and their Seed/Expanded/none tags "S/E/–". If an input gene is an alias or a gene synonym, BEERE will automatically map the queried gene to HAPPI 2.0 database gene symbols. In the case of a term list using "substring" matching, the matching table shows the matches and modified matches with the lowest Levenshtein distance (L-distance) as the best candidates. Click on any of the green "expansion" buttons of the retrieved bioentities, and it will display all the potential mismatches for an adjustment.

3. **Retrieve known relationships**. Check the quality of the relationships in the relationships-table (Fig. 3). Users can choose to refine the table by clicking on the "previous" button or clicking on any step provided at the top of the page to go back, change parameters, and match again, or to perform entity prioritization. Click on the "advanced setting," users can adjust

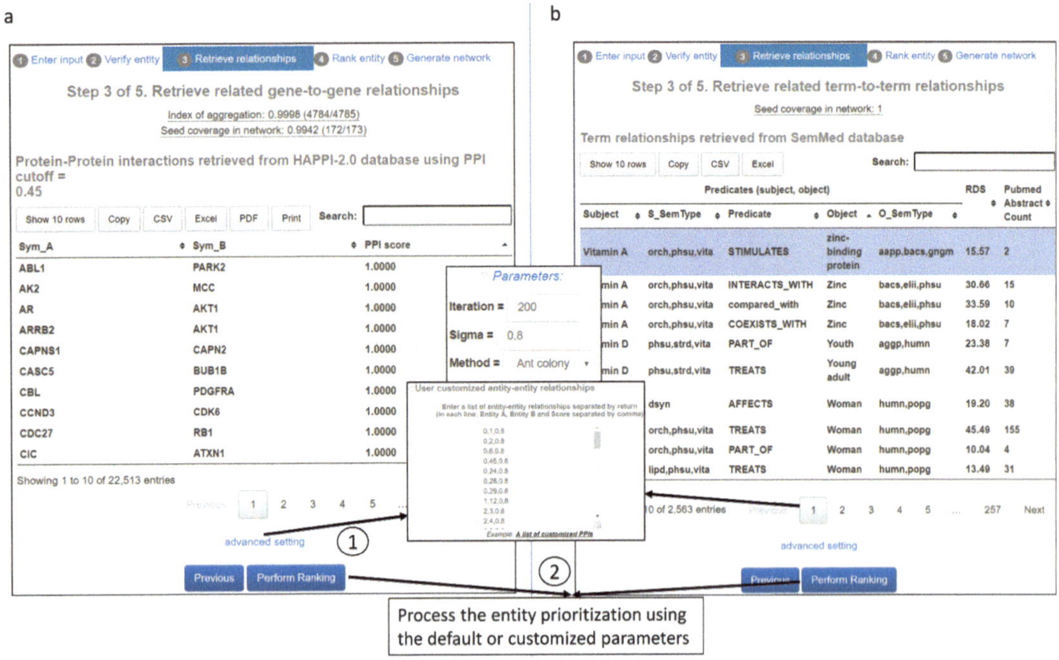

Fig. 3 The page for retrieving relationships. (1) Panel (**a**) displays the retrieved related gene-to-gene relationships and Panel (**b**) displays the retrieved related term-to-term relationships. By clicking on the "advanced setting," users can check on the parameters in performing bioentity prioritization, or add customized biomedical entity-to-entity relationships. (2) By clicking on the "perform ranking" button, users can process bioentity prioritization using the default or customized parameters

the parameters "iteration," "sigma," and "method." In the parameter "method," select either "page rank" or "ant colony" algorithm in the bioentity prioritization (*see* **Note 3**). Enter a number in the parameter "iteration" to specify the iterations run by the algorithm. Enter a float number ranging from 0 to 1 (default value is 0.8) in the damping factor parameter "sigma" (*see* **Note 4**). Optionally, users can enter a list of entity-entity relationships (one per line) in the user-customized entity-entity relationships box.

4. **Rank entities from the network**. Check the table of biomedical entity prioritization results and two visualization panels (Fig. 4). In the prioritization table, there are 6 parameters, "entity name," "in-expanded network," "ranking score," "rank," "adjust p-value," and "significance." Click on any of the bioentities in the table and it will link to an entity information page, which shows the attributes of the selected biomedical entity and the relationships specific to that entity. In the visualization panels, two visual-friendly figures help users to intuitively view the significant entities and the ranking score distribution. In particular, the word-cloud figure displays those

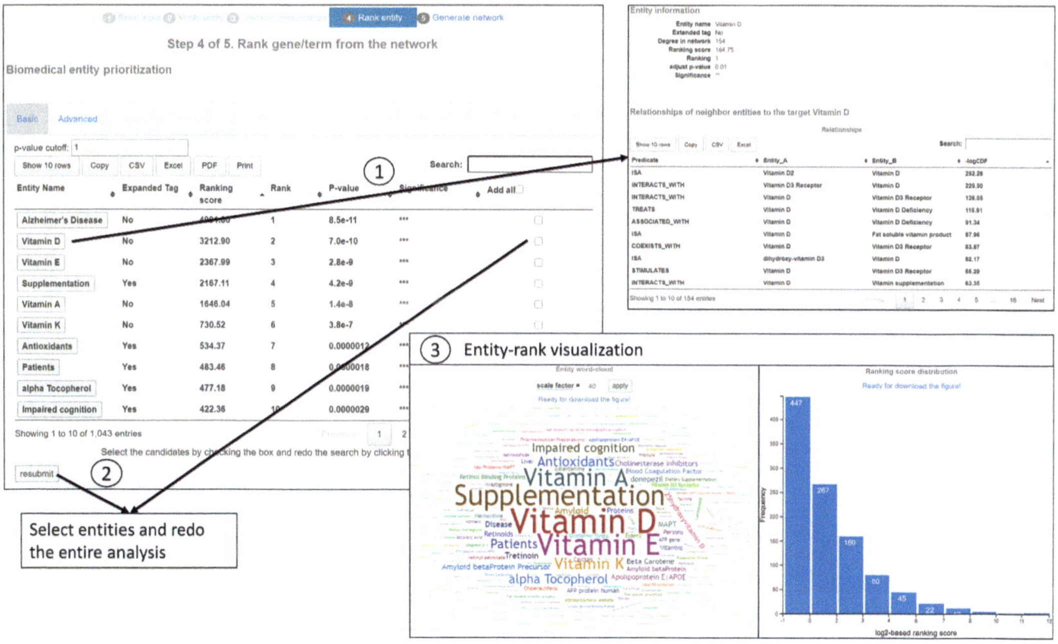

Fig. 4 The page for gene/term ranking from the network. Users can view bioentity prioritization. (1) Users can click on a bioentity name and view the bioentity information and the relationships of neighbors to the queried bioentity. (2) By checking the bioentity boxes, users can select bioentities and redo the entire analysis. (3) The bioentity-rank visualization provides a word cloud to highlight the top-ranked bioentities and a histogram to show the ranking score distribution

highly significant biomedical entities in the center with relatively larger fonts.

5. **Visually explore the network relationship data**. Check the network provided in the network visualization page. This page provides an interactive graphical panel to allow users to intuitively discover the critical entities and interactions with provenance (Fig. 5). Click on any of the three layout algorithm buttons, force-directed (default), distance-bounded energy-field minimization algorithm (DEMA), and circular to regenerate the network layout. Click on the "current view" button to export the current view of the network as a PNG image or click on the "SVG file" button to export the network's edge and node information as an SVG file. Click on the "advanced" button on the left side, enter customized entity-association information, and click on the "add" button. BEERE will visualize the color-grouped nodes in the network and show the grouping information table below the network graph. Click on any of the edges in the network to activate the panel on the right side and view a table of entity-relationship's details and the relationship provenance.

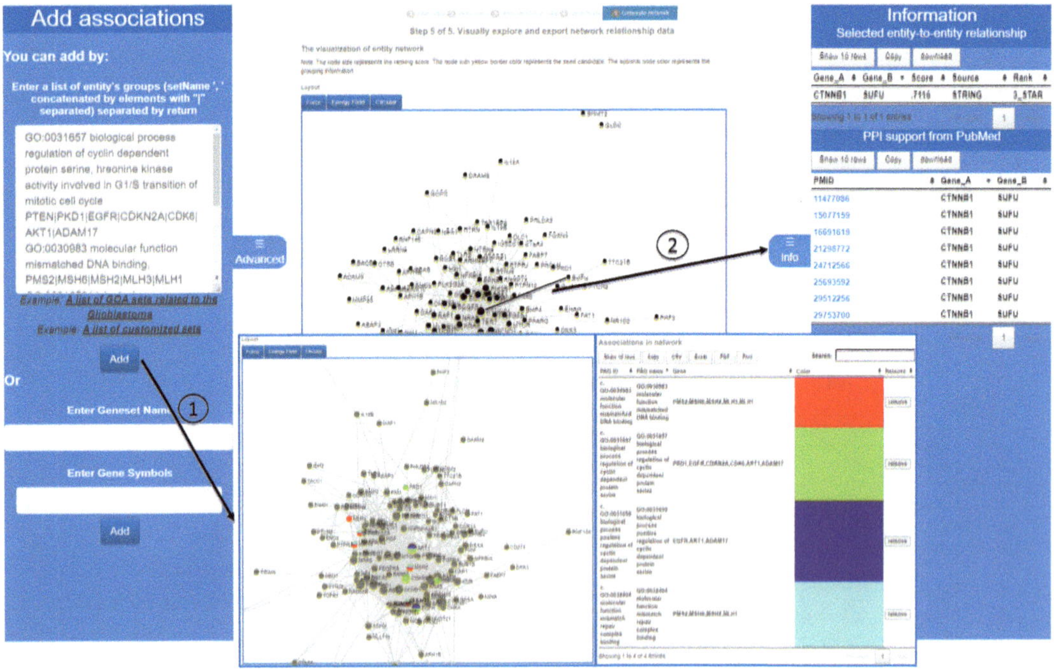

Fig. 5 The page for network visualization and relationship exploration. (1) Users can click on the "Advanced" button to add associations and visualize them in the network. (2) Users can click on an edge to activate the "Info" panel. The information panel shows detailed information about the selected biomedical entity-to-entity relationship and its provenance supported by PubMed articles

3.2 Ranking Biomedical Entity-to-Entity Associations

To perform the edge ranking on the result of BEERE, download the edge table described in **step 3** of Subheading 3.1, and convert it into a string required by WIPER API service (c.f. Subheading 2.3). Open an empty R or Python script, and copy and paste the example codes listed in the "http://discovery.informatics.uab.edu/wiper/". Replace the interaction string "a b 0.9 b c 0.9 c a 0.9" by the customized string from BEERE. BEERE also requires five additional parameters, "sigma," "PPI cutoff" (*see* **Note 5**), "path length cutoff" (*see* **Note 5**), "method," and "iteration," to run the edge prioritization. Check and revise parameters accordingly based on the detailed information of the parameters listed in WIPER's parameter table on the WIPER website. The explanation of all the output variables is provided on the WIPER website's help-page.

3.3 Performing Gene Set Enrichment Analysis Using PAGER

Given a list of ranked genes from **step 5** of Subheading 3.1, users can copy the first column (Entity Name) and paste it into the box on PAGER's advanced search page. By default, PAGER sets parameter "type of PAG" as "all," "size of PAGs" as "[2–1000]," "similarity score" $\geq$0.1 (*see* **Note 6**), "number of overlapping genes" as ">1," "cohesion score" $\geq$100 (*see* **Note 6**), "FDR" $\leq$0.05. Click on the "search PAGs" button to perform gene set enrichment analysis. Users can also update the parameters on the query result page to redo the analysis. Users can download the results by clicking the buttons above the result table.

4 Notes

1. In example A, we provide a gene list composed of 200 Glioblastoma (GBM) genetic candidates in 243 entries associated with glioblastoma from the OMIM (Online Mendelian Inheritance in Man) database (https://www.omim.org). In example B, we provide a term list of 3 different vitamins and Alzheimer's disease.
2. In the BEERE pipeline, to guarantee the quality of an expansion network, a four-step procedure involving iterative cutoff tuning and manual curation is designed (Fig. 6). The gene expansion is based on the principle of "Guilt by Association" (GBA) which says that disease-associated genes are normally located closer to each other than random pairs in the network. Given the gene candidates associated with a disease collected from statistical analysis, we follow a four-step procedure to reveal the additional gene partners coherently worked with the candidates in an expanded network. Firstly, we query genes which are matched to the seed list. Note that some genes need to be manually reviewed to improve the bioentity coverage and to avoid the error in automatic matching. Secondly, we retrieve the Protein–Protein Interactions (PPIs)

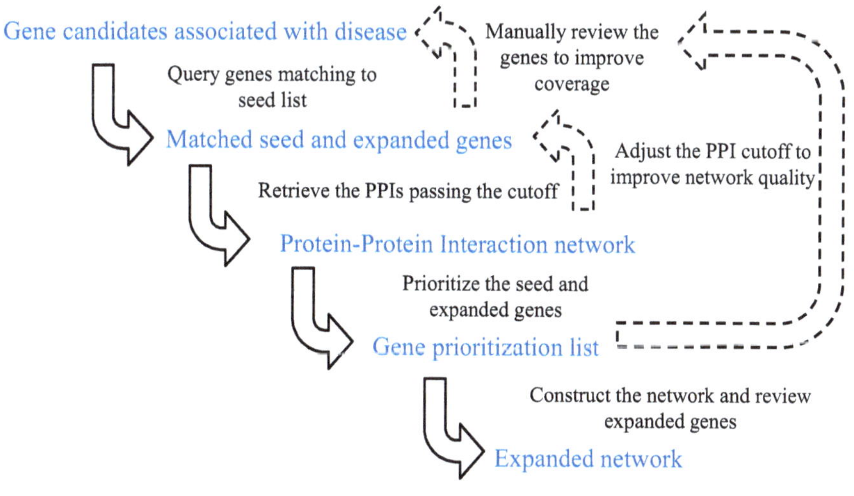

Fig. 6 The workflow for the gene expansion network using BEERE. The solid arrows are for the four steps performed to generate the expanded network. The dashed arrows are for the backtracking steps in options to refine the query

which pass the PPI confidence cutoff score. In this step, we search for a balance point where the expanded network has few isolated genes, and the quality of interactions is not too low (e.g., 3-star or below in HAPPI). To measure the network quality, two metrics, Index of Aggregation (IOA) and Seed Coverage in the Network (SCN), are introduced and displayed on the "retrieving known-relationships" page (c.f. Subheading 3.1, **step 3**). IOA assumes that a high-quality network tends to have the largest subnetwork containing as many connected nodes as possible. IOA is calculated as the number of nodes in the largest subnetwork divided by the total number of input nodes. SCN is a broader aggregation index which assumes that the higher quality a network is, the more connected nodes exist than isolated nodes. SCN is calculated as the number of connected nodes divided by the total number of input nodes. Thirdly, we prioritize the seed and expanded genes using network propagation. The gene list can be further refined by manual curation based on prior knowledge to filter out wrong candidates. The refined prioritization list can then be used as the input in another round of network expansion. Fourthly, we construct the final network and validate the expanded genes using the literature-mined predications in SemMedDB.

3. In the step of retrieving known relationships of BEERE analysis, the network propagation algorithm is preferred to the modified PageRank algorithm if we expect to measure node weights globally. Network propagation has advantages in (1) the interactive node weight updates through all possible paths between nodes and (2) a steady state of fluid can be estimated once the residue fluid is equal to the passively received incoming fluid. Therefore, the node weights at the steady state reflect node importance that optimizing the fluid distribution in the whole network. Compared to network propagation, PageRank is a partially local ranking algorithm that keeps the restart probability using the damping factor to integrate the initial local weights.
4. The damping factor "sigma" determines the probability of the click-through to prevent sinks (i.e., pages with no outgoing links) from "absorbing" the PageRanks of those pages connected to the sinks.
5. In WIPER, to reduce the cost of computing a backbone edge-to-edge traversal-path network used for edge ranking, we limit the number of edge-to-edge connections considered during the computation, by applying a maximum-hop cutoff and a traversal-path confidence-score cutoff. The default maximum-hop cutoff is infinite and the default traversal-path confidence-score cutoff is 0, but this setting is computationally very costly

Semantic type to restrict the expanded entity related to drug

Abbreviation	Unique Identifier (TUI)	Full Name
clnd	T200	Clinical Drug
chem	T103	Chemical
chvf	T120	Chemical Viewed Functionally
chvs	T104	Chemical Viewed Structurally
vita	T127	Vitamin
phsu	T121	Pharmacologic Substance

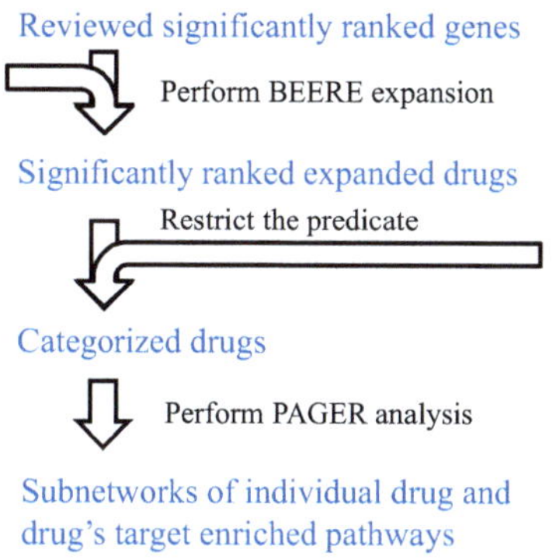

Category of the drugs according to the predicates

Predicate	Category	
TREATS	Beneficial	Conditional
PREVENTS	Beneficial	Conditional
CAUSES	Deleterious	Conditional
NEG_TREATS	Deleterious	Conditional
NEG_PREVENTS	Deleterious	Conditional
AFFECTS	Neutral	
NEG_AFFECTS	Neutral	

+

Category of the drugs by reviewing PubMed literatures

Fig. 7 The workflow of repositioned drug expansion. The arrows are the three steps to generate the subnetwork of individual drug and the drug's target enriched pathways. The table on the left lists the semantic types to restrict the expanded bioentities related to drugs using the abbreviation matching. The table on the right lists the categories of the drugs according to the predicate sentiments

especially when the network is big. Our cutoff thresholds enable the reduction of the computation workloads by allowing users to limit the number of hops during the computation, as well as removing low-confidence traversal-paths.

6. In the BEERE pipeline, given an input in the form of a human-reviewed significantly ranked gene list (which can be regarded as the potential drug targets), we can perform a three-step procedure to find repositioned drugs as illustrated by Fig. 7, which utilizes additional information such as bioentity semantic types (left-table in Fig. 7) and drug-to-bioentity predicate categories (right-table in Fig. 7). Firstly, we perform the BEERE expansion but we limit the expanded bioentity semantic types to be only drug-related on the "retrieving known-relationships" page (c.f. Subheading 3.1, **step 3**). Secondly, we can also restrict the predicates to only drug-to-bioentity predicate categories such as treats, prevents, and causes on the "searching the databases with a list of terms or genes" page (c.f. Subheading 3.1, **step 1**). In the right-table of Fig. 7, we can group the predicates into four categories (aka. predicate sentiments): beneficial, deleterious, conditional, and neutral. Thus, we can evaluate the drug sentiments by aggregating the weights of those categorized predicates obtained from the retrieved relationships-table on the "retrieving known-relationships" page (c.f. Subheading 3.1, **step 3**). Thirdly, we use PAGER described in Subheading 2.3 to perform gene set enrichment analysis to identify the functions of the target genes for each drug. The parameter "similarity score" represents the similarity between a queried list and an enriched PAG based on the combination of the overlap coefficient and the Jaccard index described in [36]. The parameter "cohesion score" represents PAG's pre-calculated quality score (called

the "nCoCo Score") that measures the biological relevance of enriched PAGs based on the scoring of intra-PAG gene–gene interaction while controlling for PAG size [33].

References

1. Guala D, Sonnhammer ELL (2017) A large-scale benchmark of gene prioritization methods. Sci Rep 7:46598. https://doi.org/10.1038/srep46598
2. Chen JY, Shen C, Sivachenko AY (2006) Mining Alzheimer disease relevant proteins from integrated protein interactome data. Pac Symp Biocomput:367–378
3. Theodosiou T, Efstathiou G, Papanikolaou N et al (2017) NAP: the network analysis profiler, a web tool for easier topological analysis and comparison of medium-scale biological networks. BMC Res Notes 10(1):278. https://doi.org/10.1186/s13104-017-2607-8
4. Wang Z, Duenas-Osorio L, Padgett JE (2015) A new mutually reinforcing network node and link ranking algorithm. Sci Rep 5:15141. https://doi.org/10.1038/srep15141
5. Wang J, Li M, Wang H et al (2012) Identification of essential proteins based on edge clustering coefficient. IEEE/ACM Trans Comput Biol Bioinform 9(4):1070–1080. https://doi.org/10.1109/TCBB.2011.147
6. Nitsch D, Tranchevent LC, Goncalves JP et al (2011) PINTA: a web server for network-based gene prioritization from expression data. Nucleic Acids Res 39(Web Server issue): W334–W338. https://doi.org/10.1093/nar/gkr289
7. Chen J, Bardes EE, Aronow BJ et al (2009) ToppGene suite for gene list enrichment analysis and candidate gene prioritization. Nucleic Acids Res 37(Web Server issue):W305–W311. https://doi.org/10.1093/nar/gkp427
8. Adie EA, Adams RR, Evans KL et al (2006) SUSPECTS: enabling fast and effective prioritization of positional candidates. Bioinformatics 22(6):773–774. https://doi.org/10.1093/bioinformatics/btk031
9. Yu W, Wulf A, Liu T et al (2008) Gene prospector: an evidence gateway for evaluating potential susceptibility genes and interacting risk factors for human diseases. BMC Bioinformatics 9:528. https://doi.org/10.1186/1471-2105-9-528
10. Tranchevent LC, Barriot R, Yu S et al (2008) ENDEAVOUR update: a web resource for gene prioritization in multiple species. Nucleic Acids Res 36(Web Server issue):W377–W384. https://doi.org/10.1093/nar/gkn325
11. Doncheva NT, Kacprowski T, Albrecht M (2012) Recent approaches to the prioritization of candidate disease genes. Wiley Interdiscip Rev Syst Biol Med 4(5):429–442. https://doi.org/10.1002/wsbm.1177
12. Bornigen D, Tranchevent LC, Bonachela-Capdevila F et al (2012) An unbiased evaluation of gene prioritization tools. Bioinformatics 28(23):3081–3088. https://doi.org/10.1093/bioinformatics/bts581
13. Moreau Y, Tranchevent LC (2012) Computational tools for prioritizing candidate genes: boosting disease gene discovery. Nat Rev Genet 13(8):523–536. https://doi.org/10.1038/nrg3253
14. Oti M, Ballouz S, Wouters MA (2011) Web tools for the prioritization of candidate disease genes. Methods Mol Biol 760:189–206. https://doi.org/10.1007/978-1-61779-176-5_12
15. Piro RM, Di Cunto F (2012) Computational approaches to disease-gene prediction: rationale, classification and successes. FEBS J 279 (5):678–696. https://doi.org/10.1111/j.1742-4658.2012.08471.x
16. Tranchevent LC, Capdevila FB, Nitsch D et al (2011) A guide to web tools to prioritize candidate genes. Brief Bioinform 12(1):22–32. https://doi.org/10.1093/bib/bbq007
17. Szklarczyk D, Morris JH, Cook H et al (2017) The STRING database in 2017: quality-controlled protein-protein association networks, made broadly accessible. Nucleic Acids Res 45(D1):D362–D368. https://doi.org/10.1093/nar/gkw937
18. Chen JY, Pandey R, Nguyen TM (2017) HAPPI-2: a comprehensive and high-quality map of human annotated and predicted protein interactions. BMC Genomics 18(1):182. https://doi.org/10.1186/s12864-017-3512-1
19. Lupski JR, Reid JG, Gonzaga-Jauregui C et al (2010) Whole-genome sequencing in a patient with Charcot-Marie-tooth neuropathy. N Engl J Med 362(13):1181–1191. https://doi.org/10.1056/NEJMoa0908094
20. Isik Z, Baldow C, Cannistraci CV et al (2015) Drug target prioritization by perturbed gene expression and network information. Sci Rep

5:17417. https://doi.org/10.1038/srep17417

21. Sivachenko AY, Yuryev A (2007) Pathway analysis software as a tool for drug target selection, prioritization and validation of drug mechanism. Expert Opin Ther Targets 11 (3):411–421. https://doi.org/10.1517/14728222.11.3.411
22. Yue Z, Arora I, Zhang EY et al (2017) Repositioning drugs by targeting network modules: a Parkinson's disease case study. BMC Bioinformatics 18(Suppl 14):532. https://doi.org/10.1186/s12859-017-1889-0
23. Denecke K (2008) Semantic structuring of and information extraction from medical documents using the UMLS. Methods Inf Med 47 (5):425–434
24. Burger G, Abu-Hanna A, de Keizer N et al (2016) Natural language processing in pathology: a scoping review. J Clin Pathol. https://doi.org/10.1136/jclinpath-2016-203872
25. Matthies F, Hahn U (2017) Scholarly information extraction is going to make a quantum leap with PubMed central (PMC). Stud Health Technol Inform 245:521–525
26. Yang H, Robinson PN, Wang K (2015) Phenolyzer: phenotype-based prioritization of candidate genes for human diseases. Nat Methods 12(9):841–843. https://doi.org/10.1038/nmeth.3484
27. Song Y, Kim E, Lee GG et al (2005) POSBIOTM-NER: a trainable biomedical named-entity recognition system. Bioinformatics 21(11):2794–2796. https://doi.org/10.1093/bioinformatics/bti414
28. Wang X, Zhang Y, Ren X et al (2018) Cross-type biomedical named entity recognition with deep multi-task learning. Bioinformatics. https://doi.org/10.1093/bioinformatics/bty869
29. Zhao Z, Yang Z, Luo L et al (2017) Disease named entity recognition from biomedical literature using a novel convolutional neural network. BMC Med Genet 10(Suppl 5):73. https://doi.org/10.1186/s12920-017-0316-8
30. Lee S, Kim D, Lee K et al (2016) BEST: next-generation biomedical entity search tool for knowledge discovery from biomedical literature. PLoS One 11(10):e0164680. https://doi.org/10.1371/journal.pone.0164680
31. Yue Z, Nguyen T, Zhang E et al (2019) WIPER: weighted in-path edge ranking for biomolecular association networks. Quant Biol 7(4):313–326. https://doi.org/10.1007/s40484-019-0180-y
32. Ivanov AA, Khuri FR, Fu H (2013) Targeting protein-protein interactions as an anticancer strategy. Trends Pharmacol Sci 34 (7):393–400. https://doi.org/10.1016/j.tips.2013.04.007
33. Yue Z, Zheng Q, Neylon MT et al (2018) PAGER 2.0: an update to the pathway, annotated-list and gene-signature electronic repository for human network biology. Nucleic Acids Res 46(D1):D668–D676. https://doi.org/10.1093/nar/gkx1040
34. Yue Z, Kshirsagar MM, Nguyen T et al (2015) PAGER: constructing PAGs and new PAG-PAG relationships for network biology. Bioinformatics 31(12):i250–i257. https://doi.org/10.1093/bioinformatics/btv265
35. Kilicoglu H, Shin D, Fiszman M et al (2012) SemMedDB: a PubMed-scale repository of biomedical semantic predications. Bioinformatics 28(23):3158–3160. https://doi.org/10.1093/bioinformatics/bts591
36. Huang H, Wu X, Sonachalam M et al (2012) PAGED: a pathway and gene-set enrichment database to enable molecular phenotype discoveries. BMC Bioinformatics 13(Suppl 15):S2. https://doi.org/10.1186/1471-2105-13-S15-S2

Chapter 9

Identification of Gene Regulatory Networks from Single-Cell Expression Data

Song Li, Haidong Yan, and Jiyoung Lee

Abstract

Single-cell RNAseq is an emerging technology that allows the quantification of gene expression in individual cells. In plants, single-cell sequencing technology has been applied to generate root cell expression maps under many experimental conditions. DAP-seq and ATAC-seq have also been used to generate genome-scale maps of protein-DNA interactions and open chromatin regions in plants. In this protocol, we describe a multistep computational pipeline for the integration of single-cell RNAseq data with DAP-seq and ATAC-seq data to predict regulatory networks and key regulatory genes. Our approach utilizes machine learning methods including feature selection and stability selection to identify candidate regulatory genes. The network generated by this pipeline can be used to provide a putative annotation of gene regulatory modules and to identify candidate transcription factors that could play a key role in specific cell types.

Key words Single-cell RNAseq, DAP-seq, ATAC-seq, Machine learning

1 Introduction

Single-cell RNAseq has been recently used to generate gene expression maps of Arabidopsis root cells under different environmental stresses or in different genetic mutants [1–5]. Using bulk RNAseq, many computational tools have been developed to predict gene regulatory networks based on expression profiles of transcription factors and their target genes [6–8]. However, such an approach has not been widely adopted in single-cell RNAseq data in plants. In addition to the transcriptome data generated by single-cell RNAseq, DAP-seq (DNA affinity purification sequencing), and ATAC-seq (Assay for Transposase-Accessible Chromatin using sequencing) are two recent technologies which have generated genome-scale maps of protein-DNA interactions and open chromatin regions in Arabidopsis [9, 10]. DAP-seq is an in vitro technology such that the interactions identified by DAP-seq are not always active in vivo. Although ATAC-seq can be applied in a single-cell level, published ATAC-seq datasets in plants were

Shahid Mukhtar (ed.), *Modeling Transcriptional Regulation: Methods and Protocols*, Methods in Molecular Biology, vol. 2328, https://doi.org/10.1007/978-1-0716-1534-8_9,

generated from the whole organ or tissue samples. Therefore, computational methods are needed to integrate published single-cell RNAseq, DAP-seq, and ATAC-seq data to predict condition-specific activities of regulatory genes and networks. We have developed a computational tool called ConSReg, which utilizes machine learning methods including feature selection and stability selection to detect regulatory networks and stable regulatory genes [11]. In this protocol, we will provide detailed instructions on how to use ConSReg to analyze single-cell RNAseq data and to predict regulatory genes using machine learning algorithms implemented in the ConSReg package.

This pipeline for data integration and regulatory network prediction will include three major steps (Fig. 1). First, we will download published single-cell RNAseq data from Arabidopsis roots. Data will be integrated and normalized across different studies. Second, we will use a published method called Identity of Cell Index (ICI) to classify cells into known cell types in Arabidopsis root [12]. Differentially expressed genes will be identified using a published method called DESingle [13]. Third, we will use ConSReg to integrate DAP-seq and ATAC-seq data with differentially expressed genes identified in the previous step. A gene regulatory network will be generated for visualization. Network visualization can be achieved using widely used programs such as Cytoscape [14], but we will not focus on the steps for visualization in this chapter. This pipeline can be used not only for single-cell data but also for bulk tissue RNAseq or time course RNAseq data. The analytical protocols for such data analysis have been established elsewhere [7, 11] and the only input for network inference will be a list of differentially expressed genes and a list of control genes that are not differentially expressed under specific conditions of interest.

2 Materials

2.1 System Requirement and Pipeline Download

All scripts used in this analysis can be obtained from github using the following command (*see* **Note 1**). The "$" means the command is executed under a Linux terminal (*see* **Note 2**).

```
$ git clone https://github.com/LiLabAtVT/SingleCellNetwork_-
MIMB.git
```

This protocol was tested under Ubuntu 20.04 (Linux) and MacOS Majove 10.14.6, which supports a command-line interface that is commonly found in UNIX-type operating system. The steps described in this protocol can be used in most Linux operating systems and MacOSX. We tested Ubuntu 20.04 using a Docker container under MacOS (https://www.docker.com). We

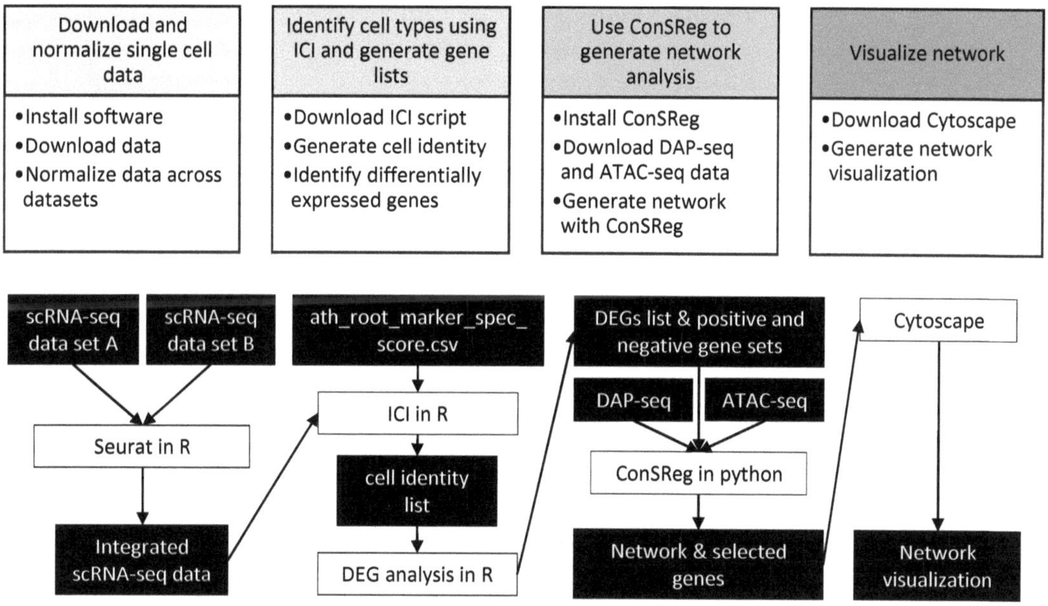

Fig. 1 The analysis pipeline for single-cell RNAseq and network analysis

recommend using Docker for Windows users to gain access to tools that are compatible with the Linux operating system. However, many operations used in this protocol can also be performed using Anaconda (*see* **Note 3**), which allows the user to install Rstudio for R and Jupyter notebook for Python. In this protocol, the first step of data download and integration will use wget, a Linux command-line tool to download data and R for data normalization (*see* **Note 4**). The second step will also be performed using R under the Rstudio environment. The final step will require an installation of Python and Jupyter notebook. Both R and Python can be installed under major operating systems (Windows, MacOS and Linux) using Anaconda platform (*see* **Note 3**).

2.2 Folder Structure

Creating a clean working folder structure is recommended practice for computational data analysis and a well-organized folder structure can help bioinformaticians to track their data and code systematically to achieve reproducible research [15]. In this project, we will perform three steps analysis and we recommend to set up three folders accordingly. Using the "git clone" command provided above will automatically create three folders under the user's working directory.

Under Linux or MacOS terminals, we can also create the folder structure using "mkdir" command. The following commands will create a working folder for the whole project, three sub-folders for each analytical step and one folder for downloading data and software packages.

```
$ mkdir SingleCellNetwork_MIMB
$ cd SingleCellNetwork_MIMB
$ mkdir step1 step2 step3
$ mkdir downloads
```

2.3 Data Download and Software Installation

2.3.1 Single-Cell Expression Data Download

In this analysis, we will download several published single-cell data sets. There are five recently published scRNA-seq datasets of Arabidopsis roots, and four of these have deposited their data matrices to the GEO database (GSE123013, GSE121619, GSE122687 and GSE123818; https://www.ncbi.nlm.nih.gov/geo/). Another dataset can be obtained from a web server (PRJNA517021; http://wanglab.sippe.ac.cn/rootatlas/), with raw data available from the SRA database. These data can be downloaded by simply click the datalink provided on the GEO database. For Linux and MacOS users, a command-line program called "wget" can be used to download these datasets. Here are some examples.

1. To download GSE123013 (*see* **Note 5** on the use of forward slash "\")

```
$ wget \
https://ftp.ncbi.nlm.nih.gov/geo/series/GSE123nnn/GSE123013/
suppl/GSE123013_5way_merge_raw.tsv.gz

$ wget \
https://ftp.ncbi.nlm.nih.gov/geo/series/GSE123nnn/
GSE123013/suppl/GSE123013_RAW_matrices.tar.gz
```

2. To download GSE121619

```
$ wget \
https://ftp.ncbi.nlm.nih.gov/geo/series/GSE121nnn/
GSE121619/suppl/GSE121619_Control_Heatshock_cds.rds.gz

$ wget \
https://ftp.ncbi.nlm.nih.gov/geo/series/GSE121nnn/
GSE121619/suppl/GSE121619_Control_Heatshock_pData.tsv.gz

$ wget \
https://ftp.ncbi.nlm.nih.gov/geo/series/GSE121nnn/GSE121619/
suppl/GSE121619_barcodes.tsv.gz

$ wget \
https://ftp.ncbi.nlm.nih.gov/geo/series/GSE121nnn/GSE121619/
suppl/GSE121619_genes.tsv.gz
```

3. To download GSE123818

```
$ wget \
https://ftp.ncbi.nlm.nih.gov/geo/series/GSE123nnn/
GSE123818/suppl/GSE123818_Root_single_cell_wt_datamatrix.
csv.gz
```

There are two types of technologies used in generating these single-cell datasets. Data from GSE122687 were generated using the drop-seq platform. The data that will be used for downstream analysis will include several matrices representing gene expression from each cell. For other published datasets, single-cell expressions were generated using the 10× genomics platform. The input data for our analysis will include three files: barcodes.tsv, genes.tsv, and matrix.mtx. We will provide specific scripts in Subheading 3 to explain the methods for handling these two different types of data.

2.3.2 Installation of R and Required R Packages

We will use Rstudio as our platform for the R interactive programming environment. Several packages will be installed using R command line, which is easily accessible from the Rstudio interface. To install Rstudio, the latest version can be downloaded from (https://rstudio.com/products/rstudio/download/). Alternatively, Rstudio can also be downloaded and installed as part of the Anaconda environment. R environment in Linux can also be invoked in the command line by typing "r" followed by hitting the "enter" key. This protocol has been tested under R version 3.6.3 and Rstudio version 1.2.5.

We will install the Seurat [16] package for single-cell data preprocessing and normalization. We will install DESingle [13] to detect differentially expressed genes from single-cell RNAseq data. We will also install ChIPSeeker [17], CoReg [18], gglasso [19], and RRF [20] which will be used for network analysis and machine learning prediction. Two types of R package ecosystem co-exist. For example, ChIPSeeker is part of the Bioconductor system, and should be installed using the installer provided by Bioconductor. Seurat and other packages can be installed using CRAN and associated installation scripts.

1. To install Seurat, gglasso, and RRF:

```
>install.packages('Seurat')
>install.packages("gglasso")
>install.packages("RRF")
```

2. To install CoReg, we will need to install from the github source:

```
> install.packages("devtools")
> library(devtools)
> install_github("LiLabAtVT/CoReg")
```

3. To install ChIPSeeker and DEsingle, we will need to install from Bioconductor system.

```
> if (!requireNamespace("BiocManager", quietly = TRUE))
   install.packages("BiocManager")
> BiocManager::install("ChIPseeker")
> BiocManager::install("DEsingle")
```

Each of these packages may depend on functions from other packages and may require the installation of other packages. This can be typically handled by R or Bioconductor package management system (*see* **Note 6** for potential complications in software installation).

2.3.3 Installation of Python and Required Python Packages

We highly recommend using Anaconda to install python and manage python scripts using Conda environment. Creating a new Conda environment for python and associated package installation will guarantee correct dependencies will be installed. Once Anaconda is downloaded and installed, a new environment can be created using the following command.

```
$ conda create --name consreg python=3.6
```

We used Python 3.6 because several other packages used in this analysis have a stable version under Python 3.6. Once the environment is created, we will need to activate the environment first.

```
$ conda activate consreg
```

The next step is to install ConSReg which will use pip to pull different dependencies from respective resources.

```
$ pip install consreg
```

ConSReg package depends on a number of other Python packages include numpy 1.16.2, scipy 1.1.0, pandas 0.21.1, joblib 0.11, rpy2 2.8.6, networkx 2, sklearn 0.19.1, intervaltree 2.1.0. All of these packages are constantly updated and may require tuning if a different version of Python is used.

2.4 Download DAP-seq and ATAC-seq Data

Preprocessing and analysis of published DAP-seq and ATAC-seq both require significant programming skills and complex computational pipelines. To use the data for downstream analysis, we have downloaded published, analyzed datasets from both DAP-seq and ATAC-seq results. DAP-seq data include hundreds of transcription factors and their binding sites which were saved as Narrow Peak files. ATAC-seq results were saved as BED files. Both results from DAP-seq and ATAC-seq data are provided in the github repository of ConSReg (https://github.com/LiLabAtVT/ConSReg.git). We also require an annotation file of Arabidopsis genomes to associate peaks from DAP-seq and ATAC-seq to neighboring genes. The genome annotation file (TAIR10_GFF3_genes.gff) is also provided in ConSReg repository.

3 Methods

3.1 Data Normalization and Integration

There are many established protocols and pipelines for RNAseq raw data preprocessing, quality control and expression quantification. Our current pipeline will start from published read count data. For the upstream analysis from reads to mapped read counts, we refer to the original publications [1, 2] or other published protocols [21].

Published single-cell RNAseq data for Arabidopsis roots have been generated using two different platforms: drop-seq and 10× genomics. We will demonstrate the protocols for importing data sets from both platforms. There are two steps in this section. First, data will be imported from their original format into an R environment as Seurat objects. Second, we will normalize and integrate data across multiple experiments.

3.1.1 R Command to Import Drop-seq Data

Step 1. Load the R package Seurat.

```
> library('Seurat')
```

Step 2. Import each data file separately.

Several data files were generated using plants treated by sucrose. We only import data from control samples in this protocol.

```
> samp_d <-
read.table(gzfile("GSM3478127_replicate_D_error_
detected_200genes.dge.txt.gz",'rt'), sep = "\t", header =
T, row.names = 1, colClasses = c("character", "numeric"))
> samp_e <-
read.table(gzfile("GSM3478128_replicate_E_error_
detected_200genes.dge.txt.gz",'rt'), sep = "\t", header =
T, row.names = 1, colClasses = c("character", "numeric"))
> samp_f <-
read.table(gzfile("GSM3478129_replicate_F_error_
detected_200genes.dge.txt.gz",'rt'), sep = "\t", header =
T, row.names = 1, colClasses = c("character", "numeric"))
> samp_g <-
read.table(gzfile("GSM3478130_replicate_G_error_
detected_200genes.dge.txt.gz",'rt'), sep = "\t", header =
T, row.names = 1, colClasses = c("character", "numeric"))
> samp_h <-
read.table(gzfile("GSM3478131_replicate_H_error_
detected_200genes.dge.txt.gz",'rt'), sep = "\t", header =
T, row.names = 1, colClasses = c("character", "numeric"))
> samp_i <-
read.table(gzfile("GSM3478132_replicate_I_error_
detected_200genes.dge.txt.gz",'rt'), sep = "\t", header =
```

```
T, row.names = 1, colClasses = c("character", "numeric"))
> samp_j <-
read.table(gzfile("GSM3478133_replicate_J_error_
detected_200genes.dge.txt.gz",'rt'), sep = "\t", header =
T, row.names = 1, colClasses = c("character", "numeric"))
```

Step 3. The above commands load each file into the workspace sequentially.

The following commands will generate Seurat objects from each file separately before merging the files.

```
> obj_d <- CreateSeuratObject(samp_d, project =
"GSE122687_WT", min.cells = 3, min.features = 200)
> obj_e <- CreateSeuratObject(samp_e, project =
"GSE122687_WT", min.cells = 3, min.features = 200)
> obj_f <- CreateSeuratObject(samp_f, project =
"GSE122687_WT", min.cells = 3, min.features = 200)
> obj_g <- CreateSeuratObject(samp_g, project =
"GSE122687_WT", min.cells = 3, min.features = 200)
> obj_h <- CreateSeuratObject(samp_h, project =
"GSE122687_WT", min.cells = 3, min.features = 200)
> obj_i <- CreateSeuratObject(samp_i, project =
"GSE122687_WT", min.cells = 3, min.features = 200)
> obj_j <- CreateSeuratObject(samp_j, project =
"GSE122687_WT", min.cells = 3, min.features = 200)
```

The parameters min.cells and min.features were used to filter cells and features by removing those features (genes) that appeared in fewer than three cells and by removing those cells with fewer than 200 features (genes).

Step 4. The following commands are to add more meta-data to each object.

```
> obj_d <- AddMetaData(obj_d, "replicate D_WT", "sample_origin")
> obj_e <- AddMetaData(obj_e, "replicate E_WT", "sample_origin")
> obj_f <- AddMetaData(obj_f, "replicate F_WT", "sample_origin")
> obj_g <- AddMetaData(obj_g, "replicate G_WT", "sample_origin")
> obj_h <- AddMetaData(obj_h, "replicate H_WT", "sample_origin")
> obj_i <- AddMetaData(obj_i, "replicate I_WT", "sample_origin")
> obj_j <- AddMetaData(obj_j, "replicate J_WT", "sample_origin")
```

Step 5. Merge data from all data files generated from wild type plants without any treatment using the merge() function.

```
> GSE122687_WT <- merge(obj_d, c(obj_e, obj_f,obj_g,obj_h,obj_i,
obj_j),
```

```
add.cell.ids = c("GSE122687_WT", "GSE122687_WT",
"GSE122687_WT", 'GSE122687_WT',
'GSE122687_WT', 'GSE122687_WT',
'GSE122687_WT'), project = "GSE122687_WT")
```

Step 6. Additional quality control steps will use mitochondrial genes to filter datasets. Data set was filtered using several features and percentage of mitochondrial genes.

```
> GSE122687_WT[["pct.mt"]] <- PercentageFeatureSet
(GSE122687_WT, pattern = "^ATMG")
> GSE122687_WT <- subset(GSE122687_WT,
subset = nFeature_RNA > 200 & nFeature_RNA < 10000 & pct.mt < 4)
```

Step 7. Normalize gene expression, and we used the default log normalization. The data will be saved as an RDS file such that it is easier to load back into R for downstream analysis.

```
> GSE122687_WT <- NormalizeData(GSE122687_WT)
> saveRDS(GSE122687_WT, "GSE122687_WT_norm.rds")
```

3.1.2 R Command to Import 10× Genomics Data

Step 1. Data generated using 10× genomics default pipeline can be imported using a build-in function in Seurat package. In this case, we use GSE123013 as one example to demonstrate how to import 10× genomics data. We only import data from wild type plants in this example.

```
> expression_matrix <- Read10X(data.dir = 'Sample_WT-WERGFP/
filtered_gene_bc_matrices/TAIR10')
> expression_matrix2 <- Read10X(data.dir = 'Sample_WT-
WERGFP_2/filtered_gene_bc_matrices/TAIR10')
> expression_matrix3 <- Read10X(data.dir = 'Sample_WT-
WERGFP_3/filtered_gene_bc_matrices/TAIR10')
```

Step 2. The function "Read10X()" can import the three files generated by 10x analysis pipeline. We will then merge the expression matrices first, before generating the Seurat object.

```
> expression_matrix_all <- cbind(expression_matrix,
expression_matrix2, expression_matrix3)
> GSE123013_seurat = CreateSeuratObject(counts =
expression_matrix_all)
> saveRDS(GSE123013_seurat, file = "GSE123013_
seurat.rds")
```

Step 3. Merge datasets into a single object for backup. Utilize the similar pipeline to process other matrix tables (GSE123818, GSE123013, GSE121619, GSE122687, and PRJNA517021) to generate GSE121619_wt_dat, GSE123818_wt_dat, GSE122687_wt_dat, and PRJNA517021_wt_dat objects. Next, merge these datasets into a combined dataset that will be integrated into the next step.

```
>merge_dat <- merge(GSE123818_wt_dat, c(GSE121619_wt_dat,
GSE123013_wt_dat, GSE122687_wt_dat, PRJNA517021_wt_dat),
merge.data = T)
>saveRDS(merge_dat, "merge_dat.rds")
```

3.1.3 R Command for Cross-Study Integration (Optional)

This step is to use the Seurat package to integrate single-cell data from five published studies. This is an optional step because a typical study does not need to analyze other published data (*see* **Note 7**).

Step 1. Split the previous combined object into multiple objects.

```
> exp_data.list <- SplitObject(merge_dat,split.by =
'orig.ident')
```

Step 2. Identify variable features individually for each object.

Seurat package uses "anchors" to perform data merge across experiments. Before identifying anchors or shared correlation structures, we identify variable features individually for each object. Note that 29,929 is the feature number in the merge_-dat matrix. We used the first 30 dimensions of data for data integration.

```
>exp_data.list <- lapply(X = exp_data.list, FUN = function(x) {
x <- FindVariableFeatures(x, selection.method = "vst",
nfeatures = 29929)
})

> exp_data.anchors <- FindIntegrationAnchors(object.
list = exp_data.list, dims = 1:30, anchor.features = 29929)
> exp_data.integrated <- IntegrateData(anchorset =
exp_data.anchors, dims = 1:30)
```

Step 3. Save the integrated data to an RDS format data for backup and future analysis.

```
> DefaultAssay(exp_data.integrated) <- "integrated"
>saveRDS(exp_data.integrated, "exp_data_integrated.rds")
```

Step 4. Convert the data into a matrix format which is more flexible for downstream statistical analysis using other packages.

```
> assay_data <- GetAssayData(object = exp_data.integrated)
> assay_data <- as.matrix(assay_data)
> saveRDS(assay_data, file = "Integrated_matrix.rds")
```

3.2 Identify Cell Types Using ICI and Generate Gene Lists

In this section, we perform analysis using a published ICI approach to assign cells into respective cell types [12]. We use only a single dataset as an example of this analysis. This is because a single dataset requires smaller memory which can be performed on a typical laptop computer (>8 Gb of memory recommended).

Step 1. Import expression data save in the previous step.

```
> GSE122687<-readRDS('GSE122687_WT_norm.rds')
> assay_data <- GetAssayData(object = GSE122687)
> ExpMat <- as.matrix(assay_data)
```

The ExpMat object contains the expression data in a matrix format with each row represents one gene (feature) and each column represents one cell.

Step 2. Import a data matrix that contains the weight (spec score) from known root cell marker genes.

```
> spec_scores <- read.csv("ath_root_marker_spec_score.csv",
row.names=1)
```

The subset of genes from the whole expression matrix will be used to determine cell-type specificity using the spec_scores.

```
> markergenes <- rownames(spec_scores)
> markergenesk <- markergenes[markergenes%in%rownames(ExpMat)]
> MarkerMat <- ExpMat[markergenesk,]
```

Step 3. Calculate SPEC identity matrix and Nt vector (number of marker gene per tissue)

```
> speck <- spec_scores[rownames(MarkerMat),]
> specI <- speck;specI[specI>0]=1;specI[specI<1]=0
> Nt <- apply(specI,2,sum)
```

Step 4. Calculate marker identity matrix, (m by C). We calculate ICI using a matrix formulation. The original implementation used nested for loops.

```
> markerI <- MarkerMat; markerI[markerI>0]=1; markerI[markerI<1]=0
> B <- t(as.matrix(markerI))%*%as.matrix(specI)
> A <- t(as.matrix(MarkerMat))%*%as.matrix(speck)
> numerator <- A * B
> ICI <- t(t(numerator)/(Nt**2))
```

Step 5. Calculate the sum of the ICI scores over the rows, and normalized ICI for cells with ICI that are none zero.

```
> rowsumk <- apply(ICI,1,sum)
> rowsumk <- rowsumk[rowsumk>0]
> ICIn <- ICI[names(rowsumk),]
> ICIn <- ICIn/rowsumk
```

Step 6. The output will be saved in a csv file. Each cell includes 15 ICI scores that indicate the probability of this cell belongs to one in 15 cell types.

```
> write.csv(ICIn,'ICIn.csv')
```

In the following analysis, we will only work with endodermis, cortex cells as examples. Other comparisons can be performed as needed. In this chapter, we focus on cell-type specificity.

Step 1. Identify cells that are defined by ICI > 0.5 and use cells that are defined as specific cell types (endodermis and cortex) to determine differentially expressed genes.

```
> ICInk<-ICIn[,c('Cortext', 'Endodermis')]
> cellN<-rownames(ICInk)
> cort<-cellN[which(ICInk[,'Cortext']>=0.5)]
> endo<-cellN[which(ICInk[,'Endodermis']>=0.5)]
> cellLabel<- rbind(cbind(cort,rep('Cortex',length(cort))),
cbind(endo,rep('Endodermis',length(endo))))
> colnames(cellLabel)<-c('cellname','celltype')
> write.csv(cellLabel,'CellLabel.csv')
```

Step 2. Use DESignle [13] to determine genes that are specifically expressed in endodermis as compared to cortex. We first reload expression data save in the previous step, and the ICI cell labels, and only use cells that are labeled as endodermis or cortex for analysis.

```
> GSE122687<-readRDS('GSE122687_WT_norm.rds')
> assay_data <- GetAssayData(object = GSE122687)
> ExpMat <- as.matrix(assay_data)
> cellLabel<-read.csv('CellLabel.csv',as.is=T,row.names=1)
> group<-factor(cellLabel [,2])
> countM<-ExpMat[,cellk[,1]]
```

Step 3. Define differentially expressed genes. We have obtained 3275 cells that are labeled as either endodermis or cortex. We first performed a filtering step to remove genes expressed <10% of cells. Because we have roughly 50% of each cell type, genes expressed in <10% of cells are unlikely to be the marker genes for the specific cell types. The output file DEsingle.csv will be used as input for next analysis to generate networks.

```
> countMK<-countM
> countMK[countMK>1]=1
> countMK[countMK<1]=0
> countMKRS<-rowSums(countMK)
> countK<-countM[countMKRS>(3275*0.1),]
> results <- DEsingle(counts = countK, group = group)
> results.classified<-DEtype(results=results, threshold=0.05)
> write.csv(results.classified,'DEsingle.csv')
```

3.3 Network Analysis Using ConSReg

To perform network analysis, the python package, ConSReg should be installed based on instructions in Subheading 2. In the installed python package folder, a Jupyter notebook is provided as a template for data analysis. The annotation files from DAP-seq, ATAC-seq and genome annotations are also provided. The user can make a copy of the single_cell_analysis.ipynb with a different name for analysis.

Step 1. Data preprocessing and constructing a feature matrix. We first import ConSReg class from the ConSReg package and specify a number of parameters using a dictionary.

```
>>> from ConSReg.main import ConSReg
>>> analysis = ConSReg()
>>> params = {
'dap_files':dap_files,
'diff_files':sc_diff_file,
```

```
'atac_file':atac_file,
'gff_file':gff_file,
'up_tss':3000,
'down_tss':500,
'n_jobs':1,
}
>>> analysis.preprocess(**params)
```

Several parameters are important for the program to run correctly. "dap_files" provides the location of dap-seq files. This can also be other narrow peak files generated by ChIP-seq or motif searching results. "diff_files" is the single-cell differential expression data generated by DEsingle or other packages. "atac_file" is a BED file representing open chromatin regions in the genome. "gff_file" is the genome annotation file in GFF format. "up_tss" and "down_tss" are two parameters that can be used to determine the length of promoter regions. This can be fine-tuned to fit different organisms. Our experiments showed that for Arabidopsis, 3000 and 500 are optimal parameter, whereas in maize, different parameters should be used. "n_jobs" is the number of threads to use for the analysis. This number should be set based on the computer used for analysis (*see* **Note 8**).

The preprocessing step performs a lot of analysis. The main time-consuming step is to match DAP-seq peaks to known genes and to map ATAC-seq peaks to known genes. The overlapping between DAP-seq and ATAC-seq peaks can also be used to set weight for the downstream analysis. Finally, the peak height signals from either DAP-seq or ATAC-seq can also be used to set weight for downstream analysis.

Step 2. Generate feature matrix and perform machine learning-based feature selection. The user can first define "negative" or "control" genes. In ConSReg analysis, we use transcription factor-target gene interactions to predict gene expression of differentially expressed genes. The "positive" genes will be a list of genes that are known to be differentially expressed between two samples. In our case, these genes will include those genes identified by DESingle in the previous step. The "negative" or "control" genes will be used as a background dataset to train machine learning models to identify the best combinations of transcription factors that predict "positive" genes. In this protocol, we use "udg" which stands for undetected genes as our negative genes. Other options are available and which set is useful can be determined based on different biological questions (*see* **Note 9**).

```
>>> analysis.gen_feature_mat(neg_type = 'udg')
```

Step 3. We performed "lrlasso" which is logistic regression with LASSO penalty. The results such as auROC (area under receiver operation characteristic curve) can be assessed.

```
>>> analysis.eval_by_cv(ml_engine = 'lrlasso',rep = 5)
>>> analysis.auroc
```

Step 4. We also perform stability selection which generated resampling based stability scores as "importance score." Transcription factors selected more often in the resampling analysis are more stable and thus are better candidate for follow up analysis.

```
>>> analysis.compute_imp_score(n_resampling = 200, n_jobs =
36, verbose = True)
>>> analysis.gen_networks(imp_cutoff = 0.5, verbose = True)
```

Step 5. Resulted auROC, auPRC and importance scores can be saved for further analysis and for selecting candidate genes.

```
>>> analysis.auroc.to_csv("auroc_result.csv")
>>> analysis.auprc.to_csv("auprc_result.csv")
>>> analysis.imp_scores_UR.to_csv("imp_score_UR.csv")
>>> analysis.imp_scores_DR.to_csv("imp_score_DR.csv")
```

Step 6. Finally, the results can be saved in a network format. In this case, we save the predicted network as an edge list which can be used for visualization using programs such as Cytoscape [14].

```
>>> for diff_name, network in zip(analysis._diff_name_list,
analysis.networks_UR):
network.to_csv("{}_UR_network.csv".format(diff_name))
>>> for diff_name, network in zip(analysis._diff_name_list,
analysis.networks_DR):
network.to_csv("{}_DR_network.csv".format(diff_name))
```

4 Notes

1. The "git" software is installed in most Linux systems by default. If git is not installed in your system, please refer to https://git-scm.com for installation instructions.
2. Code blocks: All code blocks started with "$" are command-line scripts that should be executed under a Linux or MacOS terminal. All code blocks started with ">" are command-line scripts that should be executed under an interactive R programming language console. All code blocks started with ">>>"

are command-line scripts that should be executed under an interactive Python programming language console. Python scripts can also be executed inside a Jupyter notebook command block.

3. Anaconda can be installed by downloading the installation package from the following website (https://www.anaconda.com/products/individual). We recommend the use of the Python 3 versions of the anaconda package. Once Anaconda is installed, the user can start an Anaconda Navigator, and within this program, the user can select to install Jupyter Notebook and Rstudio. Alternatively, R can be downloaded from this website (https://cran.r-project.org/). Rstudio can be downloaded from this website (https://rstudio.com/). Python can be downloaded from this website (https://www.python.org/). And Jupyter Notebook can be installed using pip after download and install python.
4. For example, MacOSX does not include "wget" by default. To download a file from the command line, the reader can use "curl -O" followed by the URL of the target package. The reader can also type the link provided in this protocol in a web-browser and the download should start automatically. After a manual download, the reader should move the downloaded package to the folder that has been created for data analysis.
5. In the following sections, a single line command usually starts with "$". The user does not need to type this symbol, which should be provided by the operating system by default. If a line does not start with "$", that means this line is a continuation of the previous line. For example, in the command "$ wget https://ftp.ncbi.nlm.nih.gov/geo...... raw.tsv.gz" appears in two rows. However, in the actual command line, this should be in the same row. This line appears in two rows because of the page width limit. In both Linux and MacOS, a forward slash ("\") indicates that the current line is not complete, the command will continue in the next line.
6. The user may experience some problems when installing other packages required for this analysis. For example, when installing ChIPSeeker package in R, dozens of other packages and the dependencies of these packages will be installed. Based on the author's experience, it is very rare that one will not experience an error during the installation of these packages. However, most of the packages can be eventually installed once the required dependencies were met. This is particularly challenging or Linux and MacOS systems because some of the shared libraries are not installed by default. To solve this problem, it is recommended to use Docker containers to install packages which allow the user to install many other dependent packages.

7. This step of cross experiment integration also requires a large memory machine. We have tested this on a Linux server with 128Gb and a Macbook Pro with 16Gb memory. However, the 16Gb computer did not have enough memory to handle data integration across all datasets.
8. In these commands, we use "n_jobs: 1". The reader should check how many cores are available. For a typical laptop, there are either 2 or 4 cores. For a typical computing server, there could be 12 to 48 or even more cores. In Linux, the command to check the number of cores available is "lscpu". Using more cores usually can speed up the analysis process, however, the reader is also advised to not use more cores than the actual available number of cores.
9. Options include: "udg"—undetected genes (UDGs), which have a mean expression value equal to zero. "leg"—low-expressed genes (LEGs), which have mean expression between 0 and 0.5. "ndeg"—non-significantly differentially expressed genes (NDEGs), which have p-value > 0.05. "high_mean"—high mean genes, ndegs which have mean expression <= 8. In Arabidopsis, UDG performed best in our benchmarking experiment. When applying to other species, different parameters should be tested and compared based on auROC and auPRC.

Acknowledgment

This work is supported by the Virginia Agricultural Experiment Station (Blacksburg) and the USDA National Institute of Food and Agriculture, US Department of Agriculture (Washington, DC).

References

1. Ryu KH, Huang L, Kang HM, Schiefelbein J (2019) Single-cell RNA sequencing resolves molecular relationships among individual plant cells. Plant Physiol 179:1444–1456. https://doi.org/10.1104/pp.18.01482
2. Shulse CN, Cole BJ, Ciobanu D et al (2019) High-throughput single-cell transcriptome profiling of plant cell types. Cell Rep 27:2241–2247.e4. https://doi.org/10.1016/j.celrep.2019.04.054
3. Denyer T, Ma X, Klesen S et al (2019) Spatiotemporal developmental trajectories in the arabidopsis root revealed using high-throughput single-cell RNA sequencing. Dev Cell 48:840–852.e5. https://doi.org/10.1016/j.devcel.2019.02.022
4. Jean-Baptiste K, McFaline-Figueroa JL, Alexandre CM et al (2019) Dynamics of gene expression in single root cells of Arabidopsis thaliana. Plant Cell 31:993–1011. https://doi.org/10.1105/tpc.18.00785
5. Zhang T-QQ, Xu Z-GG, Shang G-DD, Wang J-WW (2019) A single-cell RNA sequencing profiles the developmental landscape of Arabidopsis root. Mol Plant 12:648–660. https://doi.org/10.1016/j.molp.2019.04.004
6. Haque S, Ahmad JS, Clark NM et al (2019) Computational prediction of gene regulatory networks in plant growth and development. Curr Opin Plant Biol 47:96–105. https://doi.org/10.1016/j.pbi.2018.10.005
7. Redekar N, Pilot G, Raboy V et al (2017) Inference of transcription regulatory network in low phytic acid soybean seeds. Front Plant Sci 8:2029. https://doi.org/10.3389/fpls.2017.02029

8. Silva JCF, Teixeira RM, Silva FF et al (2019) Machine learning approaches and their current application in plant molecular biology: a systematic review. Plant Sci 284:37–47. https://doi.org/10.1016/j.plantsci.2019.03.020
9. O'Malley RC, Huang SC, Song L et al (2009) Cistrome and epicistrome features shape the regulatory DNA landscape. Cell 165:1280–1292. https://doi.org/10.1016/J.CELL.2016.04.038
10. Lu Z, Hofmeister BT, Vollmers C et al (2016) Combining ATAC-seq with nuclei sorting for discovery of cis-regulatory regions in plant genomes. Nucleic Acids Res 45. https://doi.org/10.1093/nar/gkw1179
11. Song Q, Lee J, Akter S et al (2020) Prediction of condition-specific regulatory genes using machine learning. Nucleic Acids Res. https://doi.org/10.1093/nar/gkaa264
12. Efroni I, Ip PL, Nawy T et al (2015) Quantification of cell identity from single-cell gene expression profiles. Genome Biol 16:9. https://doi.org/10.1186/s13059-015-0580-x
13. Miao Z, Deng K, Wang X, Zhang X (2018) DEsingle for detecting three types of differential expression in single-cell RNA-seq data. Bioinformatics 34:3223–3224. https://doi.org/10.1093/bioinformatics/bty332
14. Franz M, Lopes CT, Huck G et al (2015) Cytoscape.js: a graph theory library for visualisation and analysis. Bioinformatics 32:309–311. https://doi.org/10.1093/bioinformatics/btv557
15. Sandve GK, Nekrutenko A, Taylor J, Hovig E (2013) Ten simple rules for reproducible computational research. PLoS Comput Biol 9: e1003285. https://doi.org/10.1371/journal.pcbi.1003285
16. Stuart T, Butler A, Hoffman P et al (2019) Comprehensive integration of single-cell data. Cell 177:1888–1902.e21. https://doi.org/10.1016/j.cell.2019.05.031
17. Yu G, Wang LG, He QY (2015) ChIP seeker: an R/Bioconductor package for ChIP peak annotation, comparison and visualization. Bioinformatics 31:2382–2383. https://doi.org/10.1093/bioinformatics/btv145
18. Song Q, Grene R, Heath LS, Li S (2017) Identification of regulatory modules in genome scale transcription regulatory networks. BMC Syst Biol 11:140. https://doi.org/10.1186/s12918-017-0493-2
19. Yang Y, Zou H (2015) A fast unified algorithm for solving group-lasso penalize learning problems. Stat Comput 25:1129–1141. https://doi.org/10.1007/s11222-014-9498-5
20. Deng H, Runger G (2013) Gene selection with guided regularized random forest. Pattern Recogn 46:3483–3489. https://doi.org/10.1016/j.patcog.2013.05.018
21. Qi S, Akter S, Li S (2019) Identification of novel lincRNA and co-expression network analysis using RNA-sequencing data in plants. Plant Long Non-Coding RNAs:207–221

Chapter 10

Inference of Gene Regulatory Network from Single-Cell Transcriptomic Data Using pySCENIC

Nilesh Kumar, Bharat Mishra, Mohammad Athar, and Shahid Mukhtar

Abstract

With the advent of recent next-generation sequencing (NGS) technologies in genomics, transcriptomics, and epigenomics, profiling single-cell sequencing became possible. The single-cell RNA sequencing (scRNA-seq) is widely used to characterize diverse cell populations and ascertain cell type-specific regulatory mechanisms. The gene regulatory network (GRN) mainly consists of genes and their regulators—transcription factors (TF). Here, we describe the lightning-fast Python implementation of the SCENIC (Single-Cell reEgulatory Network Inference and Clustering) pipeline called pySCENIC [28]. Using single-cell RNA-seq data, it maps TFs onto gene regulatory networks and integrates various cell types to infer cell-specific GRNs. There are two fast and efficient GRN inference algorithms, GRNBoost2 and GENIE3, optionally available with pySCENIC. The pipeline has three steps: (1) identification of potential TF targets based on co-expression; (2) TF-motif enrichment analysis to identify the direct targets (regulons); and (3) scoring the activity of regulons (or other gene sets) on single cell types.

Key words Gene co-expression network, Gene regulatory network, scRNA-seq, RNA-Seq count data

1 Introduction

The RNA-seq technologies have been widely used for the discovery and relative quantification of transcripts in a diverse organisms including animals and plants [1–6]. Such transcriptomics data provides an opportunity for co-expression network construction to understand the correlation among gene expression patterns [7]. Furthermore, the regulatory nodes within co-expressed genes can be highlighted using a wide variety of indicators of centralities for the co-expression network [2, 8, 9]. The most prominent indicators of centralities applicable to biological networks, including co-expression networks, are degree, closeness centrality,

The original version of this chapter was revised. The correction to this chapter is available at https://doi.org/ https://doi.org/10.1007/978-1-0716-1534-8_20

Shahid Mukhtar (ed.), *Modeling Transcriptional Regulation: Methods and Protocols*, Methods in Molecular Biology, vol. 2328, https://doi.org/10.1007/978-1-0716-1534-8_10, © Springer Science+Business Media, LLC, part of Springer Nature 2021, Corrected Publication 2021 betweenness centrality, eigenvector centrality, PageRank centrality, etc. [10, 11]. The main purpose of using these indicators of centralities is to rank the nodes that are often overlooked such as genes not currently known as TFs, but playing significant regulatory roles. It has been shown that most of the genes or proteins that play crucial roles in diverse biological processes possess high centrality [7, 10–14]. Recently, a similar approach has been applied to bracket regulatory genes involved in host modifying processes such as protein translation, ubiquitination, interferon signaling, and the cytokine storm in SARS-CoV-2 pathogenesis [15].

In recent years, single-cell RNA sequencing (scRNA-seq) [16] has become the preferred choice. The key difference between conventional RNA-seq and scRNA-seq technologies is their resolution and precision at the cellular level rather than tissue level [16–19]. Each cell has a unique transcriptome and thus requires an individualized snapshot of its transcription profiles that can be provided by the scRNA-seq sequencing technique. While RNA-seq is only able to highlight expression patterns of large numbers of cells together, scRNA-seq provides a typical window of opportunity to obtain an expression profile at the single-cell level [19]. Additionally, the bulk analysis techniques are prone to incorrect assumptions of homogeneity of the pooled cells and usually underrate the relevant expression patterns in rare cell populations [20]. To be able to establish an understanding of the regulatory network of cells in different conditions, a wide variety of gene regulatory networks (GRNs) mapping algorithms have been published [21]. Some of the algorithms were developed for bulk sequencing data such as RNA-seq GRN but they can also be utilized for scRNA-seq GRN inference.

1.1 GENIE3

GENIE3 was the best performer algorithm in the DREAM4; *In Silico Multifactorial challenge*, for the inference of GRNs of bulk transcriptional data [22]. GENIE3 can recover regulatory networks from gene multivariate expression data. The resulting regulatory networks are directed graphs with "V" genes connected through unsigned directed edges. GRN is reconstructed using "V" regression task to obtain a subset of genes whose expression profiles are the most predictive of a target gene's expression profile.

1.2 PPCOR

PPCOR is an R package that can be used to compute the pairwise partial correlations and *p*-value for each pair of genes with respect to all the other genes [23]. After partial correlation computation, an undirected regulatory network is constructed where a sign of the correlation is used to infer negative and positive regulation.

1.3 GRNBoost2

GRNBoost2 is an algorithm faster than GENIE3, and a possible alternative when the datasets have a high volume of observations. The key difference between both methods is the use of stochastic gradient boosting machine regression with early stopping regularization to infer the network in GRNBoost2, while GENIE3 computes regression for each gene [24].

1.4 MICRAT

MICRAT (Maximal Information coefficient with Conditional Relative Average entropy and Time-series mutual information) is used to infer GRNs from time-series expression data. It uses the maximal information coefficient to construct an undirected network of genes. Subsequently, the conditional relative average entropies of each pair of genes are used to interpret the potential regulators and their targets in the network [25].

1.5 PIDC

The partial information decomposition (PIDC) method is based upon information theory to calculate statistical dependencies between triplets of genes in expression datasets. It measures the average ratio of unique information between two genes across all of the third genes in the remainder of the network [26].

2 Materials

1. The SCENIC [29] workflow can be set up on a standard UNIX (Linux or MacOS recommended) or Windows 10 workstation. For efficient analysis, High Performance Computing (HPC) cluster is recommended. For Windows operation system, it is recommended to install Windows Subsystem for Linux.
2. The workflow is available in both R (version 4.0.3) and Python (version 3) environments. The Python implementation (*pySCENIC* [28, 30]) is recommended over R implementation because of its efficient multiprocessing capability since it uses Dask parallel processing.
3. Conda or Miniconda virtual environments for Python and BiocManager package manager for Bioconductor in R, respectively [27].
4. HPC cluster access, for faster processing 24 core and at least 64 gigabytes (Gb).
5. Create a virtual environment and install conda and pip packages needed for the analysis (*see* **Note 1**).
6. Create working directory and subdirectories (Bash Shell).

```
$ mkdir pySCENIC
$ $wd = pySCENIC
$ cd $wd
$ mkdir pySCENIC_data
$ mkdir pySCENIC_data/auxilliaries
$ mkdir pySCENIC_data/figures
$ mkdir pySCENIC_data/resources
$ mkdir pySCENIC_data/results
```

7. Download meta and supporting files (Bash Shell).

```
$ cd pySCENIC_data/auxilliaries
$ wget
https://raw.githubusercontent.aertslab/aertslab/pySCENIC/ma
ster/resources/lambert2018.txt
$ wget
https://resources.aertslab.org/cistarget/databases/hom
o_sapiens/hg19/refseq_r45/mc9nr/gene_based/hg19-500bp-
upstream-10species.mc9nr.feather
$ wget
https://resources.aertslab.org/cistarget/databases/hom
o_sapiens/hg19/refseq_r45/mc9nr/gene_based/hg19-tss-
centered-5kb-10species.mc9nr.feather
$ wget
https://resources.aertslab.org/cistarget/databases/hom
o_sapiens/hg19/refseq_r45/mc9nr/gene_based/hg19-tss-
centered-10kb-10species.mc9nr.feather
$ wget
https://resources.aertslab.org/cistarget/motif2tf/
motifs-v9-nr.hgnc-m0.001-o0.0.tbl
```

Keep the cell annotations, cell count, Transcripts per Kilobase Million (TMP) and metadata files in the "resources" directory (*see* **Note 2**).

3 Methods

For efficient processing, some of the analysis steps should be done in command-line mode, and the rest of the steps can be done using Python scripts. The whole workflow has four major steps: (1) preprocessing, (2) GRN analysis using expression data [31], (3) TF-motif enrichment and regulons-cell enrichment, and (4) scoring.

3.1 Preprocessing and Cleaning of the Data (Python Shell)

3.1.1 Import Python Libraries

```
import os, glob, re, pickle
from functools import partial
from collections import OrderedDict
import operator as op
from cytoolz import compose
import pandas as pd
import seaborn as sns
import numpy as np
import scanpy as sc
import anndata as ad
import matplotlib as mpl
import matplotlib.pyplot as plt
from pyscenic.export import export2loom,
```

```
add_scenic_metadata
from pyscenic.utils import load_motifs
from pyscenic.transform import df2regulons
from pyscenic.aucell import aucell
```

3.1.2 Setup the File and Folder Variables (Python Shell)

```
wd = "pySCENIC_data/" #working directory
RESOURCES_FOLDERNAME = wd + "resources/"
AUXILLIARIES_FOLDERNAME = wd + "auxilliaries/"
RESULTS_FOLDERNAME = wd + "results/"
FIGURES_FOLDERNAME = wd + "figures/"
BASE_URL = "http://motifcollections.aertslab.
org/v9/logos/"
COLUMN_NAME_LOGO = "MotifLogo"
COLUMN_NAME_MOTIF_ID = "MotifID"
COLUMN_NAME_TARGETS = "TargetGenes"
HUMAN_TFS_FNAME =
os.path.join(AUXILLIARIES_FOLDERNAME,'lambert2018.txt')
        RANKING_DBS_FNAMES = list(map(lambda fn:
        os.path.join(AUXILLIARIES_FOLDERNAME, fn),

['hg19-500bp-upstream-
        10species.mc9nr.feather',

'hg19-tss-centered-5kb-
        10species.mc9nr.feather',

'hg19-tss-centered-10kb-
        10species.mc9nr.feather']))
        MOTIF_ANNOTATIONS_FNAME =
      os.path.join(AUXILLIARIES_FOLDERNAME, 'motifs-
v9-
        nr.hgnc-m0.001-o0.0.tbl')
        DATASET_ID = "Experiment_name" # for example
GEO ID
        TCGA_CODE = 'CANCERTYPE' # TCGA cancer code.
        CELL_ANNOTATIONS_FNAME =
os.path.join(RESOURCES_FOLDERNAME, " cell.annotations.csv")
        SAMPLE_METADATA_FNAME =
      os.path.join(RESOURCES_FOLDERNAME, "meta_data.
xlsx")
        EXP_MTX_TPM_FNAME =
        os.path.join(RESOURCES_FOLDERNAME,
        'Expression_tpm.csv')
        EXP_MTX_COUNTS_FNAME =
        os.path.join(RESOURCES_FOLDERNAME,
```

```
                                'Expression_counts.csv')
                                # Output files and folders.
                                METADATA_FNAME = os.path.join(RESULTS_FOLDE
RNAME,
                                '{}.metadata.csv'.format(DATASET_ID))
                                  EXP_MTX_QC_FNAME = os.path.join(RESULTS_
FOLDERNAME,
                                '{}.qc.tpm.csv'.format(DATASET_ID))
                                  ADJACENCIES_FNAME = os.path.join(RESULTS_
FOLDERNAME,
                                '{}.adjacencies.tsv'.format(DATASET_ID))
                              MOTIFS_FNAME = os.path.join(RESULTS_FOLDERNAME,
                                '{}.motifs.csv'.format(DATASET_ID))
                                 REGULONS_DAT_FNAME = os.path.join(RESULTS_
FOLDERNAME,
                                '{}.regulons.dat'.format(DATASET_ID))
                                  AUCELL_MTX_FNAME = os.path.join(RESULTS_
FOLDERNAME,
                                '{}.auc.csv'.format(DATASET_ID))
                              BIN_MTX_FNAME = os.path.join(RESULTS_FOLDERNAME
,
                                '{}.bin.csv'.format(DATASET_ID))
                                THR_FNAME = os.path.join(RESULTS_FOLDERNAME,
                                '{}.thresholds.csv'.format(DATASET_ID))
                              ANNDATA_FNAME = os.path.join(RESULTS_FOLDERNAME
,
                                '{}.h5ad'.format(DATASET_ID))
                               LOOM_FNAME = os.path.join(RESULTS_FOLDERNAME,
                                '{}_{}.loom'.format(TCGA_CODE, DATASET_ID))
```

3.1.3 Cleaning the Data (Python Shell)

```
df_annotations = pd.read_csv(CELL_ANNOTATIONS_FNAME)
df_annotations['samples'] =
df_annotations['samples'].str.upper()
df_annotations.rename(columns={'cell.types':
'cell_type', 'cells': 'cell_id', 'samples':
'sample_id',
                              'treatment.group':
'treatment_group', 'Cohort': 'cohort'}, inplace=True)
          df_annotations['cell_type'] =
          df_annotations.cell_type.replace({
                 'Mal': 'Malignant Cell',
                 'Endo.': 'Endothelial Cell',
                 'T.CD4': 'Thelper Cell',
                 'CAF': 'Fibroblast',
                 'T.CD8': 'Tcytotoxic Cell',
                 'T.cell': 'T Cell',
```

```
                    'NK': 'NK Cell',
                    'B.cell': 'B Cell'})
df_samples = pd.read_excel(SAMPLE_METADA TA_FNAME,
header=2)
df_samples = df_samples.drop(['Cohort'], axis=1)
df_samples['Sample'] = df_samples.Sample.str.upper()
df_metadata = pd.merge(df_annotations, df_samples,
left_on='sample_id', right_on='Sample')
df_metadata = df_metadata.drop(['Sample',
'treatment_group'], axis=1)
                    df_metadata.rename(columns={'Patient':
'patient_id',
```

3.1.4 Quality Control of TPM Data (Python Shell)

```
df_tpm = pd.read_csv(EXP_MTX_TPM_FNAME, index_col=0)
df_tpm.shape
adata = sc.AnnData(X=df_tpm.T.sort_index())
df_obs = df_metadata[['cell_id', 'sample_id',
'cell_type', 'cohort', 'patient_id', 'age', 'sex',
treatment',
'treatment_group', 'lesion_type',
'site']].set_index('cell_id').sort_index()
adata.obs = df_obs
adata.var_names_make_unique()
sc.pp.filter_cells(adata, min_genes=200)
sc.pp.filter_genes(adata, min_cells=3)
adata.raw = adata
sc.pp.log1p(adata)
adata.write_h5ad(ANNDATA_FNAME)
adata.to_df().to_csv(EXP_MTX_QC_FNAME) # Write csv
file on the disk.
```

3.2 Gene Regulatory Network (GRN) Analysis Using grnboost2 (See Note 3)

For this step use command-line version of pySCENIC to use multiple processors (Bash Shell).

```
$ pyscenic grn
pySCENIC_data/results/GSE115978.qc.tpm.csv \
pySCENIC_data/auxilliaries/lambert2018.txt \
-o pySCENIC_data/results/GSE115978.adjacencies.tsv \
--num_workers 16
```

3.3 Create Regulons

Using RcisTarget then identify modules for which the regulator's binding motif is significantly enriched across the target genes and creates regulons with direct targets (*see* **Note 4**) (Python Shell).

```
$ pyscenic ctx
pySCENIC_data/results/GSE115978.adjacencies.tsv
pySCENIC_data/auxilliaries/hg19-500bp-upstream-
```

```
10species.mc9nr.feather
pySCENIC_data/auxilliaries/hg19-tss-centered-5kb-
10species.mc9nr.feather
pySCENIC_data/auxilliaries/hg19-tss-centered-10kb-
10species.mc9nr.feather \
--annotations_fname pySCENIC_data/auxilliaries/motifs-
v9-nr.hgnc-m0.001-o0.0.tbl \
--expression_mtx_fname
pySCENIC_data/results/GSE115978.qc.tpm.csv \
--output pySCENIC_data/results/GSE115978.motifs.csv \
--mask_dropouts \
--num_workers 10
```

3.4 Identify Cells with Active Gene Sets

Using AUCell method identify cells with active gene regulatory networks. For this part of the analysis use pyscenic.aucell Python library (Python Shell).

```
df_motifs = load_motifs(MOTIFS_FNAME)
regulons = derive_regulons(df_motifs)
auc_mtx = aucell(df_tpm.T, regulons, num_workers=26)
auc_mtx.to_csv(AUCELL_MTX_FNAME)
```

3.5 Save Loom File on Disk (Python Shell).

```
export2loom(adata.to_df(), regulons, LOOM_FNAME,
cell_annotations=adata.obs['cell_type'].to_dict(),
                    embeddings=OrderedDict([('AUCell + tSNE',
embedding_aucell_tsne), ('PCA + tSNE', embedding_pca_tsne)]),
                 auc_mtx = auc_mtx,
                 tree_structure=(),
                 title='{}_{}'.format(TCGA_CODE,
DATASET_ID),
                 nomenclature="HGNC",
auc_thresholds=thresholds,
                 compress=True)
```

4 Notes

1. pySCENIC Installation
   (a) Create and set up a new conda environment.

   ```
   $ conda create -n scenic python=3.6
   ```

   (b) Activate the conda environment.

   ```
   $ conda activate scenic
   ```

   (c) Install conda packages

   ```
   $ conda install numpy pandas matplotlib seaborn
   ```

```
$ conda install -c anaconda xlrd
$ conda install -c anaconda cytoolz
$ conda install seaborn scikit-learn statsmodels
numba pytables
$ conda install -c conda-forge python-igraph
louvain
$ conda install -c conda-forge multicore-tsne
$ conda install dask==2.9
```

(d) Install pip packages

```
$ pip install scanpy
$ pip install pyscenic
```

2. Resource directory file format.
   (a) Cell annotation csv file (comma delimited).

Cells	Samples	Cell. types	Treatment. group	Cohort	No.of. genes	No.of. reads
Cell 1	Mel78	Mal	post.treatment	Tirosh	8258	357,919
Cell 2	Mel79	Mal	treatment. naive	Tirosh	2047	5727
Cell 3	Mel88	Mal	post.treatment	Tirosh	5375	139,218
Cell 4	Mel79	Mal	treatment. naive	Tirosh	5648	73,996
Cell *n*	Mel78	Mal	post.treatment	Tirosh	7409	380,341

   (b) The cell count file (CSV format). The first column should contain the gene names whereas the first rows should be named after the individual cell types. Keep the first cell of the file empty.

	Cell 1	Cell 2	Cell 3	Cell 4	Cell *n*
C9orf152	0	0	0	0	0
RPS11	370	1	75	15	345
ELMO2	43	0	2	18	43
CREB3L1	0	0	0	0	0
PNMA1	68	0	1	8	103

(c) The TPM normalized expression CSV file. The first column should contain the gene names whereas the first rows should be named after the individual cell types. Keep the first cell of the file empty.

	Cell 1	Cell 2	Cell 3	Cell 4	Cell *n*
C9orf152	0	0	0	0	0
RPS11	8.144184	5.915091	7.243164	6.019502	7.974753
ELMO2	2.639232	0	0.732052	3.687956	2.599318
CREB3L1	0	0	0	0	0
PNMA1	3.656496	0	0.536053	3.041418	4.132741

(d) The excel sheet of the metadata file.

Sample	Patient	Age	Sex	Treatment	Treatment group	Lesion type	Site
Mel53	Mel53	77	F	None	Untreated	Metastasis	Subcutaneous back lesion
Mel58	Mel58	83	M	Ipilimumab	Post-ICI (resistant)	Metastasis	Subcutaneous leg lesion
Mel60	Mel60	60	M	Trametinib, ipilimumab	Post-ICI (resistant)	Metastasis	Spleen
Mel71	Mel71	79	M	None	Untreated	Metastasis	Transverse colon
Mel72	Mel72	57	F	IL-2, nivolumab, ipilimumab + anti-KIR-Ab	Post-ICI (resistant)	Metastasis	External iliac lymph node

3. *Choose GRN algorithm.* In pySCENIC there are two available algorithms genie3 and grnboost2 out of these two method grnboost2 is the default. To choose genie3 use:

```
-m or --method [genie3]
```

4. Multiprocessing mode setup. Depending upon the Dask and pySCENIC version and HPC facility choose appropriate multiprocessing mode.

```
--mode "dask_multiprocessing" # Fast and default
--mode "custom_multiprocessing" or
```

Acknowledgments

This work was supported by the National Science Foundation (IOS-1557796 and IOS-2038872) to M.S.M., and U54 ES 030246 from NIH/NIEHS to M.A. and M.S.M. We also acknowledge Dr. Karolina Mukhtar for critical read of this book chapter.

References

1. Sengupta U, Ukil S, Dimitrova N et al (2009) Expression-based network biology identifies alteration in key regulatory pathways of type 2 diabetes and associated risk/complications. PLoS One 4(12):e8100. https://doi.org/10.1371/journal.pone.0008100
2. Garbutt CC, Bangalore PV, Kannar P et al (2014) Getting to the edge: protein dynamical networks as a new frontier in plant-microbe interactions. Front Plant Sci 5:312. https://doi.org/10.3389/fpls.2014.00312
3. Tully JP, Hill AE, Ahmed HM et al (2014) Expression-based network biology identifies immune-related functional modules involved in plant defense. BMC Genomics 15:421. https://doi.org/10.1186/1471-2164-15-421
4. Naqvi RZ, Zaidi SS, Akhtar KP et al (2017) Transcriptomics reveals multiple resistance mechanisms against cotton leaf curl disease in a naturally immune cotton species, Gossypium arboreum. Sci Rep 7(1):15880. https://doi.org/10.1038/s41598-017-15963-9
5. Naqvi RZ, Zaidi SS, Mukhtar MS et al (2019) Transcriptomic analysis of cultivated cotton Gossypium hirsutum provides insights into host responses upon whitefly-mediated transmission of cotton leaf curl disease. PLoS One 14(2):e0210011. https://doi.org/10.1371/journal.pone.0210011
6. Zaidi SS, Naqvi RZ, Asif M et al (2020) Molecular insight into cotton leaf curl geminivirus disease resistance in cultivated cotton (Gossypium hirsutum). Plant Biotechnol J 18 (3):691–706. https://doi.org/10.1111/pbi.13236
7. Mishra B, Sun Y, Howton TC et al (2018) Dynamic modeling of transcriptional gene regulatory network uncovers distinct pathways during the onset of Arabidopsis leaf senescence. NPJ Syst Biol Appl 4:35. https://doi.org/10.1038/s41540-018-0071-2
8. Mishra B, Sun Y, Ahmed H et al (2017) Global temporal dynamic landscape of pathogen-mediated subversion of Arabidopsis innate immunity. Sci Rep 7(1):7849. https://doi.org/10.1038/s41598-017-08073-z
9. McCormack ME, Lopez JA, Crocker TH et al (2016) Making the right connections: network biology and plant immune system dynamics. Curr Plant Biol 5:2–12
10. Mishra B, Kumar N, Mukhtar MS (2019) Systems biology and machine learning in plant-pathogen interactions. Mol Plant-Microbe Interact 32(1):45–55. https://doi.org/10.1094/MPMI-08-18-0221-FI
11. Ahmed H, Howton TC, Sun Y et al (2018) Network biology discovers pathogen contact points in host protein-protein interactomes. Nat Commun 9(1):2312. https://doi.org/10.1038/s41467-018-04632-8
12. Jeong H, Mason SP, Barabasi AL et al (2001) Lethality and centrality in protein networks. Nature 411(6833):41–42. https://doi.org/10.1038/35075138
13. Li X, Li W, Zeng M et al (2020) Network-based methods for predicting essential genes or proteins: a survey. Brief Bioinform 21 (2):566–583. https://doi.org/10.1093/bib/bbz017
14. van Leeuwen J, Pons C, Tan G et al (2020) Systematic analysis of bypass suppression of essential genes. Mol Syst Biol 16(9):e9828. https://doi.org/10.15252/msb.20209828
15. Kumar N, Mishra B, Mehmood A et al (2020) Integrative network biology framework elucidates molecular mechanisms of SARS-CoV-2 pathogenesis. iScience 23(9):101526. https://doi.org/10.1016/j.isci.2020.101526
16. Haque A, Engel J, Teichmann SA et al (2017) A practical guide to single-cell RNA-sequencing for biomedical research and clinical applications. Genome Med 9(1):75. https://doi.org/10.1186/s13073-017-0467-4
17. Olsen TK, Baryawno N (2018) Introduction to single-cell RNA sequencing. Curr Protoc Mol Biol 122(1):e57. https://doi.org/10.1002/cpmb.57
18. Hwang B, Lee JH, Bang D (2018) Single-cell RNA sequencing technologies and bioinformatics pipelines. Exp Mol Med 50(8):96. https://doi.org/10.1038/s12276-018-0071-8

19. Chen G, Ning B, Shi T (2019) Single-cell RNA-Seq technologies and related computational data analysis. Front Genet 10:317. https://doi.org/10.3389/fgene.2019.00317
20. Todorov H, Cannoodt R, Saelens W et al (2019) Network inference from single-cell transcriptomic data. Methods Mol Biol 1883:235–249. https://doi.org/10.1007/978-1-4939-8882-2_10
21. Chen S, Mar JC (2018) Evaluating methods of inferring gene regulatory networks highlights their lack of performance for single cell gene expression data. BMC Bioinformatics 19 (1):232. https://doi.org/10.1186/s12859-018-2217-z
22. Huynh-Thu VA, Irrthum A, Wehenkel L et al (2010) Inferring regulatory networks from expression data using tree-based methods. PLoS One 5(9). https://doi.org/10.1371/journal.pone.0012776
23. Kim S (2015) ppcor: an R package for a fast calculation to semi-partial correlation coefficients. Commun Stat Appl Methods 22 (6):665–674. https://doi.org/10.5351/CSAM.2015.22.6.665
24. Moerman T, Aibar Santos S, Bravo Gonzalez-Blas C et al (2019) GRNBoost2 and Arboreto: efficient and scalable inference of gene regulatory networks. Bioinformatics 35 (12):2159–2161. https://doi.org/10.1093/bioinformatics/bty916
25. Yang B, Xu Y, Maxwell A et al (2018) MICRAT: a novel algorithm for inferring gene regulatory networks using time series gene expression data. BMC Syst Biol 12 (Suppl 7):115. https://doi.org/10.1186/s12918-018-0635-1
26. Chan TE, Stumpf MPH, Babtie AC (2017) Gene regulatory network inference from single-cell data using multivariate information measures. Cell Syst 5(3):251–267.e253. https://doi.org/10.1016/j.cels.2017.08.014
27. Gentleman RC, Carey VJ, Bates DM et al (2004) Bioconductor: open software development for computational biology and bioinformatics. Genome Biol 5(10):R80. https://doi.org/10.1186/gb-2004-5-10-r80
28. Van de Sande B, Flerin C, Davie K et al (2020) A scalable SCENIC workflow for single-cell gene regulatory network analysis. Nat Protoc 15(7):2247–2276. https://doi.org/10.1038/s41596-020-0336-2
29. Aibar S, Gonzalez-Blas CB, Moerman T et al (2017) SCENIC: single-cell regulatory network inference and clustering. Nat Methods 14(11):1083–1086. https://doi.org/10.1038/nmeth.4463
30. Aertslab (2021) pySCENIC-GitHub. GitHub. https://github.com/aertslab/pySCENIC. 2021
31. Jerby-Arnon L, Shah P, Cuoco MS et al (2018) A Cancer Cell Program Promotes T Cell Exclusion and Resistance to Checkpoint Blockade. Cell 175(4):984–997 e924. https://doi.org/10.1016/j.cell.2018.09.006

Chapter 11

Modeling Immune Dynamics in Plants Using JIMENA-Package

Özge Osmanoglu, Shabana Shams, Thomas Dandekar, and Muhammad Naseem

Abstract

Plant immunity is a highly dynamic process and requires dynamic modeling to capture the events of complexity mediated by the interaction between plant host and the attacking pathogen. The events of recognition are invoked by pathogen-based epitopes, while the subversion of host defenses are orchestrated by pathogen-originated effector molecules. The pathogen constitutes an immune signaling network inside the host cells. We model plant immune dynamics by using JIMENA-package, which is a java-based genetic regulatory network (GRN) simulation framework. It can efficiently compute network behavior and system states mediated by pathogenic perturbations. Here, we describe a step-by-step protocol to introduce the application of JIMENA-package to quantify immune dynamics in plant–pathogen interaction networks.

Key words Dynamic modeling, Plant immunity, Boolean logic, PAMPs, NLRs, Effectors

1 Introduction

Plants are challenged by a plethora of harmful pathogens that are broadly categorized into biotrophs, hemibiotrophs, and necrotrophs. To cause infection in the host, necrotrophs kill plant cells whereas biotrophs keep the host cells alive and hemibiotrophs are the pathogens with a dual lifestyle [1, 2]. A dynamic relationship of evolvability exists between microbial pathogens and their cognate hosts, formerly called gene-for-gene interaction latter called as the ZIGZAG model of plant immunity [3]. This model is a didactic and intuitive representation of the events that lead to recognition between host and pathogen and subsequent processes that culminate in mechanisms of resistance or susceptibility in plants. The plant pathogen *Pseudomonas syringae* pv. tomato DC3000 (*Pst* DC3000) is the most well-studied hemibiotroph [4], and the zigzag model of plant immunity [3] is predominantly grounded on experimental observations that have been acquired from the interaction of this model pathogen with *Arabidopsis*. Plants have

Shahid Mukhtar (ed.), *Modeling Transcriptional Regulation: Methods and Protocols*, Methods in Molecular Biology, vol. 2328, https://doi.org/10.1007/978-1-0716-1534-8_11,

attained multilayer, robust immune systems [1–3] to defend against the ever-evolving virulence of the pathogen. Plant surface cellular receptors, such as FLS2 and EFR are receptor-like kinases (RLKs) contain ectodomains that perceive pathogen-associated molecular patterns (PAMPs), such as flagellin, chitin, glycoproteins and lipopolysaccharides and mount an innate immune response known as PAMPs-triggered immunity (PTI) [5, 6]. PTI is transient but broadly effective against many microbes and is often associated with rapid production of reactive oxygen species (ROS), membrane depolarization, calcium signatures, activation of mitogen protein kinases (MAPKs) as well as the induction of defense genes such as WRKY29, WRKY30, FRK1, and PR1 [7, 8]. To cope with PTI, virulent plant pathogens deliver effectors into the host cell. Virulent effectors redundantly inhibit the functions of several host targets and thus mitigate PTI, which culminates in susceptibility (effectors mediated susceptibility) of the host plant. However, the effectors of avirulent pathogens are recognized by mobile receptors with Nucleotide binding and Leucine Rich repeat domains (NLRs) proteins [3, 9]. This recognition leads to a robust immune response known as effector-triggered immunity (ETI) [10, 11], which normally results in cell death and is aliased as Hypersensitive Response (HR).

Although the ZIGZAG model is neither quantitative nor predictive, the biochemical interactions among the downstream components of PTI and ETI constitute an immune signaling network (*see* **Note 1**: [12, 13]). The topology of this network should represent interacting nodes such as immune receptors, regulatory proteins and important transcription factors crosslinking with pathogenic factors such as PAMPs and effectors. Whereas edges of this network depict functional attributes of node-to-node activation and/or inhibition [13]. The assembled Boolean based network can be analyzed to model the effect of perturbations (effector-based interventions) mediated by the attacking pathogen. The JIMENA-package [14] performs steady-state analyses and dynamic simulations, recognizing nodes and interacting edges automatically in a given Gene Regulatory Network (GRN). To analyze the network for steady-states, JIMENA initiates the processing by calculating the discrete stable states as per a differential equation developed by Di Cara et al. [15]. Depending on the network topology and the existing connections the Boolean characteristics are calculated which culminates in a discrete version of the dynamical system representing the network. Based on the resulting discrete dynamical model the framework can efficiently predict stable system states (SSS) with minimal required computing power. For the discrete dynamic analysis, JIMENA uses generic assumptions: The presence of any one of the activators can effectively activate a target node, and the presence of any one of the inhibitors can successfully inactivate the target node. It identifies the number of stable steady-states located in the network and approximates activity values of each node at a given steady-state. Here, we performed JIMENA based simulations, steady-state analyses and perturbations taking *Pst* DC3000 as an input signal to the system in our study.

2 Materials

1. Java 7 or above.
2. yED. Software to generate and modify diagrams. https://www.yworks.com/products/yed [16].
3. JIMENA Software-package for simulation of genetic regulatory networks (GRN). https://www.biozentrum.uni-wuerzburg.de/bioinfo/computing/jimena/
4. A GRN of directed interactions generated in graphml format (*see* **Note 1**).
5. A personal computer running on any of the current operating systems.

3 Methods

1. In order to reconstruct a GRN for simulation using JIMENA (*see* **Note 1**), a detailed collection of interaction data is necessary (*see* **Note 2**). Literature survey and previously curated pathway databases are the main attributes to the network reconstruction phase. In our research, we collected literature and consulted various protein–protein interaction databases for the verification of plant hormone immune interactions (*see* **Note 3**). For instance, databases such as Kyoto Encyclopedia for Genes and Genomes (KEGG) [17, 18], STRING based search tool for the retrieval of interacting Genes Proteins [19] and PMN [20] can be screened to build hormone immune network for reference plant Arabidopsis (*see* **Notes 1–6**).
2. Import the network reconstructed in graphml format in JIMENA [14, 15] (Network → Import yEd file). The network can be reconstructed using yED Graph Editor [16]. For the reconstruction of the network, interactions from several databases including KEGG [17, 18], STRING [19], and from the literature should be collected.
3. Analyze the stable states of the imported network (Analysis→Find Stable States). The default calculation time is 10 time units and can be modified if desired. Stable states can be exported to a text file for later use (Fig. 1).

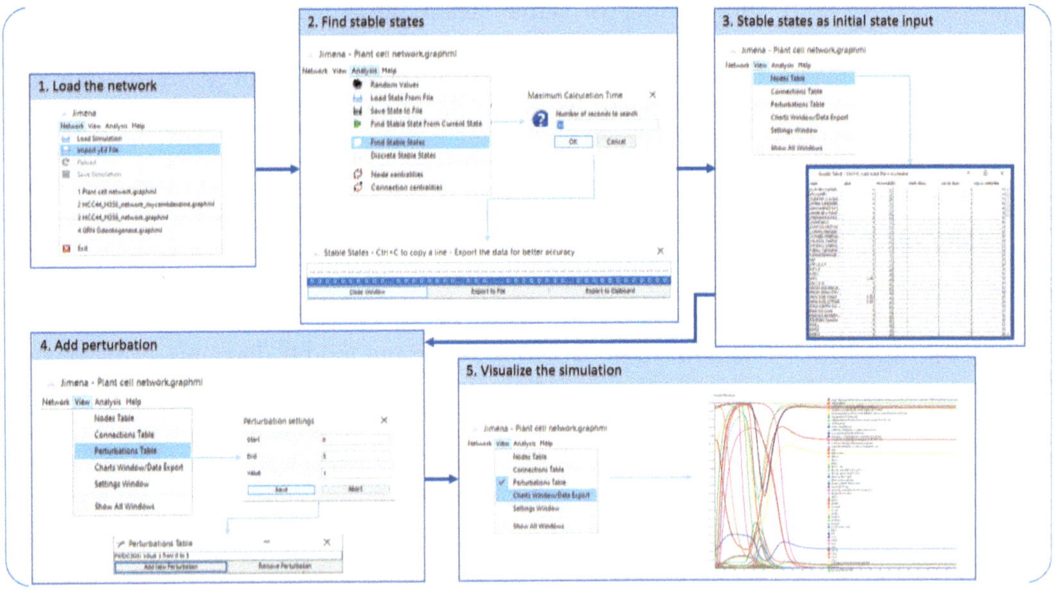

Fig. 1 Key steps in running simulations on a previously assembled plant hormone immune signaling network. Simulation steps (1–5) showing the dynamic readout regarding the activity of various nodes in a previously assembled plant hormone immune network (adopted from Naseem et al. 2012). Uploading of the pre-assembled network on the JIMENA portal by importing the yEd file from the local computer (step 1). JIMENA identifies stable states in the network (step 2) and prepares the simulation environment by defining stable states as initial input to the system (step 3). Next, perturbations are added to the system by selecting the nodes the impact of which is supposed to be determined (step 4). The visualization of the activity of immune network nodes is the final simulation step (step 5). Besides other nodes, visualization of the PR1 (pathogenesis-related protein-1) in response to *Pst* DC3000 infection serves as an index of plant immunity

4. Open "Nodes Table" (View→Nodes Table). Here, activation levels of individual nodes can be set or stable states obtained from **step 2** can be used as a starting point. To use stable states, copy (Ctrl + C) the line stable state of interest from stable states table and paste (Ctrl + V) in Nodes Table.
5. For a simulation of overexpression or knockout of individual nodes, open the Perturbations Table (View→Perturbations table). click on "Add New Perturbation" and specify the value and the time period that the given value is applicable to the selected node. For Knockout/Overexpression perturbations, "On/Off Perturbation" should be chosen. A short-lasting overexpression of *Pst* DC3000 can be given as perturbation for five time units. A value of 1 represents overexpression, while the value of 0 is used to simulate complete knockout of a node (Fig. 1; *see* **Note 5**).
6. Start the simulation by clicking on the green double arrow in IMIENA window. The default simulation time is 50 time units and can be changed to shorter simulation time in case the network reaches a new stable state faster than 50 time units.

7. Visualize the simulation results by opening the Charts Table (View→Charts Table). In the charts table, activation states of all nodes during the simulation can be viewed as well as individual nodes of interest. The activation values of chosen nodes during the simulation can be exported by "Export Selected" and saved.
8. Return to the initial state of the network by clicking on the yellow circular arrow and run further simulations with activation/inhibition of other nodes of interest (Fig. 1).

4 Notes

1. The reconstruction of the model is based upon Boolean logic, where each key unit of the involved immune system action (from genes to proteins) is designated as a node of the network. The simulations exemplified here are based on a previously assembled hormone immune network [13, 20, 21] where pathogen-based regulatory epitopes such as PAMPs, effectors as well as important host proteins for instance immune receptor RLKs, NB-LRR containing R proteins, transcription factors and host hormone receptors were all considered. To model in detail the host–pathogen interaction, we present here a refined and detailed model of plant signaling, extending our previous work [13]. Moreover, non-protein nodes such as COR toxin secreted by *Pst* DC3000 or host-mediated micro RNAs (such as miR393) as well as immune relevant molecules such as reactive oxygen species (ROS) and deposition of callose were also considered for assembling the network.
2. The connectivity among nodes in the network is maintained through edges. They either show activation or inhibition. Nodes of the network are turned out to be either linear as well as sparsely connected in metabolic pathways such as those representing hormone biosynthesis or to be densely connected such as hormone signaling crosstalking points. The latter serve as hub-nodes that have central positions in the network and highly influence the immune response of the system. The resulting Boolean model was established in the network biology portal CellDesigner and was then converted into yED format to seek compatibility by using JIMENA-package.
3. Plant immunity is a highly dynamic process and should ideally be dealt with in terms of quantitative dynamic modeling. The Boolean (1 or 0) based activation or inhibition experienced by nodes from their respective edges reveals the logic encoded in the network. We converted the static network model into a continuous dynamic system by using the JIMENA-package that interpolates concatenated e-functions between fully "ON" and "OFF" states, taking the logical connectivity into account (OR, AND, or NOT gates) [13, 15]

4. We parameterized our model for *Pst* DC3000 and *Arabidopsis* interactions such that full activation state was assigned to the attacking pathogen *Pst* DC3000 (capable of injecting all mentioned pathogenicity determinants into *Arabidopsis*) as the input signal. Likewise, the effect of *Pst* mutants can be modeled by removing corresponding nodes of pathogenicity determinants such as PAMPs, effectors or COR toxin from the network. Similarly, the effects of individual PAMPs, virulent and non-virulent effectors can also be predicted with JIMENA package.
5. The simulation shows how a change from minimum activation "0" to maximum "1" overtime is represented as gain (sigmoid curve) in signaling activity and the opposite change as decay in the activity. Furthermore, the dynamics of the activation of marker node PR-1 is a good indicator [13, 21] of the overall immune response strength of the system for this plant–pathogen interaction. Network reconstruction encompassing host immune regulatory proteins and pathogenic factors associated with disease development.
6. The dynamics of the activation of marker node pathogenesis-related-1 (PR1) is a good indicator of the overall immune response strength of the system for this plant–pathogen interaction [11]. The connectivity among nodes in the network is maintained through edges; they either show activation or inhibition. Densely nodes of the network serve as hub-nodes that have central positions in the network and orchestrate the ultimate immune response of the system. The JIMENA based dynamic simulation and prediction of the activity of the involved proteins can be extended to other plant–pathogen interaction systems as well as other plant processes such as growth, development as well as abiotic stresses in plants.

Acknowledgments

We thank the Land Bavaria (contribution to DFG Project number 324392634/TR221 to T.D.) for funding and are thankful for a grant from the Research Incentive Fund (R19073) and a CLUSTER grant (R20141) by Zayed University to M.N.

References

1. Glazebrook J (2005) Contrasting mechanisms of defense against biotrophic and necrotrophic pathogens. Annu Rev Phytopathol 43:205–227
2. Pieterse CMJ, Van der Does D, Zamioudis C, Leon-Reyes A, Van Wees SCM (2012) Hormonal modulation of plant immunity. Annu Rev Cell Dev Biol 28:489–521
3. Jones JDG, Dangl JL (2006) The plant immune system. Nature 444:323–329
4. Xin X-F, He SY (2013) Pseudomonas syringae pv. tomato DC3000: a model pathogen for

probing disease susceptibility and hormone signaling in plants. Annu Rev Phytopathol 51:473–498
5. Monaghan J, Zipfel C (2012) Plant pattern recognition receptor complexes at the plasma membrane. Curr Opin Plant Biol 15:349–357
6. Zipfel C, Kunze G, Chinchilla D, Caniard A, Jones JDG, Boller T, Felix G (2006) Perception of the bacterial PAMP EF-Tu by the receptor EFR restricts agrobacterium-mediated transformation. Cell 125:749–760
7. Chinchilla D, Zipfel C, Robatzek S, Kemmerling B, Nürnberger T, Jones JDG, Felix G, Boller T (2007) A flagellin-induced complex of the receptor FLS2 and BAK1 initiates plant defence. Nature 448:497–500
8. Asai T, Tena G, Plotnikova J, Willmann MR, Chiu W-L, Gomez-Gomez L, Boller T, Ausubel FM, Sheen J (2002) MAP kinase signalling cascade in Arabidopsis innate immunity. Nature 415:977–983
9. Sarris PF et al (2015) A plant immune receptor detects pathogen effectors that target WRKY transcription factors. Cell 161:1089–1100
10. Göhre V, Robatzek S (2008) Breaking the barriers: microbial effector molecules subvert plant immunity. Annu Rev Phytopathol 46:189–215
11. Cui H, Tsuda K, Parker JE (2014) Effector-triggered immunity: from pathogen perception to robust defense. Annu Rev Plant Biol 66:1–25
12. Dong X, Jiang Z, Peng YL, Zhang Z (2015) Revealing shared and distinct gene network organization in Arabidopsis immune responses by integrative analysis. Plant Physiol 167 (3):1186–1203
13. Naseem M, Philippi N, Hussain a, Wangorsch G, Ahmed N, Dandekar T (2012) Integrated systems view on networking by hormones in arabidopsis immunity reveals multiple crosstalk for cytokinin. Plant Cell 24:1793–1814
14. Karl S, Dandekar T (2015) Jimena: efficient computing and system state identification for genetic regulatory networks. BMC Bioinformatics *14*:306. https://doi.org/10.1186/1471-2105-14-306
15. Di Cara A, Garg A, De Micheli G, Xenarios I, Mendoza L (2007) Dynamic simulation of regulatory networks using SQUAD. BMC Bioinformatics *8*(1):462
16. https://www.yworks.com/products/yed
17. Kanehisa M, Goto S (2000) KEGG: kyoto encyclopedia of genes and genomes. Nucleic Acids Res 28:27–30
18. Kanehisa M, Sato Y, Furumichi M, Morishima K, Tanabe M (2019) New approach for understanding genome variations in KEGG. Nucleic Acids Res 47:D590–D595
19. Szklarczyk D, Gable AL, Lyon D, Junge A, Wyder S, Huerta-Cepas J, Simonovic M, Doncheva NT, Morris JH, Bork P, Jensen LJ, Mering CV (2019) STRING v11: protein-protein association networks with increased coverage, supporting functional discovery in genome-wide experimental datasets. Nucleic Acids Res *47*(D1):D607–D613. https://doi.org/10.1093/nar/gky1131
20. Schläpfer P, Zhang P, Wang C, Kim T, Banf M, Chae L et al (2017) Genome-wide prediction of metabolic enzymes, pathways, and gene clusters in plants. *Plant Physiol 173* (4):2041–2059
21. Kaltdorf M, Dandekar T, Naseem M (2017) Reconstruction of an immune dynamic model to simulate the contrasting role of auxin and cytokinin in plant immunity. In: Hurny A, Benková E (eds) Auxins and cytokinins in plant biology. Humana Press, New York, NY, pp 83–92

Chapter 12

Dynamic Regulatory Event Mining by iDREM in Large-Scale Multi-omics Datasets During Biotic and Abiotic Stress in Plants

Bharat Mishra, Nilesh Kumar, Jinbao Liu, and Karolina M. Pajerowska-Mukhtar

Abstract

The system-wide complexity of genome regulation encoding the organism phenotypic diversity is well understood. However, a major challenge persists about the appropriate method to describe the systematic dynamic genome regulation event utilizing enormous multi-omics datasets. Here, we describe Interactive Dynamic Regulatory Events Miner (iDREM) which reconstructs gene-regulatory networks from temporal transcriptome, proteome, and epigenome datasets during stress to envisage "master" regulators by simulating cascades of temporal transcription-regulatory and interactome events. The iDREM is a Java-based software that integrates static and time-series transcriptomics and proteomics datasets, transcription factor (TF)–target interactions, microRNA (miRNA)–target interaction, and protein–protein interactions to reconstruct temporal regulatory network and identify significant regulators in an unsupervised manner. The hidden Markov model detects specialized manipulated pathways as well as genes to recognize statistically significant regulators (TFs/miRNAs) that diverge in temporal activity. This method can be translated to any biotic or abiotic stress in plants and animals to predict the master regulators from condition-specific multi-omics datasets including host–pathogen interactions for comprehensive understanding of manipulated biological pathways.

Key words Gene-regulatory network, Temporal transcriptome, RNA-Seq, Gene regulation, Visualization, Plant–pathogen interactions

1 Introduction

The overwhelming complexity of biological, physical, and social networks have been represented and comprehensively analyzed by graph theory [1–3]. Networks are collections of nodes connected by edges in a defined pattern; similarly, biological networks are connections of several biomolecules [4–7]. Some of the highly studied biological networks gene coexpression, gene-regulatory network (GRN), protein–protein interaction (PPI), and metabolic networks participate in the system-wide functional pathways [8–

Shahid Mukhtar (ed.), *Modeling Transcriptional Regulation: Methods and Protocols*, Methods in Molecular Biology, vol. 2328, https://doi.org/10.1007/978-1-0716-1534-8_12,

12]. Among these networks, GRNs represent the pan regulatory circuits of numerous biological functions [13]. Regulators such as transcription factors (TFs) and microRNAs (miRNAs) are the controllers of activation or inhibition of certain components of gene-regulatory networks of signaling pathways following any changes in basal conditions. Many studies of biotic/abiotic stress in plants have exposed the prominence of TFs in controlling the regulatory mechanisms to cope with or fight off the stress [5, 13–17]. Plant–pathogen interactions represent a robust system for studies on the dynamicity of host gene-regulatory associations owing to the coevolution of host and pathogen proteins [18–22]. These evolutionary relationships between pathogen and its host to counter each-other lead to a perpetual tug-of-war and molecular arms race, which has been summarized in the ZIGZAG model of plant–microbe interactions [23, 24]. The ZIGZAG model is a series of complex dynamic events that enforces the manipulation of regulatory mechanisms subsequently leading to plant host resistance or susceptibility (Fig. 1) [8, 25–29]. While additional, case-specific dynamic models of plant-pathogen interactions have been proposed [24, 25, 30, 31], the ZIGZAG model describes an accurate representation of the molecular events during a phytopathogen infection and can be used to guide the discovery of system-wide regulatory dynamic mechanisms to reveal the intricate layers of various plant pathosystems [32, 33].

The recent developments in high-throughput genome-wide experimental techniques aided in the accumulation of massive-omics datasets across different species of eukaryotes and prokaryotes [8, 34]. Development of methodologies for utilization of different datasets to comprehensively explore the biological system is one of the top research priorities of the current times [35]. The genetic expression is partially controlled by the TF proteins that bind to the promoter regions of DNA to activate or repress their downstream target genes [15, 36]. These genetic regulations are ultimately responsible for the organismal phenotypic diversity and have been accumulated by the scientific community for the understanding of the pan regulatory repertoire [13]. However, these regulatory networks provide only a partial picture of condition-specific regulomics and fail to elaborate on the other-omics layers, limiting our systematic understanding of the biological processes involved [4, 8]. Previously, many tools have been developed to study the dynamics of regulatory networks based on temporal transcriptomics (microarray and RNA-Seq) to identify master/significant regulators (TFs and/or miRNAs) controlling the activation or inhibition of diverse biological pathways [24, 37–40]. Recently, a new tool, Interactive Dynamic Regulatory Events Miner (iDREM), was developed by Ding et al., 2018 that integrates several high-throughput multi-omics layers including transcriptomics, miRNA-omics, epigenomics, proteomics, and single-cell RNA-Seq datasets (Fig. 2)

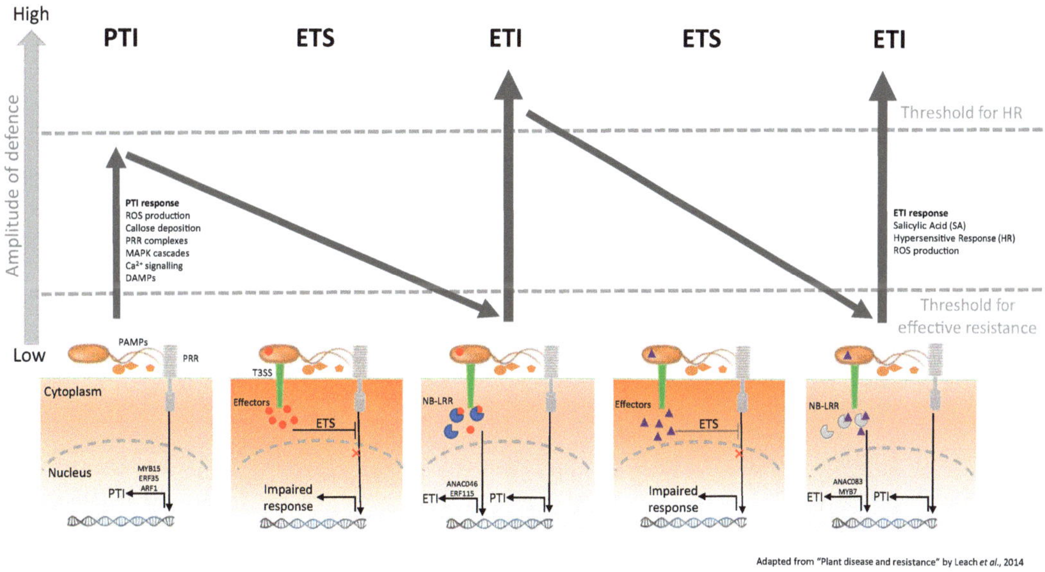

Fig. 1 ZIGZAG model of plant–pathogen interaction. A visual representation of the ZIGZAG model between pathogens and plant immune systems (adapted from Leach et al. 2014). In phase 1, pathogen-associated molecular patterns (PAMPs) are detected by plant pattern recognition receptors (PRRs), leading to a series of PAMP-associated immunity (PTI) responses such as Reactive Oxygen Species (ROS) production and callose deposition. In order to even up the disadvantaged scenario, surviving pathogens in phase 2 deliver effectors into plant cells by Type III secretion systems (T3SSs) to interfere with the PTI pathway, resulting in effector-triggered susceptibility (ETS). ETS breaks through the threshold of plant effective resistance to induce corresponding disease progression. Then, in phase 3, nucleotide-binding site leucine-rich repeat (NBS-LRR) proteins encoded by plant resistance genes directly or indirectly recognize pathogen effectors to proceed with effector-triggered immunity (ETI). Typically, ETI passes the threshold of hypersensitive response (HR), inducing various responses such as Salicylic Acid (SA) production and HR. This tug-of-war goes on in phase 4 when a new type of effectors secreted by pathogens helps the pathogen fight off ETI, followed by the counterattack of the host plant with the novel corresponding NB-LRR proteins in phase 5. The competition can continue for several rounds, forming a ZIGZAG model between ETS and ETI

[41]. iDREM utilizes time-series datasets and static biological networks (GRNs and PPI) to merge and identify bifurcation points of expression paths by learning from an Input-Output Hidden Markov Model (IOHMM) [42]. These bifurcation points at a specific time points determine the divergence in the co-expression of downstream target genes due to changes in the expression patterns [13]. The significant regulators (TFs/miRNAs) are predicted for each bifurcation based on the changes in target genes expression and enriched functional pathways are assigned to each path [13].

iDREM has implementation support for multi-omics datasets with impressive interactive visualization and analyses of reconstructed dynamic network and event mining. The analysis also illustrates the significant regulators-associated path as well as their functional annotations [43, 44]. Additionally, iDREM can be used for single-cell RNA-Seq datasets to model the regulatory

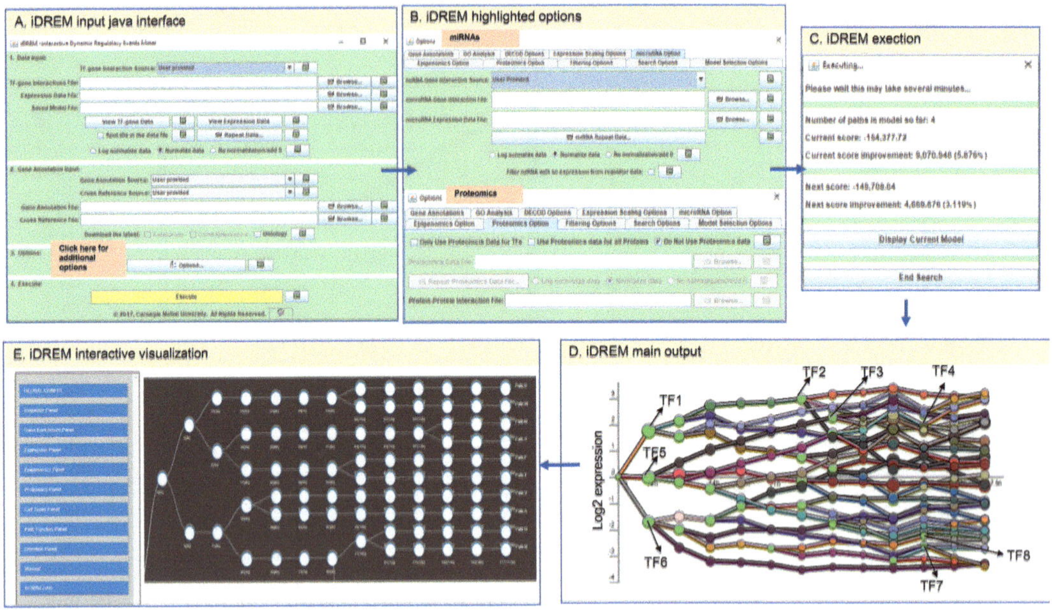

Fig. 2 iDREM interface and execution. (**a**) The input interface for basic iDREM analysis. At-least transcription factor (TF)—gene interaction file, gene expression file, and gene annotation file is required for this minimalistic modeling. (**b**) iDREM highlighted options including miRNAs, proteomics, and others. Multi-omics datasets can be integrated through options for better system-wide understanding of dynamic gene regulation and event mining. (**c**) iDREM execution window will pop-up after click of execution button on the iDREM main interface. The execution may take extended periods of time to finish based on the size of input dataset. If the expression dataset has multiple time points and more than 10,000 genes, the modeling might take couple of hours to execute. (**d**) iDREM main output interface. The colored lines are different regulated event paths modeled by the iDREM modeler. The light green nodes on the paths represent the significant regulatory bifurcation points by regulators (TFs/miRNAs). (**e**) iDREM interactive visualization interface. Please note that the representation on the right side removes the actual regulatory levels and only offers a graphic view of the bifurcation in the model and the paths. The actual expression levels of genes, TFs, proteins, miRNAs, and numerous additional features (epigenomic, proteomics, path functions, and cell types for single-cell) of the model can be interactively inspected through several panels available on the left side of interface

associations among diverse cell types and predict the master regulators in cell fate transitions. This integrative multi-omics systems biology platform reconstructs the dynamics of the regulatory network, which may indicate the evolution of innovative hypotheses on the dynamic genetic mechanisms altered during stress in any model system including the plant-pathosystem. Here, we explore the implementation of iDREM to reconstruct the dynamic regulatory networks to mine the dynamic events and predict master regulators during bacterial pathogen *Pseudomonas syringae* pv. *tomato* DC3000 (*Pst* DC3000) infection on *Arabidopsis thaliana* (Arabidopsis) based on *in planta* temporal transcriptome (Fig. 2) [36], TF-target network, PPI, miRNA–target interactions and gene annotation from The Arabidopsis Information Resource (TAIR) as initial datasets for our comprehensive study [45].

2 Materials

1. A personal computer/workstation running on supported operating systems; macOS, Windows 10 or Linux (Ubuntu 20.04 or Fedora 32).
2. 4 Gigabytes RAM for most experiments (8 Gigabytes RAM preferred for large datasets).
3. Java: A version of Java 7 or later must be installed to use iDREM.
4. Internet: To enable the interactive visualization powered by JavaScript, an internet connection is required.
5. To download iDREM, simply download and extract iDREM package. https://github.com/phoenixding/idrem
6. Tab-delimited TF-gene Interaction File with the format; 1st row: header, should be “TF Gene Input” (*see* **Note 1**).
7. Tab-delimited mRNA Expression Data File with the format; header row (1st row): 1st column—Gene, remaining columns—time point IDs (*see* **Note 2**).
8. Gene Annotation File from iDREM Source for the organism of interest.
9. Cross Reference File from iDREM Source.
10. Options (Optional, iDREM can run without these options).
    (a) MicroRNA-gene Interaction File (*see* **Note 3**).
    (b) MicroRNA Expression Data File (*see* **Note 4**).
    (c) Epigenomic data File in BED6 format (*see* **Note 5**).
    (d) GTF File of a given organism.
    (e) Proteomics Data File (*see* **Note 6**).
    (f) Protein–Protein Interaction File (PPI) (*see* **Note 7**).

3 Methods

1. In order to execute the dynamic GRN events mining, open the iDREM jar package (idrem.jar) by JAVA(TM) Platform SE binary inside the iDREM unzipped directory or execute from the command line after changing to iDREM directory and then type:

```
> java -Xmx8g -jar idrem.jar
```

(to use more available memory (16GB) change -Xmx8g to -Xmx16g)

2. To model the dynamic GRN events using iDREM, TF–gene interactions, transcriptome/mRNAs expression data, and Gene Annotation File is required for any minimalistic experiment (Fig. 2). Model given TF-gene interactions or retrieve the open-source network from ENCODE database and upload to TF–gene Interactions File (*see* **Note 1**). Get a temporal transcriptome expression dataset and upload it to Expression Data File (*see* **Note 2**). Verify that the files are tab-delimited and in the correct format by the View tab for TF-gene Data and Expression Data. Normalize expression data if needed. If the expression data is already in log-space, use "normalize data," otherwise use "log normalize data." If using a fold change data against control/zero-time scale, use the "No normalization/ add 0" method to correctly represent the data. Use Gene Ontology (GO) annotation files provided by iDREM team for well-studied organisms. A complete list of organisms with GO annotation files can be found in the iDREM manual. Additionally, download the latest Annotation, Cross Reference, and Ontology files from the iDREM server.
3. Click the "Execute" tab for the minimalistic modeling of dynamic regulatory events mining with a transcriptome dataset.
4. Otherwise, click the "Options" tab to add multi-omics layers and extra filters in dynamic event mining experiment. A dialog box will appear with additional options involving gene annotations, GO analysis, expression scaling, microRNA, epigenomic, proteomics, gene filtering, search, and model selection options.
5. Gene Annotations: select the types of annotation that need to be identified, which are associated with the genes' attributes such as "Biological Process," "Molecular Function," and "Cellular Component."
6. GO Analysis: Select parameters for GO analysis option of a minimum GO level, a minimum number of genes, and a multiple hypothesis correction either by "Bonferroni" or "Randomization" method. Keep in mind that a Bonferroni correction is faster, but a randomization test leads to lower p-values.
7. Expression Scaling Options: The user can elect to use activity scores for transcription factors. This will scale the regulator interaction values by its expression. By default, TF expression scaling is off.
8. MicroRNA Options: In this option, upload the modeled or iDREM provided miRNA–gene interaction file and miRNA expression data. Normalize the expression data per your need. If the expression already in log-space, use "normalize data," otherwise use "log normalize data." If using a fold change data against control/zero-time scale use "No normalization/add 0" method to correctly represent the data.

9. Epigenomic Option: Upload the epigenomic data including DNA and histone methylation, etc. The epigenomic score is used to denote the repression of the region. Therefore, different types of epigenomic data need to be pre-processed differently. Also, upload the GTF file associated with the given organism.
10. Proteomics Option: Upload the temporal protein level information supporting the model prediction. Choose the way the proteomics data should be used. Opt to either proteomics data for TFs, or for all proteins., or do not use proteomics data at all. Upload time-series proteomics dataset with gene names and data values for each time point. The header row specifies the name and time points. Normalize the expression data per your need. If the expression already in log-space, use "normalize data," otherwise use "log normalize data." If using a fold change data against the control/zero-time scale, use "No normalization/add 0" method to correctly represent your data. Upload the gene name-annotated protein–protein interactions file after downloading from STRING or BioGRID databases with the interaction strength.
11. Gene Filtering Option: Filter the genes that are not part of the transcription factor–gene interactions. Upload a list of genes to be excluded from the analysis.
12. Search Options: Use this option to merge the paths based on a common prior split that is modeled to merge. If the paths are not needed to be merged, uncheck the selection.
13. Model Selection Option: Select from any of the two frameworks: "Penalized Likelihood" or "Train-Test" for model selection. In the "Penalized Likelihood" model, all the genes are used to both train the parameters of the model during search and select the model and a regularization parameter, "Penalized Likelihood Node Penalty," is the penalty subtracted for each state to prevent overfitting. Whereas, in the Train-Test model only a subset of genes is used to train the parameters of the model under consideration, while the remainder is used to score the model based on likelihood. A second phase is then executed under this option where the data is re-split and only changes that reduce the number of states are considered. "Penalized Likelihood" is the default model framework used for dynamic event mining.
14. Click the "Execute" tab for the multi-omics modeling of dynamic regulatory events mining with transcriptome, proteomics, miRNAs, and Epigenomic datasets. If the data file has two or more time points, then the iDREM algorithm will execute. When the algorithm completes, a new interface will appear showing an annotated dynamic map.

15. The main window displays the time series of all the genes that were not filtered and are represented through the iDREM map.
16. Interactive Visualization: iDREM provides an interactive visualization of the predicted model besides the iDREM direct output. Visualize output results about significant regulators in paths, their activation or inhibition, presence of gene(s) in paths, activation/inhibition of paths based on gene expression, explore regulator specific targets and their average expression, and path functional annotations.

4 Notes

1. Arrange the TF-gene interaction file format as follows. The first column should be "TF," the second column should be "Gene" and the third column should be "Input," classified as: Positive interaction=1, negative interaction=0. The first row is the header row.

TF	Gene	Input
AT1G09530	AT1G02340	1
AT1G09530	AT1G02400	1
AT1G09530	AT1G06570	1

2. Gene expression data file format should be as follows. The first row is the header row with "gene" and time points names. Data starts from the second row.

Gene	0 h	2 h	3 h	4 h	6 h
AT1G01320	0.0176305	−1.96454	−2.20406	−1.4432	−0.49346
AT1G01390	0.774516	−0.56299	−0.96363	−1.00316	−0.7135
AT1G01790	1.14254	1.1893	0.852441	−0.15885	−0.09075
AT1G02120	−1.51476	−0.27666	−0.68379	0.779072	2.16978
AT1G02380	−0.986297	0.272888	0.595829	0.674299	−0.2216

7 h	8 h	10 h	11 h	12 h	14 h	16 h	17.5 h
−0.21562	0.466248	0.464839	0.19669	0.583211	0.430028	0.711834	0.558346
−1.07499	−0.8813	−0.78099	−0.59221	−0.82	−0.08411	−0.31392	−0.52591
−0.9054	−0.73754	−1.31293	−1.37237	−1.67218	−1.40142	−1.55822	−1.63705
1.01665	0.832013	1.26951	0.818967	2.17296	1.01995	0.587493	0.013703
0.371551	−0.69524	−0.14241	1.40945	1.81864	0.488978	1.6268	2.69634

3. miRNA–gene interaction file format should be as follows. The first column should be "miRNA" and the second column should be "Gene." The first row is the header row.

miRNA	Gene
miR156a	AT3G15270
miR157b	AT4G36770
miR161	AT1G15125

4. miRNA expression data file format should be as follows. The first row is the header row with miRNAs and time-points names. Data starts from second row.

miRNA	0 h	2 h	3 h	4 h	6 h	7 h	8 h	10 h	11 h	12 h	14 h	16 h	17.5 h
miR156a	−1.51476	−0.27666	−0.68379	0.779072	2.16978	1.01665	0.832013	1.26951	0.818967	2.17296	1.01995	0.587493	0.013703
miR157b	−0.986297	0.272888	0.595829	0.674299	−0.2216	0.371551	−0.69524	−0.14241	1.40945	1.81864	0.488978	1.6268	2.69634
miR161	−1.36035	1.86776	1.35452	0.979324	0.303071	0.420461	0.058817	−0.21339	0.322971	−0.03466	−0.12024	0.030595	−0.24147

5. Epigenomic data file should be in BED6 format as follows. There is no header row.

chr7 28372162 28373662 0h_AT3G15270 0.21 −
chr12 76532560 76534060 4h_AT1G15125 0.25 +
chr10 3739377 3740877 12h_AT1G77770 0.56 +
chr6 125380004 125381504 14h_AT1G77770 0.41 −

6. Proteomics data format should be formatted as follows. The first row is the header row with protein names and time points. Data starts from the second row.

Name	0 h	2 h	3 h	4 h	6 h	7 h	8 h	10 h	11 h	12 h	14 h	16 h	17.5 h
AT1G02120	−0.536791	−0.83126	−0.46058	−0.14563	0.226025	0.594995	1.15956	2.07151	1.70604	3.2238	1.84686	1.35978	1.8473
AT1G02380	1.19751	0.462478	−0.47047	−0.75643	−0.39834	−0.66956	−0.25331	−0.70168	−1.26553	−1.91462	−1.37707	−1.65306	−2.21151
AT1G02400	0.66193	1.57124	−0.51933	−0.09221	−0.53158	−1.01267	−1.00703	−1.5903	−1.31549	−0.83629	−1.5072	−1.74244	−1.68207

7. Protein–protein interaction should be formatted as follows. The first column should be protein1, the second column should be protein2 and the third column should be the strength of interaction. If strength is unavailable, put 1 in place of numbers. There is no header row.

AT1G02120	AT1G02750	0.813
AT1G02380	AT1G03160	0.902
AT1G02400	AT1G03410	0.904

Acknowledgments

N.K. and J.L. were supported by the National Science Foundation award (IOS-1557796 and IOS-2038872). Work in K.P.M. laboratory is supported by a NSF-CAREER (IOS-1350244 and IOS-2038872) award.

References

1. Brandes U, Robins G, McCranie ANN et al (2013) What is network science? Netw Sci 1 (1):1–15. https://doi.org/10.1017/nws.2013.2
2. Boccaletti S, Latora V, Moreno Y et al (2006) Complex networks: structure and dynamics. Phys Rep 424(4):175–308. https://doi.org/10.1016/j.physrep.2005.10.009
3. Kumar N, Mishra B, Mehmood A et al (2020) Integrative network biology framework elucidates molecular mechanisms of SARS-CoV-2 pathogenesis. iScience 23(9):101526. https://doi.org/10.1016/j.isci.2020.101526
4. McCormack ME, Lopez JA, Crocker TH et al (2016) Making the right connections: network biology and plant immune system dynamics. Curr Plant Biol 5:2–12. https://doi.org/10.1016/j.cpb.2015.10.002
5. Garbutt CC, Bangalore PV, Kannar P et al (2014) Getting to the edge: protein dynamical networks as a new frontier in plant-microbe interactions. Front Plant Sci 5:312. https://doi.org/10.3389/fpls.2014.00312
6. Lopez J, Mukhtar MS (2017) Mapping protein-protein interaction using high-throughput yeast 2-hybrid. Methods Mol Biol 1610:217–230. https://doi.org/10.1007/978-1-4939-7003-2_14
7. Mott GA, Smakowska-Luzan E, Pasha A et al (2019) Map of physical interactions between extracellular domains of Arabidopsis leucine-rich repeat receptor kinases. Sci Data 6 (1):190025. https://doi.org/10.1038/sdata.2019.25
8. Mishra B, Kumar N, Mukhtar MS (2019) Systems biology and machine learning in plant-pathogen interactions. Mol Plant-Microbe Interact 32(1):45–55. https://doi.org/10.1094/MPMI-08-18-0221-FI
9. Gao J, Barzel B, Barabasi AL (2016) Universal resilience patterns in complex networks. Nature 536(7615):238. https://doi.org/10.1038/nature18019
10. Cho DY, Kim YA, Przytycka TM (2012) Chapter 5: network biology approach to complex diseases. PLoS Comput Biol 8(12): e1002820. https://doi.org/10.1371/journal.pcbi.1002820
11. Naqvi RZ, Zaidi SS, Akhtar KP et al (2017) Transcriptomics reveals multiple resistance mechanisms against cotton leaf curl disease in a naturally immune cotton species, Gossypium arboreum. Sci Rep 7(1):15880. https://doi.org/10.1038/s41598-017-15963-9
12. Naqvi RZ, SS-e-A Z, Mukhtar MS et al (2019) Transcriptomic analysis of cultivated cotton Gossypium hirsutum provides insights into host responses upon whitefly-mediated transmission of cotton leaf curl disease. PLoS One 14(2):e0210011. https://doi.org/10.1371/journal.pone.0210011
13. Mishra B, Sun Y, Howton TC et al (2018) Dynamic modeling of transcriptional gene regulatory network uncovers distinct pathways during the onset of Arabidopsis leaf senescence. NPJ Syst Biol Appl 4:35. https://doi.org/10.1038/s41540-018-0071-2
14. Mishra B, Sun Y, Ahmed H et al (2017) Global temporal dynamic landscape of pathogen-mediated subversion of Arabidopsis innate immunity. Sci Rep 7(1):7849. https://doi.org/10.1038/s41598-017-08073-z
15. de Luis Balaguer MA, Fisher AP, Clark NM et al (2017) Predicting gene regulatory networks by combining spatial and temporal gene expression data in Arabidopsis root stem cells. Proc Natl Acad Sci U S A 114(36): E7632–E7640. https://doi.org/10.1073/pnas.1707566114
16. Baltrus DA, Nishimura MT, Romanchuk A et al (2011) Dynamic evolution of pathogenicity revealed by sequencing and comparative genomics of 19 Pseudomonas syringae isolates. PLoS Pathog 7(7):e1002132. https://doi.org/10.1371/journal.ppat.1002132
17. Ahmed H, Howton TC, Sun Y et al (2018) Network biology discovers pathogen contact points in host protein-protein interactomes. Nat Commun 9(1):2312. https://doi.org/10.1038/s41467-018-04632-8
18. Mukhtar MS, Carvunis AR, Dreze M et al (2011) Independently evolved virulence

effectors converge onto hubs in a plant immune system network. Science 333 (6042):596–601. https://doi.org/10.1126/science.1203659

19. Arabidopsis Interactome Mapping C (2011) Evidence for network evolution in an Arabidopsis interactome map. Science 333 (6042):601–607. https://doi.org/10.1126/science.1203877
20. Klopffleisch K, Phan N, Augustin K et al (2011) Arabidopsis G-protein interactome reveals connections to cell wall carbohydrates and morphogenesis. Mol Syst Biol 7:532. https://doi.org/10.1038/msb.2011.66
21. Smakowska-Luzan E, Mott GA, Parys K et al (2018) An extracellular network of Arabidopsis leucine-rich repeat receptor kinases. Nature 553(7688):342–346. https://doi.org/10.1038/nature25184
22. Washington EJ, Mukhtar MS, Finkel OM et al (2016) Pseudomonas syringae type III effector HopAF1 suppresses plant immunity by targeting methionine recycling to block ethylene induction. Proc Natl Acad Sci U S A 113(25):E3577–E3586. https://doi.org/10.1073/pnas.1606322113
23. Jones JD, Dangl JL (2006) The plant immune system. Nature 444(7117):323–329. https://doi.org/10.1038/nature05286
24. Cook DE, Mesarich CH, Thomma BP (2015) Understanding plant immunity as a surveillance system to detect invasion. Annu Rev Phytopathol 53:541–563. https://doi.org/10.1146/annurev-phyto-080614-120114
25. Pritchard L, Birch PR (2014) The zigzag model of plant-microbe interactions: is it time to move on? Mol Plant Pathol 15(9):865–870. https://doi.org/10.1111/mpp.12210
26. Sun Y, Detchemendy TW, Pajerowska-Mukhtar KM et al (2018) NPR1 in JazzSet with pathogen effectors. Trends Plant Sci 23 (6):469–472. https://doi.org/10.1016/j.tplants.2018.04.007
27. Mukhtar MS, McCormack ME, Argueso CT et al (2016) Pathogen tactics to manipulate plant cell death. Curr Biol 26(13):R608–R619. https://doi.org/10.1016/j.cub.2016.02.051
28. Leach J, Leung H, Tisserat N (2014) Plant disease and resistance. Encyclopedia of Agriculture and Food Systems 2014:360–374. https://doi.org/10.1016/B978-0-444-52512-3.00165-0
29. Tully JP, Hill AE, Ahmed HM et al (2014) Expression-based network biology identifies immune-related functional modules involved in plant defense. BMC Genomics 15:421. https://doi.org/10.1186/1471-2164-15-421
30. Thordal-Christensen H (2020) A holistic view on plant effector-triggered immunity presented as an iceberg model. Cell Mol Life Sci 77(20):3963–3976. https://doi.org/10.1007/s00018-020-03515-w
31. Fei Q, Zhang Y, Xia R et al (2016) Small RNAs add zing to the zig-zag-zig model of plant defenses. Mol Plant-Microbe Interact 29 (3):165–169. https://doi.org/10.1094/MPMI-09-15-0212-FI
32. Zaidi SS, Mukhtar MS, Mansoor S (2018) Genome editing: targeting susceptibility genes for plant disease resistance. Trends Biotechnol 36(9):898–906. https://doi.org/10.1016/j.tibtech.2018.04.005
33. Zaidi SS, Naqvi RZ, Asif M et al (2020) Molecular insight into cotton leaf curl geminivirus disease resistance in cultivated cotton (Gossypium hirsutum). Plant Biotechnol J 18 (3):691–706. https://doi.org/10.1111/pbi.13236
34. Liu Z, Miller D, Li F et al (2020) A large accessory protein interactome is rewired across environments. elife 9:e62365. https://doi.org/10.7554/eLife.62365
35. Matcovitch-Natan O, Winter DR, Giladi A et al (2016) Microglia development follows a stepwise program to regulate brain homeostasis. Science 353(6301):aad8670. https://doi.org/10.1126/science.aad8670
36. Lewis LA, Polanski K, de Torres-Zabala M et al (2015) Transcriptional dynamics driving MAMP-triggered immunity and pathogen effector-mediated immunosuppression in Arabidopsis leaves following infection with Pseudomonas syringae pv tomato DC3000. Plant Cell 27(11):3038–3064. https://doi.org/10.1105/tpc.15.00471
37. Lachmann A, Xu H, Krishnan J et al (2010) ChEA: transcription factor regulation inferred from integrating genome-wide ChIP-X experiments. Bioinformatics 26(19):2438–2444. https://doi.org/10.1093/bioinformatics/btq466
38. Jones CJ, Newsom D, Kelly B et al (2014) ChIP-seq and RNA-seq reveal an AmrZ-mediated mechanism for cyclic di-GMP synthesis and biofilm development by Pseudomonas aeruginosa. PLoS Pathog 10(3):e1003984. https://doi.org/10.1371/journal.ppat.1003984
39. Ideker T, Thorsson V, Ranish JA et al (2001) Integrated genomic and proteomic analyses of a systematically perturbed metabolic network.

Science 292(5518):929–934. https://doi.org/10.1126/science.292.5518.929

40. Ernst J, Vainas O, Harbison CT et al (2007) Reconstructing dynamic regulatory maps. Mol Syst Biol 3:74. https://doi.org/10.1038/msb4100115

41. Ding J, Hagood JS, Ambalavanan N et al (2018) iDREM: interactive visualization of dynamic regulatory networks. PLoS Comput Biol 14(3):e1006019. https://doi.org/10.1371/journal.pcbi.1006019

42. Bengio Y, Frasconi P (1995) An input-output HMM architecture. In G. Tesauro, D. S. Touretzky, & T. K. Leen (Eds.), Advances in neural information processing systems, 7, pp 427–434. Cambridge, MA: MIT Press

43. Song L, Huang SC, Wise A et al (2016) A transcription factor hierarchy defines an environmental stress response network. Science 354(6312). https://doi.org/10.1126/science.aag1550

44. Ciofani M, Madar A, Galan C et al (2012) A validated regulatory network for Th17 cell specification. Cell 151(2):289–303. https://doi.org/10.1016/j.cell.2012.09.016

45. Berardini TZ, Reiser L, Li D et al (2015) The Arabidopsis information resource: making and mining the "gold standard" annotated reference plant genome. Genesis 53(8):474–485. https://doi.org/10.1002/dvg.22877

Chapter 13

A Semi-In Vivo Transcriptional Assay to Dissect Plant Defense Regulatory Modules

Fatimah Aljedaani, Naganand Rayapuram, and Ikram Blilou

Abstract

Plants use different regulatory modules in response to changes in their surroundings. With the transcriptomic approaches governing all research areas, an integrative, fast, and sensitive approach toward validating genes of interest becomes a critical step prior to functional studies in planta. This chapter describes a detailed method for a quantitative analysis of transcriptional readouts of defense response genes using tobacco leaves as a transient system. The method uses Luciferase reporter assays to monitor activities of defense pathway promoters. Under normal conditions, the JASMONATE ZIM-DOMAIN (JAZ) proteins repress defense genes by preventing their expression. Here, we will provide a detailed protocol on the use of a dual-luciferase system to analyze activities of various defense response promoters simultaneously. We will use two well-characterized modules from the Jasmonic acid (JA) defense pathway; the JAZ3 repressor protein and the promoters of three of JA responsive genes, MYC2, 3 and 4. This assay revealed not only differences in promoter strength but also provided quantitative insights on the JAZ3 repression of *MYCs* in a quantitative manner.

Key words Promoter, Transcriptional regulation, MYCs, JAZ3, DLR, Luciferase, Tobacco, Luminescence, Fluorescence

1 Introduction

The transcriptomic era has been advantageous and instrumental in many research areas allowing high-throughput analysis of differentially expressed genes in different organisms and under various environmental conditions or diseases [1]. Consequently, many regulatory networks have been identified, and their dynamic behaviors have been established, and core components of signaling pathways have been identified [2–5]. In plants, especially when dealing with non-model systems such as crops, gene expression analysis has become a valuable tool to study physiological responses and to identify key genes that might be used for improving crop fitness [1, 6–8]. However, these approaches usually generate a large number of target genes that need to be validated before conducting

Shahid Mukhtar (ed.), *Modeling Transcriptional Regulation: Methods and Protocols*, Methods in Molecular Biology, vol. 2328, https://doi.org/10.1007/978-1-0716-1534-8_13,

functional assays in planta. This becomes important and challenging when studying crop plants, especially when selecting candidates' genes to be used for genetic modifications. Among the available techniques that can be used for transcriptome data validation are: (1) qRT-PCR (quantitative but does not provide the spatial resolution); (2) in situ hybridization (provides the expression at the cell and tissue types but: it is tedious, needs special skills, is challenging, needs to be optimized for almost every plant species; moreover, it is not quantitative); (3) reporter fusions (needs an established transformation protocol, generation time can be very long, only a few reporters can be used for expression analysis because of the tissue thickness in most crops). Also, these techniques cannot determine whether the effector protein acts as a single entity or requires binding to other protein partners to regulate the expression of a conjoint target. Therefore, having transient expression systems to study transcriptional regulation of genes of interest precisely and in a timely manner becomes highly solicited. Luciferase reporter assays offer an excellent system for assaying promoter activities transiently under different conditions, allowing thus the study of transcriptional regulation of one or multiple proteins on a given promoter [9].

The assay described in this protocol is based on the Dual-Luciferase Reporter (DLR) system (https://worldwide.promega.com/products/luciferase-assays/reporter-assays/dual_luciferase-reporter-assay-system). It relies on the use of multiple components (Fig. 1) and requires three principal elements:

1. The promoter of interest fused to a Firefly Luciferase reporter gene.
2. The regulator also called effector: the protein that activates or represses the promoter of interest.
3. An internal control (Renilla Luciferase), used to normalize the transformation efficiency.

This dual reporter system is based on the use of two luciferase enzymes that have different origins and produce luminescence in the presence of the corresponding substrates [10–14]. The Firefly luciferase (fLUC) originated from North American Firefly (*Photinuspyralis*); it produces light in the 550–570 nm wavelength using luciferin as a substrate in the presence of Magnesium, oxygen, and ATP [15]. The Firefly luciferase (fLUC) is usually used to measure the promoter response to either the regulator or to evaluate changes in promoter activity under different conditions. The other reporter enzyme is used as a control to evaluate differences resulting from transformation efficiency among the samples, lysis efficacy, and cell fitness. The second enzyme is derived from the Renilla sea pansy (*Renilla reniformis*); it emits blue light at 480 nm in the presence of the substrate coelenterazine and oxygen

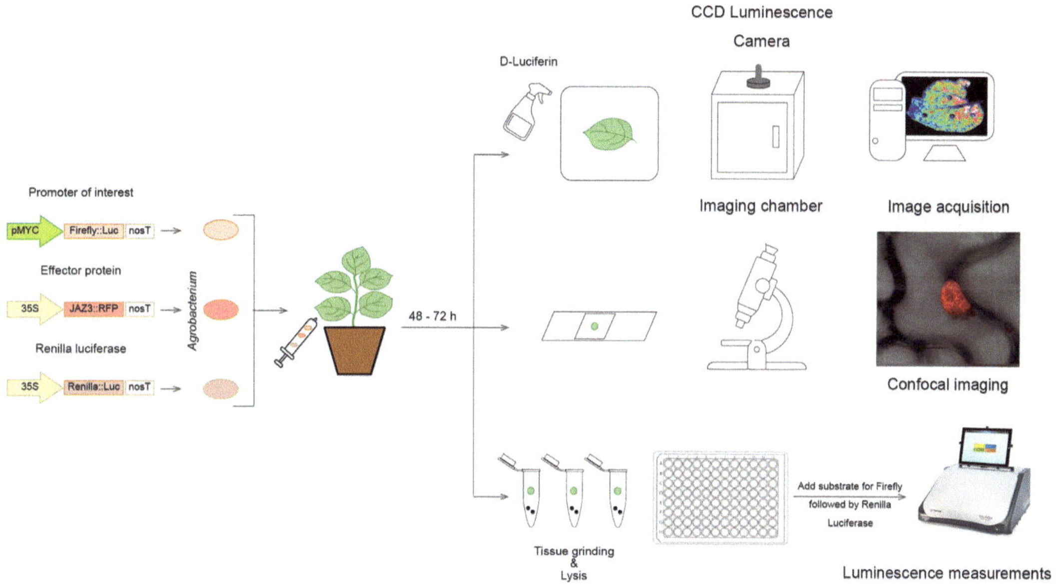

Fig. 1 Protocol workflow illustrating the essential steps for this protocol

[16]. Having distinct substrates, cofactors, and light requisite allows consecutive measurements of the activity of these two enzymes. First, the substrate for the Firefly enzyme is added, and its activity is measured, once the signal is acquired and recorded, the addition of the second substrate leads to quenching of the Firefly signal and at the same time initiates the reaction of Renilla luciferase [15].

Here we will provide a detailed protocol of the DLR assay using a well-described regulatory network controlling the plant hormones jasmonate (JA) signaling pathway in Arabidopsis, known to play fundamental roles in plant response to wounding and defense against attack by herbivores and necrotrophic fungi in addition to growth and development [17–22]. We will focus on two central regulators; the nuclear-localized protein called JASMO NATE-ZIM-DOMAIN3 (JAZ3) known to repress the activity of JA responsive genes [23–30]. The primary and the most characterized target regulated by the JAZ repressors are the MYC proteins, belonging to a bHLH transcription factor family [31–34]. Here we will exploit this regulation and use it as an example to show how to acquire qualitative and quantitative measurements on JAZ3 transcriptional regulation of the three MYC promoters, MYC2, 3, and 4. The protocol provides detailed guidelines that can be used to evaluate the transcriptional regulation of a promoter by its upstream effector protein using tobacco as a transient system (Fig. 1).

We will cover all steps; these include cloning, tissue infiltration, effector expression using confocal microscopy, expression analysis and imaging, tissue collection, protein extraction, and luciferase activity measurement analysis.

2 Materials

1. Growth chamber/greenhouse bench space under long-day condition (16 h light, 8 h dark) at 25 °C.
2. 4–5 weeks old *Nicotiana benthamiana* plants (*see* **Note 1**).
3. A 28 °C shaking incubator.
4. Competent cells of *Agrobacterium tumefaciens.*
5. Spectrophotometer (Eppendorf BioPhotometer plus 6132 UV/vis).
6. Plasmids containing the promoters fused to Firefly (*pMYC2*, *pMYC3*, and *pMYC4*).
7. Renilla Luciferase construct.
8. The effector protein (JAZ3) (*see* **Note 2**).
9. p19 helper plasmid.
10. 0.5 M MES solution, pH 5.7 adjusted with KOH and sterilized by autoclaving.
11. Acetosyringone (200 mM stock solution in DMSO) and stored at −20 °C.
12. $MgCl_2$ (1 M stock solution, sterilized by autoclaving).
13. 15 and 50 mL Falcon tubes.
14. 1 mL disposable sterile syringes.
15. 2 mL safe-lock Eppendorf tubes.
16. Tissue lyser/tissue homogenizers (Qiagen, Tissuelyser II; Cat No./ID: 85300).
17. Stainless Steel Beads (Qiagen; Cat No./ID: 69989).
18. Benchtop and bucket centrifuges.
19. Fluorescence stereo microscope, we used a Leica M205 FA.
20. A confocal microscope, we used a Zeiss LSM 880 confocal microscope. RFP fluorescence was excited at 543 nm and images were acquired at 600–660 nm.
21. D-Luciferin (D-Luciferin, Monosodium Salt; 100 mg; Thermofisher; EA 88291).
22. Biopsy punch with a plunger (Integra Miltex, Disposable Biopsy punch with plunger, 4 mm, 33-34-P/25) if not available see alternative below.

23. Luminometer with double injectors, we used GloMax Navigator 96 microplate reader (Promega).
24. Dual-Luciferase® Reporter Assay System kit (Promega; E1960) containing:
    (a) 5× Passive Lysis Buffer (PLB).
    (b) Luciferase Assay Buffer II (LARII, green cap) with dissolved lyophilized Luciferase Assay Substrate. Always keep in the dark and store at −20 °C.
    (c) 1× Stop & Glo® Reagent (Blue cap): mix 50× Stop & Glo® Substrate with Stop & Glo® Buffer. Always keep in the dark and store at −20 °C.
25. DTT stock solution 1 M stored at −20 °C.
26. 96-well white opaque microplates (Thermofisher, Catalog number: 15042).
27. Ice.
28. Liquid nitrogen.
29. Long forceps.
30. Floating microtubes racks/floaters.
31. Nitrile gloves.
32. Thermal protection gloves for low temperature.
33. Vortex.
34. Orbital shaker.
35. Benchtop centrifuge for 2 mL microtubes.
36. Bioluminescence in vivo imaging system consisting of:
    (a) Imaging chamber.
    (b) CCD camera (we used an Andor camera).
    (c) A camera controller.
    (d) Refrigeration unit.
    (e) Acquisition computer and a monitor in addition to the Andor Solis (i) software platform or any bioluminescence acquisition software.

3 Methods

3.1 Cloning

1. Clone the promoters of interest into a binary vector containing the Firefly Luciferase reporter.

   In this chapter we used, *pMYC2*, 2.3 kb, *pMYC3*, 2 kb, and *pMYC4* 1 kb promoter regions upstream of the start codon.
2. Amplify the promoter of interest from genomic DNA and subclone them into a pDONR vector (we used pGEMTeasy

Table 1
Primer sequences used in this study

pMYC2- pGEMTeasy 221F	GGGGACAAGTTTGTACAAAAAAGCAGGCTTA atagattgaggcgcttctacaaggt
pMYC2- pGEMTeasy 221R	GGGGACCACTTTGTACAAGAAAGCTGGGTT tccataaaccggtgaccggtaaaaa
pMYC3- pGEMTeasy 221F	GGGGACAAGTTTGTACAAAAAAGCAGGCTTA cttgttattagcgcaaagaggatcg
pMYC3- pGEMTeasy 221R	GGGGACCACTTTGTACAAGAAAGCTGGGTTgtgaacatacgccggttgaaaag
pMYC4- pGEMTeasy 221F	GGGGACAAGTTTGTACAAAAAAGCAGGCTTActacccaaaatgtgtgaggccc
pMYC4- pGEMTeasy 221R	GGGGACCACTTTGTACAAGAAAGCTGGGTT aacagttctctgacgtagttataaaagagaagact

221) to generate an entry clone. Primer sequences are provided in Table 1.

3. Generate the expression clones by performing an LR recombination reaction between the entry clone and a Gateway destination vector (we used pGreen II) containing the Firefly reporter gene as described in [9].
4. Clone the effectors or regulators into binary vectors. In our case, we used *35S::JAZ3:RFP* (*see* **Note 3**).
5. Use the *35S::Renilla:LUC* Luciferase [9] for normalization and a p19 helper plasmid to inhibit silencing.
6. Transform each construct independently into *Agrobacterium* (Fig. 1).

3.2 Infiltration by Agrobacterium and Confirmation of Transformation

1. Transform each construct into *Agrobacterium tumefaciens* (GV3101) and select by antibiotic resistance. After 2–3 days, inoculate a single colony into 5 mL LB medium supplemented with the appropriate antibiotics.
2. Transfer 100 μL of 48 h grown Agrobacterium culture from each construct into 15 mL falcon tubes containing new LB medium supplemented with sterile 10 mM 2-(N-morpholine)-ethanesulfonic acid (MES; pH 5.6) and 40 mM acetosyringone. Grow the agrobacteria for an additional 16 h at 28 °C.
3. When growth reaches an OD600 of approximately 3.0, centrifuge the cultures at 3200 × *g* for 10 min and re-suspend the pellets in 10 mM $MgCl_2$. Each construct should have the following final ODs: For Agrobacterium cultures containing the promoter-of-interest, effector and p19 helper plasmid; the

Table 2
Constructs generated in this study

Constructs	Vector	Resistance in bacteria
35S-JAZ3mRFP	pH7m34GW	Spectinomycin
pMYC2:FireflyLUC	pGIILUC	Kan
pMYC3:FireflyLUC	pGIILUC	Kan
pMYC4:FireflyLUC	pGIILUC	Kan

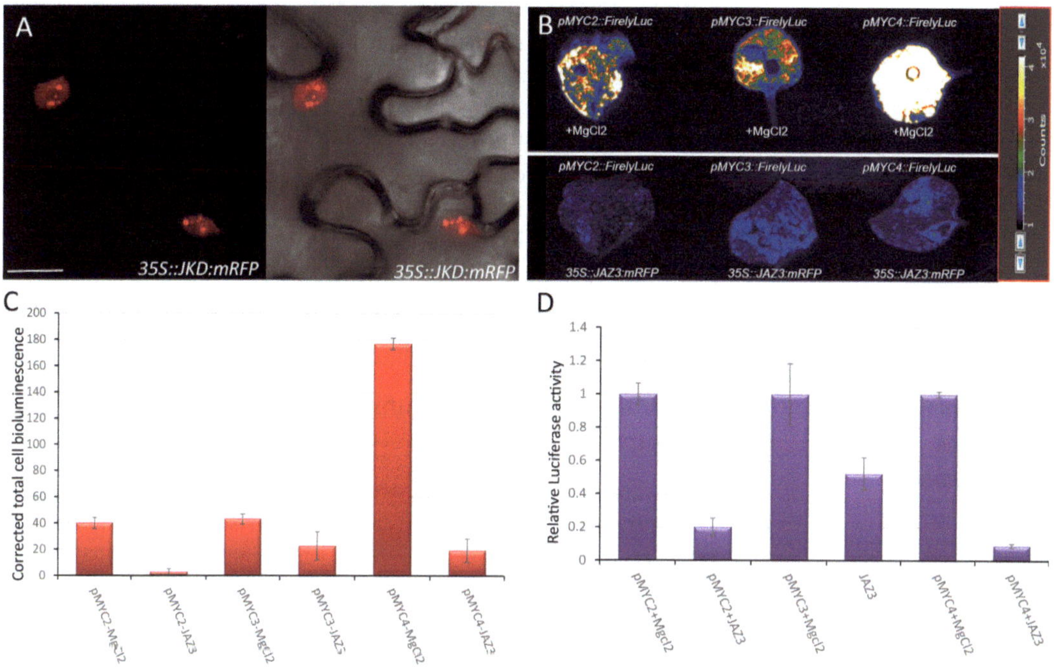

Fig. 2 JAZ3 effector protein represses *MYC* genes activity. (**a**) Confocal image of tobacco cells expressing *35S::JAZ3:RFP*. The left image is the RFP channel, the right image is an overlay of transmission and RFP channels. Scale bar: 20 μm. (**b**) Tobacco leaves expressing *pMYC2::FireflyLuc*; *pMYC3::FireflyLuc*; *pMYC4::FireflyLuc* alone (upper panel) or in the presence of *35S::JAZ3:RFP* (Lower panel). Color bar indicates bioluminescence intensity from low (dark blue) to high (white). (**c**) Quantification of bioluminescence intensity using Image J. (**d**) the Promoter activity of *pMYC2::FireflyLuc*; *pMYC3::FireflyLuc*; *pMYC4::FireflyLuc* measured by Dual-Luciferase in the presence of *35S::JAZ3:RFP*

OD should be adjusted to OD600 = 0.5; the *35S::Renilla:LUC* control should have an OD600 = 0.2.

4. Add Acetosyringone to the $MgCl_2$ solution containing Agrobacterium resuspensions to a final concentration of 200 μM and keep at room temperature for at least 3 h without shaking.
5. Use a volume ratio of Agrobacterium suspension of 1:1:0.2:3 for the promoter-of-interest:p19 helper plasmid:Renilla

control:Effector. First pre-mix Suspensions of different Agrobacterium strains carrying common constructs (promoter, p19, and Renilla control), the effector is added separately (*see* **Note 4**).

6. Before infiltrating the leaves make a scheme with the combinations to be used and amounts of bacteria to be added and label carefully each leaf/pot (Table 2).
7. Infiltrate the leaf gently using a 1 mL disposable syringe containing the bacterial mixture. The infiltration should be done from the abaxial side of the leaf. The infiltrated leaves will have a water-soaked appearance (*see* **Note 5**).
8. Grow the infiltrated samples for a maximum of 3 days.

   Three days after infiltration, the maximal expression of effectors should be observed. Effector expression can be checked using a fluorescence binocular microscope. Once the signal is observed, then a leaf disk can be cut using a biopsy punch with a plunger; if not available, then use a 15 mL falcon tube and cut the leaf disks by pressing a tube into the infiltrated leaf. The leaf disk can be mounted in water, and expression should be imaged using a confocal microscope (Fig. 1). The resulting image can be observed in Fig. 2a.

3.3 Noninvasive Luciferase Measurement

This assay involves spraying infiltrated tobacco leaves with a D-luciferin substrate (*see* **Note 6**).

1. Substrate preparation: Prepare a 200× Luciferin stock solution (30 mg/mL) in sterile water. Dilute in sterile water (1:200) prior to use.
2. In a square petri dish collect carefully one infiltrated tobacco leaf from each combination.
3. Spray with D-luciferin. This step should be done prior to imaging.
4. Incubate for 10 min at room temperature in the dark.
5. Collect the signal using the Bioluminescence in vivo imaging system following the manufacturer's instructions.
6. Save the images using the software format as well as a Tiff or JPEG format. The signal can be observed in Fig. 2b.
7. Image analysis: quantify the Luminescence signal from the image as follows:
   (a) Download image J: https://fiji.sc/#download.
   (b) Open your image with ImageJ.
   (c) In your image, select three ROIs as disks using a drawing/selection tool (i.e., circle).

(d) From the Analyze menu, select "set measurements." Make sure you have area integrated intensity selected (the rest can be ignored).

(e) Select "Measure" from the analysis menu. A popup box with the measured values will appear.

(f) For background, select a region within the leaf that has no signal.

(g) Repeat this step for all the samples to be measured.

(h) Select the data in the Results window and copy into a new spreadsheet (in excel).

(i) Use the following formula to calculate the corrected total cell Bioluminescence (CTCB).

(j) CTCB = Integrated Density − (Area of selected cell × Mean Bioluminescence of background readings).

(k) Make a graph as in Fig. 2c.

3.4 Luciferase Measurement Using DLR System

Before starting the experiment, make sure that the glomax device is functional.

First, use the Light plate to check the device performance then set the settings as follows: inject 25 μL of LAR II and Stop & Glo® Reagent sequentially into each sample for independent measurement of fLUC and rLUC activities. Each injection should be followed by a 10 s integration period and a 0.4 s delay. Prime the injectors with the substrate solutions as advised by the manufacturer [15].

3.5 Sample Collection

1. Label 2 mL safe-lock tubes and add one to two stainless steel beads per tube. Use three tubes per infiltrated leaf.
2. Isolate three leaf disks per infiltrated leaf and transfer each disk independently to separate tubes. Freeze in liquid nitrogen (*see* **Note** 7).
3. Grind the tissue using the TissueLyser.

3.6 Luciferase Measurement

Before lysing the tissue, thaw the substrate solution

4. Add 200 μL of 1× PLB buffer containing DTT into the ground tissue, and put back in liquid nitrogen.
5. Thaw on ice for 10–20 min.
6. Vortex the samples and centrifuge at maximum speed for 3 min. Carefully transfer 75 μL of the supernatant into the luminometer plate.

7. Measure luciferase activity with a luminometer. Luminescence for both rLUC and fLUC is then automatically recorded in an excel file.
8. Data analysis:
   (a) Calculate the relative ratio of fLUC/rLUC activity. This is to normalize variations introduced by different transformation efficiencies.
   (b) Average the ratios from the three technical replicates for each combination.
   (c) Set the promoter activity control with $MgCl_2$ and without effectors arbitrarily at 1, and normalize the rest of the samples against it. Calculate and normalize the standard deviations (Fig. 2d).

Repeat each experiment for at least three times to generate independent biological replicates. Each experiment should have three technical replicates per sample.

4 Notes

1. The age of the plants used for transformation is important; leaves should be used before the plants initiate flowers. Also, the plants should look green and healthy, and any stress will affect the transformation efficiency and the resulting outcome.
2. All constructs should be fused into binary vectors (primers used for cloning are listed in Table 1) of the analysis.
3. We advise to always fuse the effector with a fluorescent tag in order to monitor the expression and the efficiency of the infiltration.
4. Higher Renilla LUC saturates the luciferase activity. It is important to keep it at its lowest concentration. Promoter and p19 should be the same amount, but the ratio with the effector can be changed depending on its transactivation efficiency.
5. Prior to Agrobacterium transformation, grow *Nicotiana benthamiana* plants for 3–4 weeks in the Greenhouse and use the youngest leaves for infiltration. Infiltrate the leaf gently, do not push the bacterial suspension excessively inside the tissue to avoid damaging the leaf area and induce stress responses; ideally, the infiltration should be done in one go, if it does not work use another leaf.

6. Before starting the reaction, make sure that the imaging system is running properly, we advise turning the system on a few hours before starting the experiment and to test the functionality of the system.
7. Samples can be stored at −80 °C for a few weeks.

Acknowledgments

This study was supported by KAUST baseline research funding to I.B. We are grateful to Vinicius Lube for technical assistance when performing Luciferase measurements using the Andor camera. The authors declare no competing financial interests.

References

1. Lowe R, Shirley N, Bleackley M, Dolan S, Shafee T (2017) Transcriptomics technologies. PLoS Comput Biol 13:1–23
2. Brady SM, Long TA, Benfey PN (2006) Unraveling the dynamic transcriptome. Plant Cell 18:2101–2111
3. Shulse CN et al (2019) High-throughput single-cell transcriptome profiling of plant cell types. Cell Rep 27:2241–2247.e4
4. Li S, Yamada M, Han X, Ohler U, Benfey P (2016) High resolution expression map of the Arabidopsis root reveals alternative splicing and lincRNA regulation. Dev Cell 39:508–522
5. Rasmussen RN, Christensen KV, Holm R, Nielsen CU (2019) Transcriptome analysis identifies activated signaling pathways and regulated ABC transporters and solute carriers after hyperosmotic stress in renal MDCK I cells. Genomics 111:1557–1565
6. Wang H, Zhou P, Zhu W, Wang F (2019) De novo comparative transcriptome analysis of genes differentially expressed in the scion of homografted and heterografted tomato seedlings. Sci Rep 9:1–12
7. Iquebal MA et al (2019) RNAseq analysis reveals drought-responsive molecular pathways with candidate genes and putative molecular markers in root tissue of wheat. Sci Rep 9:1–18
8. Li JR, Liu CC, Sun CH, Chen YT (2018) Plant stress RNA-seq Nexus: a stress-specific transcriptome database in plant cells. BMC Genomics 19:1–8
9. Díaz-Triviño S, Long Y, Scheres B, Blilou I (2017) Analysis of a plant transcriptional regulatory network using transient expression systems. In: Kaufmann K, Mueller-Roeber B (eds) Plant gene regulatory networks: methods and protocols, methods in molecular biology. Humana Press, New York, pp 83–103
10. Wood KV (1994) Luciferase assay method. United States Patent
11. Williams TM, Burlein JE, Ogden S, Kricka LJ, Kant JA (1989) Advantages of firefly luciferase as a reporter gene: application to the interleukin-2 gene promoter. Anal Biochem 176:28–32
12. Allard STM, Kopish K (2008) Luciferase reporter assays: powerful , adaptable tools for cell biology research. Cell Notes 21:23–26
13. Ar B, Je T, Jf H (1988) Optimized use of the firefly luciferase assay as a reporter gene in mammalian cell lines. BioTechniques 7:1116–1122
14. Ow DW et al (1986) Transient and stable expression of the firefly luciferase gene in plant cells and transgenic plants. Science 234:856–859
15. (2006) http://kirschner.med.harvard.edu/files/protocols/Promega_dualluciferase. Dual-Luciferase® reporter assay system. 1–26
16. Bhaumik S, Gambhir SS (2002) Optical imaging of Renilla luciferase reporter gene expression in living mice. Proc Natl Acad Sci 99:377–382
17. Qi T et al (2014) Arabidopsis DELLA and JAZ proteins bind the WD-repeat/bHLH/MYB complex to modulate gibberellin and jasmonate signaling synergy. Plant Cell 26:1118–1133
18. Yang D-L et al Plant hormone jasmonate prioritizes defense over growth by interfering with gibberellin signaling cascade. Proc Natl Acad Sci 109(19):E1192–E1200. https://doi.org/10.1073/pnas.1201616109

19. Kazan K, Manners JM (2011) The interplay between light and jasmonate signalling during defence and development. J Exp Bot 62:4087–4100
20. Hou X, Lee LYC, Xia K, Yan Y, Yu H (2010) DELLAs modulate jasmonate signaling via competitive binding to JAZs. Dev Cell 19:884–894
21. Campos ML et al (2016) Rewiring of jasmonate and phytochrome B signalling uncouples plant growth-defense tradeoffs. Nat Commun 7:1–10
22. Song S, Qi T, Wasternack C, Xie D (2014) Jasmonate signaling and crosstalk with gibberellin and ethylene. Curr Opin Plant Biol 21:112–119
23. Chini A, Fonseca S, Chico JM, Fernandez-Calvo P, Solano R (2009) The ZIM domain mediates homo- and heteromeric interactions between Arabidopsis JAZ proteins. Plant J 59:77–87
24. Pauwels L, Goossens A (2011) The JAZ proteins: a crucial interface in the jasmonate signaling cascade. Plant Cell 23:3089–3100
25. Wager A, Browse J, Wang X, Arita M (2012) Social network: JAZ protein interactions expand our knowledge of jasmonate signaling. Front Plant Sci 3:41. https://doi.org/10.3389/fpls.2012.00041
26. Santino A et al (2013) Jasmonate signaling in plant development and defense response to multiple (a)biotic stresses. Plant Cell Rep 32:1085–1098
27. Adie B, Chico JM, Lorenzo O, Garcı G (2007) The JAZ family of repressors is the missing link in jasmonate signalling. Nature 448:666
28. McConn M, Creelman RA, Bell E, Mullet JE, Browse J (1997) Jasmonate is essential for insect defense in Arabidopsis. Proc Natl Acad Sci USA 94:5473–5477
29. Santner A, Estelle M (2007) The JAZ proteins link jasmonate perception with transcriptional changes. Plant Cell 19:3839–3842
30. Pieterse CMJ, Pierik R, Van Wees SCM (2014) Different shades of JAZ during plant growth and defense. New Phytol 204:261–264
31. Fernandez-Calvo P et al (2011) The Arabidopsis bHLH transcription factors MYC3 and MYC4 are targets of JAZ repressors and act additively with MYC2 in the activation of jasmonate responses. Plant Cell 23:701–715
32. Schweizer F et al (2013) Arabidopsis basic helix-loop-helix transcription factors MYC2, MYC3, and MYC4 regulate glucosinolate biosynthesis, insect performance, and feeding behavior. Plant Cell 25:3117–3132
33. Goossens J, Swinnen G, Vanden Bossche R, Pauwels L, Goossens A (2015) Change of a conserved amino acid in the MYC2 and MYC3 transcription factors leads to release of JAZ repression and increased activity. New Phytol 206:1229–1237
34. Gasperini D et al (2015) Multilayered organization of jasmonate signalling in the regulation of root growth. PLoS Genet 11:1–27

Chapter 14

Assessing Global Circadian Rhythm Through Single-Time-Point Transcriptomic Analysis

Xingwei Wang, Yufeng Xu, Mian Zhou, and Wei Wang

Abstract

Plant circadian clock has emerged as a central hub integrating various endogenous signals and exogenous stimuli to coordinate diverse plant physiological processes. The intimate relationship between crop circadian clock and key agronomic traits has been increasingly appreciated. However, due to the lack of fundamental genetic resources, more complex genome structures and the high cost of large-scale time-course circadian expression profiling, our understanding of crop circadian clock is still very limited. To study plant circadian clock, conventional methods rely on time-course experiments, which can be expensive and time-consuming. Different from these conventional approaches, the molecular timetable method can estimate the global rhythm using single-time-point transcriptome datasets, which has shown great promises in accelerating studies of crop circadian clock. Here we describe the application of the molecular timetable method in soybean and provide key technical caveats as well as related R Markdown scripts.

Key words Circadian clock, Molecular timetable, Soybean, Submergence, Transcriptome

1 Introduction

The rotation of the Earth has generated days and nights. Consequently, life on Earth exhibits periodicities. The behavioral changes of plants between days and nights are easily observed; e.g., tulips blossom during the day and close during the night. However, plants do not simply respond to these diurnal environmental changes. They anticipate and prepare for them before they actually happen by their internal clock, called the circadian clock. As an autonomous system, the circadian clock maintains oscillation even under constant light/temperature conditions, one of the key defining features of the circadian clock.

Plant circadian clock controls diverse plant physiological processes including germination, hypocotyl elongation, leaf

Supplementary Information: The online version of this chapter (https://doi.org/10.1007/978-1-0716-1534-8_14) contains supplementary material, which is available to authorized users.

Shahid Mukhtar (ed.), *Modeling Transcriptional Regulation: Methods and Protocols*, Methods in Molecular Biology, vol. 2328, https://doi.org/10.1007/978-1-0716-1534-8_14,

movement, shade avoidance, flowering time, flower opening, winter dormancy, etc. [1]. The plant circadian clock also has a profound impact on key agronomic traits [2]. Investigation of hybrid vigor identified the circadian clock as a critical contributor to the growth vigor and increased biomass [3]. Based on 43 field experiments spanning three growing seasons, Monsanto reported a consistent yield increase by modulating the soybean circadian clock [4]. Specific selection of the circadian clock traits during crop domestication was also shown to promote crop adaptation to growth areas with drastically different latitudes from their origins in both tomato [5] and soybean [6].

While the plant circadian clock has widespread regulations on many aspects of plant physiology, it also integrates diverse endogenous signals and exogenous stimuli to optimize energy allocation [7]. The expression of central oscillator genes of plant circadian clock has been shown to respond to biotic stresses like oomycete [8], fungal [9] and bacterial pathogens [10] and abiotic stresses including cold stress [11], drought [12], nitrogen supply [13], and iron deficiency [14–16]. However, all these findings are based on the studies of the model plant, *Arabidopsis thaliana*. How crop circadian clock respond and integrate environmental stimuli remains to be further elucidated.

Our lack of knowledge in most crop circadian clock is partly due to the lack of cost-effective methods to assess crop circadian rhythms. Noninvasive methods rely on imaging systems to record clock-controlled physiological processes including leaf movement [17], hypocotyl elongation [18], photosystem II operating efficiency [19], delayed fluorescence [20], chlorophyll content [21], volume changes of pulvinar motor cells [22], and floral stem elongation [23]. However, these methods require specialized imaging systems that are usually too expensive for most labs. Aside from these noninvasive methods, destructive approaches like quantification of clock gene expression through qPCR, microarray or RNA-seq have been widely used. Since a time-course experiment with 2–4 h sampling frequency spanning 2–3 days is needed to assess the circadian rhythm of the gene expression profile, the large number of samples make these methods relatively expensive.

With the prevalence of the application of microarray and next-generation sequencing technologies, we have accumulated a huge amount of transcriptome information. NCBI Gene Expression Omnibus database has archived the transcriptome information of over 13,000 rice samples, 8000 maize samples, 7000 soybean samples and 4000 wheat samples. However, the sampling schemes of these transcriptome datasets rarely fulfill the standard required for conventional circadian rhythm analysis. Therefore, this rich collection of crop transcriptome datasets has not been exploited for circadian clock studies yet.

The molecular timetable method can reveal the global rhythm using single-time-point transcriptome datasets rather than time-course datasets [24]. Therefore, the molecular timetable method makes the exploitation of existing crop transcriptome datasets possible [25]. The key to the establishment of the molecular timetable method is the availability of time-indicating genes. Time-indicating genes have a robust circadian expression profile with relatively fixed peak expression time. The normalized average expression level of time-indicating genes can therefore indicate the phase of the sample. The molecular timetable was first developed in mice [24] and later adapted in *Arabidopsis* [26], tomato [27], and soybean [25]. By applying this method on the publicly available soybean transcriptome datasets, we were able to survey the effect of various abiotic stresses on the soybean circadian clock comprehensively [25]. Compared to other methods used for the circadian analysis, the molecular timetable method is certainly the most cost-effective approach, once the time-indicating genes are identified. However, as a computational heuristics, the effectiveness and accuracy of the molecular timetable method are inherently constrained by the quality of the original transcriptome datasets. Therefore, the conclusions drawn by the molecular timetable analysis need to be further verified experimentally [25].

2 Materials

1. List of time-indicating genes (*see* **Note 1**).
2. Expression matrix of the transcriptome dataset (*see* **Note 2**).
3. R version 3.6.0 or newer.
4. R packages: ggplot2, gridExtra, stringr.

3 Methods

To demonstrate the application of the molecular timetable method as well as related statistical analysis, here we use the soybean time-indicating genes previously identified by Li et al. [25] (Data 1) and the soybean transcriptome related with submergence treatment reported by Tamang et al. [28] (Data 2). The relevant R Markdown scripts for the molecular timetable analysis and related statistical analysis are provided in Data 3.

1. Depending on the type of technology used to obtain the transcriptome dataset (microarray or RNA-seq), normalize the expression matrix of the input data using corresponding algorithms before being used for the molecular timetable analysis.

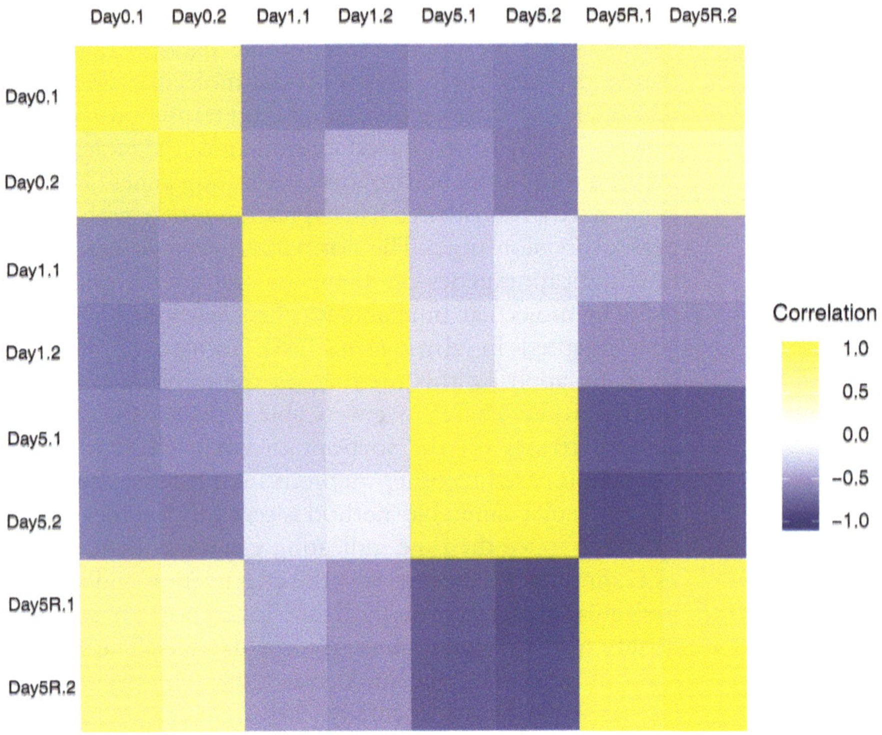

Fig. 1 Heatmap showing a pairwise correlation between all the samples based on the Z-scores of soybean time-indicating genes. The two replicates of each condition is labeled as ".1" and ".2" respectively. Day0, Day1, and Day5 are samples with 0, 1, and 5 days' submergence treatment respectively. Day5R are samples with 5-days' submergence treatment followed by 1-day's reoxygenation

2. Extract the time-indicating genes from the input expression matrix based on the list of the time-indicating genes previously identified.

3. Perform standardization across all the samples and all the time-indicating genes to generate Z-scores of the expression matrix. This procedure is the most critical step of the molecular timetable method (*see* **Note 3**).

4. Most transcriptome datasets have biological replicates. Presumably, the biological replicates should be sampled at a similar time-of-a-day. Replicated samples with drastically different sampling time can severely jeopardize the reliability of the molecular timetable analysis. Use heatmaps of the pairwise correlation of each sample for a quick check of the relative reproducibility of the replicates (Fig. 1). While heatmaps are very useful for a quick check of the reproducibility of a large number of samples, line plots of Z-score of each sample vs. their corresponding circadian time (CT) groups can reveal a more detailed difference between replicates (Fig. 2). The demo data showed acceptable reproducibility.

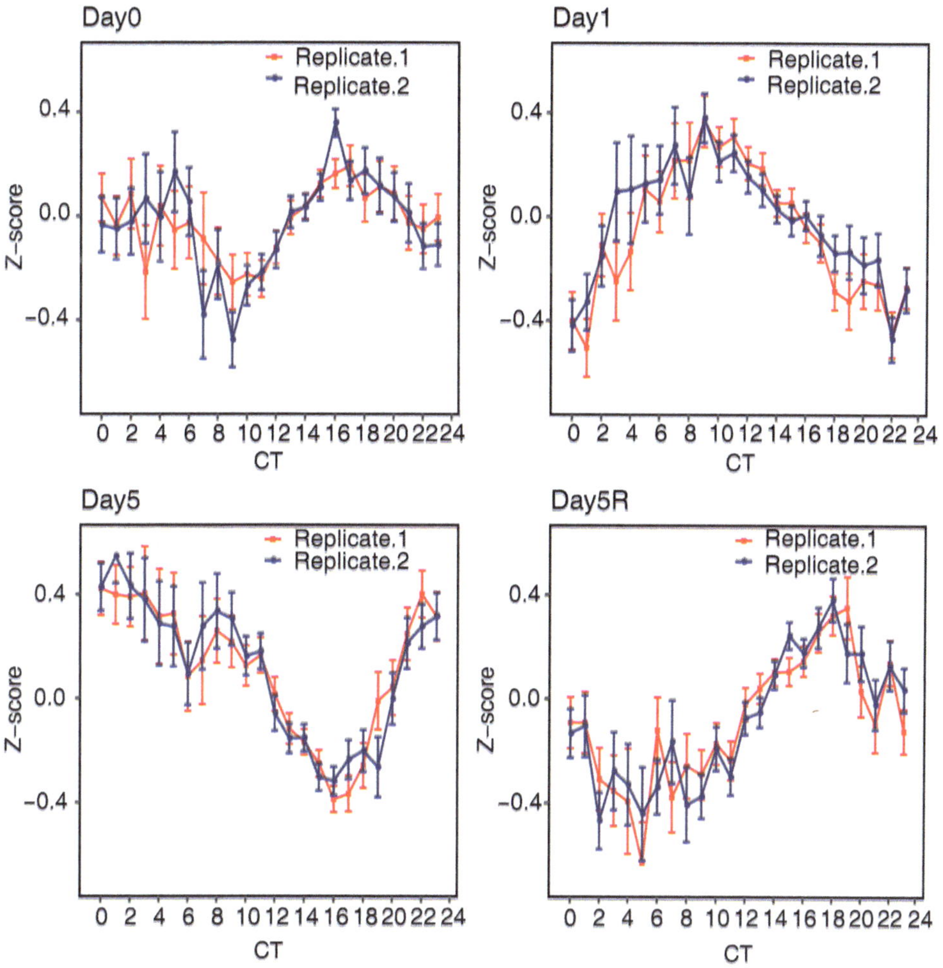

Fig. 2 Line plots of the Z-scores of soybean time-indicating genes in each sample. Mean ± SEM is shown. *CT* circadian time. Day0, Day1, and Day5 are samples with 0, 1, and 5 days' submergence treatment respectively. Day5R are samples with 5-days' submergence treatment followed by 1-day's reoxygenation

5. Once the reproducibility among replicates is confirmed, combine the biological replicates to generate the line plot (Fig. 3a) and the heatmap (Fig. 3b) of the molecular timetable. At this step, the line plot can reveal the global rhythm of the sample. In this example, the comparisons between Day0 and Day1, Day5 suggest that submergence may change the phase of the circadian rhythm while reoxygenation (Day5R) may have partially restored the phase (Fig. 3a). When a large number of samples are considered, the heatmap can be a more effective visualization method to show the molecular timetable (Fig. 3b).
6. Amplitude, period, phase and robustness of oscillation are four key parameters to characterize the inherent features of the circadian rhythm. Estimate period-compensated estimation of

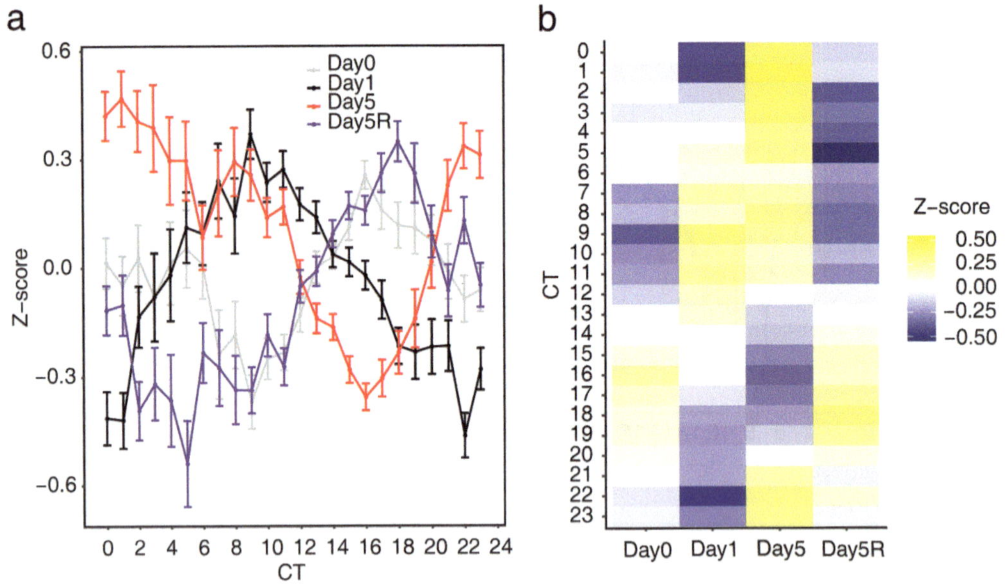

Fig. 3 Line plot (**a**) and heatmap (**b**) of the Z-scores of soybean time-indicating genes with replicates combined. Mean ± SEM is shown. *CT* circadian time. Day0, Day1, and Day5 are samples with 0, 1, and 5 days' submergence treatment respectively. Day5R are samples with 5-days' submergence treatment followed by 1-day's reoxygenation

the circadian body time, Phase24, through nonlinear regression (*see* the relevant R scripts in Data **3** for more details on the nonlinear regression). Evaluate the robustness of oscillation by estimating the linear fit of the measured Z-score versus the predicted Z-score by the nonlinear regression model. The $-\log_{10}(p)$ of this linear regression is used as the indicator of oscillation robustness (Fig. 4). Due to several features of the molecular timetable method, we do not try to perform any analysis on the amplitude and the period derived by the nonlinear regression (*see* **Note 4**).

7. In some cases, the line plot of the Z-score of the time-indicating genes may not show the ideal oscillatory pattern. However, it is still possible to compare different samples through the line plots (Fig. 5a) or heatmaps (Fig. 5b) of the moving correlation (*see* **Note 5**).

4 Notes

1. For demonstration, soybean time-indicating genes are used [25]. Time-indicating genes can be identified through circadian time-course transcriptomics analysis. Time-indicating genes have robust circadian expression rhythm and relatively fixed peak expression time-of-a-day. We have compared various

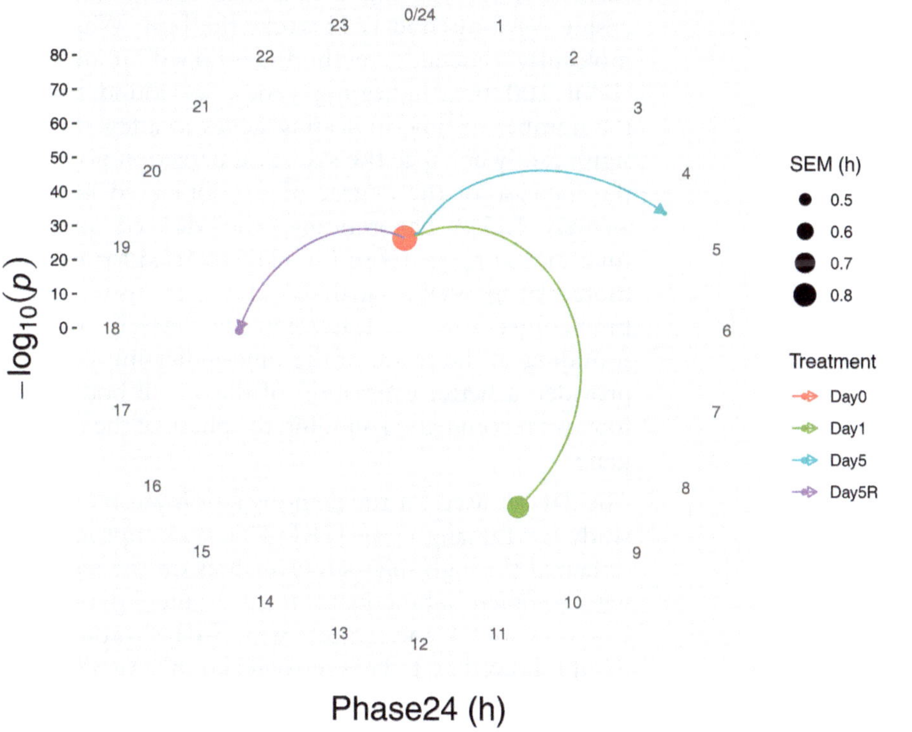

Fig. 4 Polar plot showing the Phase24 and robustness of oscillation. The angular coordinates represent Phase24. The $-\log_{10}$-transformed oscillation p values represent the robustness of the oscillation and are plotted as radial coordinates. The size of the symbols is proportional to the SEM of Phase24

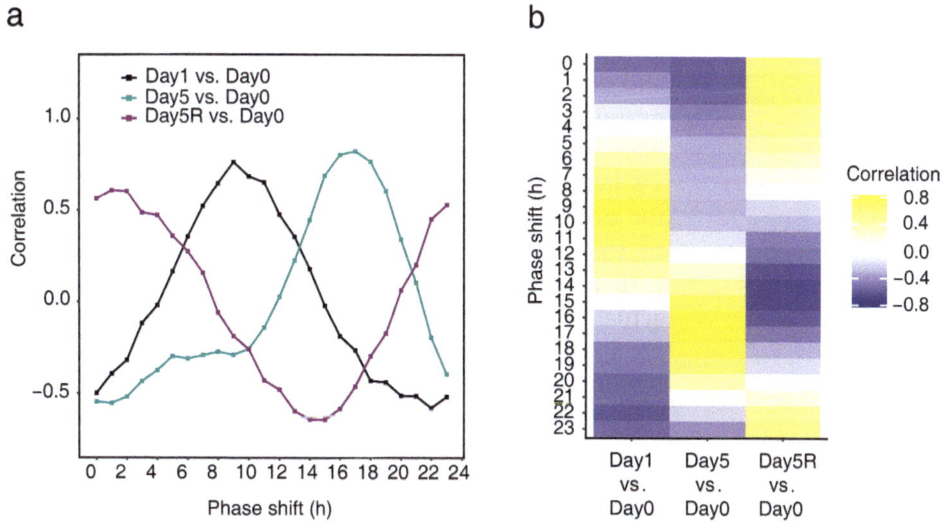

Fig. 5 Line plot (**a**) and heatmap (**b**) of moving correlation calculated using the data from Day0 as the reference

oscillation analysis algorithms and found that the original cosine wave method [24] works the best. While the original molecular timetable method developed in mice used only about 100 time-indicating genes, we found that increasing the number of time-indicating genes to a few thousand could significantly enhance the statistical detection power because of the increase of the degree of freedom used in the statistical analysis [25]. Theoretically, the derived phase of each time-indicating genes can be directly used for the subsequent molecular timetable analysis. However, using our circadian time-course soybean transcriptome datasets, we found that rounding of the phase of the time-indicating genes to integer provided a better estimation of the actual body time. Therefore, we recommend rounding the phase of the time-indicating genes.

2. The dataset used for the demonstration was retrieved from the study by Tamang et al. [28]. This transcriptome dataset was obtained through microarray analysis on the soybean samples with complete submergence for 0, 1, and 5 days (Day0, Day1, Day5) as well as the samples with 5-days' submergence treatment followed by 1-day's reoxygenation (Day5R). Each condition has two replicates.

3. The molecular timetable method relies on the standardized expression level (Z-score transformation of the non-normalized data) of the time-indicating genes in each CT group to derive the global rhythm of each sample. Different time-indicating genes have drastically different average expression levels computed across all the samples. For example, the expression pattern of soybean time-indicating genes in CT group 6 and CT group 14 showed robust circadian rhythms and the peak expression time matches their corresponding CT group assignment (Fig. 6a). However, the average expression level of CT group 6 genes is 43 fold of that of CT group 14 genes. Therefore, without the normalization procedure, the peak time of the global rhythm curve will be dominated by the CT group that is composed of time-indicating genes with a very high average expression level, regardless of the actual sampling time of the sample. As shown in Fig. 6b, CT groups 5 and 6 genes have a very high average expression level (computed across all samples). As a result, the peak times of almost all the global rhythm curves generated using non-normalized expression levels are dominated by CT groups 5 and 6 genes (Fig. 6c). Moreover, the overall pattern of these global rhythm curves is largely similar to Fig. 6b due to the dominance of CT groups 5 and 6 genes. When the Z-score transformation is performed, the molecular timetable method can precisely reflect the actual sampling time (Fig. 6d).

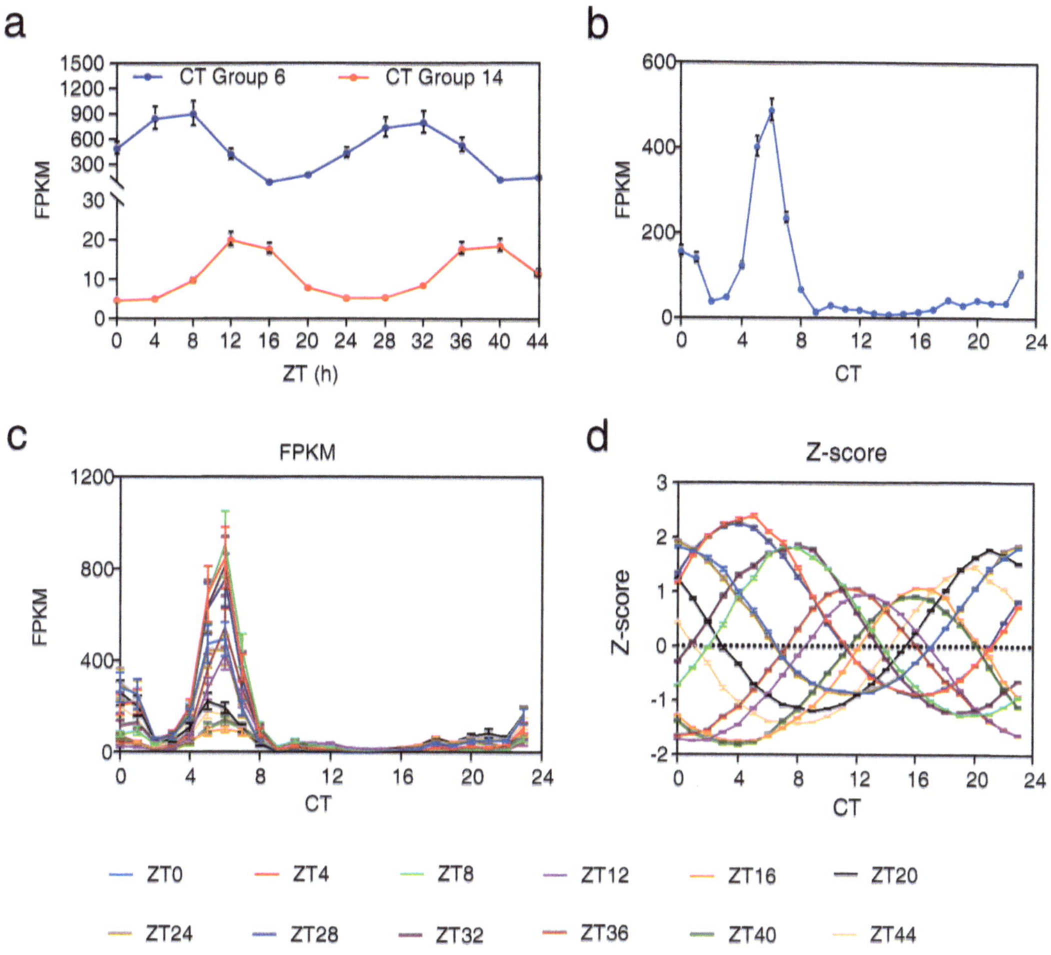

Fig. 6 Significance of Z-score transformation. (**a**) Average FPKM of soybean CT group 6 and 14 time-indicating genes computed at each sampling time. (**b**) Average FPKM of soybean time-indicating genes of every CT group computed across all the sampling time. (**c**) Average FPKM of 24 CT groups of soybean time-indicating genes. (**d**) Z-scores of 24 CT groups of soybean time-indicating genes. Mean $\pm$ SEM is shown in all figures. Standard errors of some data points are smaller than the size of the data symbol. Therefore, they may appear invisible. *ZT* zeitgeber time, *CT* circadian time. These figures were based on the re-analysis of the soybean circadian time-course transcriptome dataset we generated previously [25]

4. The Z-score transformation scales the mean and standard deviation. Therefore, it is not meaningful to compare the amplitudes of different conditions as these amplitudes are also scaled. The molecular timetable analysis uses single-time-point transcriptome datasets. For a single-time-point sample, it is not meaningful to study its period, as only one-time point rather than time-series information is involved.
5. By shifting the Z-scores of one sample by 0 through 23 h in a recursive manner, we can calculate the correlations of these phase-shifted profiles with another sample and generate the

moving correlation curve (Fig. 5a). If the original Z-score profiles of two samples show a similar pattern, the maximum correlation will be achieved at phase shift 0. If the original Z-score profiles of two samples show a reversed pattern, the maximum correlation will be achieved at phase shift 12. In this example, submergence treatment causes drastic changes in phase (Day1 vs. Day0 and Day5 vs. Day0) while reoxygenation restores the rhythm (Day5R vs. Day0).

Acknowledgment

This work was supported by the funds from State Key Laboratory for Protein and Plant Gene Research, Peking University, Center for Life Sciences and the National Natural Science Foundation of China (31970641) to W.W. and the funds from Beijing Nova Program of Science and Technology (Z191100001119027) and the National Natural Science Foundation of China (31970283) to M. Z.

References

1. Yakir E, Hilman D, Harir Y, Green RM (2007) Regulation of output from the plant circadian clock. FEBS J 274(2):335–345. https://doi.org/10.1111/j.1742-4658.2006.05616.x
2. Bendix C, Marshall CM, Harmon FG (2015) Circadian clock genes universally control key agricultural traits. Mol Plant 8(8):1135–1152. https://doi.org/10.1016/j.molp.2015.03.003
3. Ni Z, Kim ED, Ha M, Lackey E, Liu J, Zhang Y, Sun Q, Chen ZJ (2009) Altered circadian rhythms regulate growth vigour in hybrids and allopolyploids. Nature 457 (7227):327–331. https://doi.org/10.1038/nature07523
4. Preuss SB, Meister R, Xu Q, Urwin CP, Tripodi FA, Screen SE, Anil VS, Zhu S, Morrell JA, Liu G, Ratcliffe OJ, Reuber TL, Khanna R, Goldman BS, Bell E, Ziegler TE, McClerren AL, Ruff TG, Petracek ME (2012) Expression of the Arabidopsis thaliana BBX32 gene in soybean increases grain yield. PLoS One 7(2): e30717. https://doi.org/10.1371/journal.pone.0030717
5. Muller NA, Wijnen CL, Srinivasan A, Ryngajllo M, Ofner I, Lin T, Ranjan A, West D, Maloof JN, Sinha NR, Huang S, Zamir D, Jimenez-Gomez JM (2016) Domestication selected for deceleration of the circadian clock in cultivated tomato. Nat Genet 48 (1):89–93. https://doi.org/10.1038/ng.3447
6. Lu SJ, Zhao XH, Hu YL, Liu SL, Nan HY, Li XM, Fang C, Cao D, Shi XY, Kong LP, Su T, Zhang FG, Li SC, Wang Z, Yuan XH, Cober ER, Weller JL, Liu BH, Hou XL, Tian ZX, Kong FJ (2017) Natural variation at the soybean J locus improves adaptation to the tropics and enhances yield. Nat Genet 49(5):773–779. https://doi.org/10.1038/ng.3819
7. Greenham K, McClung CR (2015) Integrating circadian dynamics with physiological processes in plants. Nat Rev Genet 16(10):598–610. https://doi.org/10.1038/nrg3976
8. Wang W, Barnaby JY, Tada Y, Li H, Tor M, Caldelari D, Lee DU, Fu XD, Dong X (2011) Timing of plant immune responses by a central circadian regulator. Nature 470 (7332):110–114. https://doi.org/10.1038/nature09766
9. Windram O, Madhou P, McHattie S, Hill C, Hickman R, Cooke E, Jenkins DJ, Penfold CA, Baxter L, Breeze E, Kiddle SJ, Rhodes J, Atwell S, Kliebenstein DJ, Kim YS, Stegle O, Borgwardt K, Zhang C, Tabrett A, Legaie R, Moore J, Finkenstadt B, Wild DL, Mead A, Rand D, Beynon J, Ott S, Buchanan-Wollaston V, Denby KJ (2012) Arabidopsis defense against Botrytis cinerea: chronology and regulation deciphered by high-resolution temporal transcriptomic analysis. Plant Cell 24 (9):3530–3557. https://doi.org/10.1105/tpc.112.102046

10. Zhang C, Xie QG, Anderson RG, Ng GN, Seitz NC, Peterson T, McClung CR, McDowell JM, Kong DD, Kwak JM, Lu H (2013) Crosstalk between the circadian clock and innate immunity in Arabidopsis. Plos Pathog 9(6):e1003370. https://doi.org/10.1371/journal.ppat.1003370. ARTN e1003370
11. Bieniawska Z, Espinoza C, Schlereth A, Sulpice R, Hincha DK, Hannah MA (2008) Disruption of the Arabidopsis circadian clock is responsible for extensive variation in the cold-responsive transcriptome. Plant Physiol 147(1):263–279. https://doi.org/10.1104/pp.108.118059
12. Pokhilko A, Mas P, Millar AJ (2013) Modelling the widespread effects of TOC1 signalling on the plant circadian clock and its outputs. BMC Syst Biol 7:23. https://doi.org/10.1186/1752-0509-7-23
13. Gutierrez RA, Stokes TL, Thum K, Xu X, Obertello M, Katari MS, Tanurdzic M, Dean A, Nero DC, McClung CR, Coruzzi GM (2008) Systems approach identifies an organic nitrogen-responsive gene network that is regulated by the master clock control gene CCA1. Proc Natl Acad Sci U S A 105 (12):4939–4944. https://doi.org/10.1073/pnas.0800211105
14. Salome PA, Oliva M, Weigel D, Kramer U (2013) Circadian clock adjustment to plant iron status depends on chloroplast and phytochrome function. EMBO J 32(4):511–523. https://doi.org/10.1038/emboj.2012.330
15. Hong S, Kim SA, Guerinot ML, McClung CR (2013) Reciprocal interaction of the circadian clock with the iron homeostasis network in Arabidopsis. Plant Physiol 161(2):893–903. https://doi.org/10.1104/pp.112.208603
16. Chen YY, Wang Y, Shin LJ, Wu JF, Shanmugam V, Tsednee M, Lo JC, Chen CC, Wu SH, Yeh KC (2013) Iron is involved in the maintenance of circadian period length in Arabidopsis. Plant Physiol 161(3):1409–1420. https://doi.org/10.1104/pp.112.212068
17. Greenham K, Lou P, Remsen SE, Farid H, McClung CR (2015) TRiP: Tracking Rhythms in Plants, an automated leaf movement analysis program for circadian period estimation. Plant Methods 11:33. https://doi.org/10.1186/s13007-015-0075 5
18. Dowson-Day MJ, Millar AJ (1999) Circadian dysfunction causes aberrant hypocotyl elongation patterns in Arabidopsis. Plant J 17 (1):63–71. https://doi.org/10.1046/j.1365-313X.1999.00353.x
19. Litthauer S, Battle MW, Lawson T, Jones MA (2015) Phototropins maintain robust circadian oscillation of PSII operating efficiency under blue light. Plant J 83(6):1034–1045. https://doi.org/10.1111/tpj.12947
20. Gould PD, Diaz P, Hogben C, Kusakina J, Salem R, Hartwell J, Hall A (2009) Delayed fluorescence as a universal tool for the measurement of circadian rhythms in higher plants. Plant J 58(5):893–901. https://doi.org/10.1111/j.1365-313X.2009.03819.x
21. Pan WJ, Wang X, Deng YR, Li JH, Chen W, Chiang JY, Yang JB, Zheng L (2015) Nondestructive and intuitive determination of circadian chlorophyll rhythms in soybean leaves using multispectral imaging. Sci Rep 5:11108. https://doi.org/10.1038/srep11108
22. Mayer WE, Fischer C (1994) Protoplasts from Phaseolus-Coccineus L pulvinar motor cells show circadian volume oscillations. Chronobiol Int 11(3):156–164. https://doi.org/10.3109/07420529409057235
23. Jouve L, Greppin H, Agosti RD (1998) Arabidopsis thaliana floral stem elongation: evidence for an endogenous circadian rhythm. Plant Physiol Bioch 36(6):469–472. https://doi.org/10.1016/S0981-9428(98)80212-X
24. Ueda HR, Chen W, Minami Y, Honma S, Honma K, Iino M, Hashimoto S (2004) Molecular-timetable methods for detection of body time and rhythm disorders from single-time-point genome-wide expression profiles. Proc Natl Acad Sci U S A 101 (31):11227–11232. https://doi.org/10.1073/pnas.0401882101
25. Li M, Cao L, Mwimba M, Zhou Y, Li L, Zhou M, Schnable PS, O'Rourke JA, Dong X, Wang W (2019) Comprehensive mapping of abiotic stress inputs into the soybean circadian clock. Proc Natl Acad Sci U S A 116(47):23840–23849. https://doi.org/10.1073/pnas.1708508116
26. Kerwin RE, Jimenez-Gomez JM, Fulop D, Harmer SL, Maloof JN, Kliebenstein DJ (2011) Network quantitative trait loci mapping of circadian clock outputs identifies metabolic pathway-to-clock linkages in Arabidopsis. Plant Cell 23(2):471–485. https://doi.org/10.1105/tpc.110.082065
27. Higashi T, Tanigaki Y, Takayama K, Nagano AJ, Honjo MN, Fukuda H (2016) Detection of diurnal variation of tomato transcriptome through the molecular timetable method in a sunlight-type plant factory. Front Plant Sci 7:87. https://doi.org/10.3389/fpls.2016.00087
28. Tamang BG, Magliozzi JO, Maroof MA, Fukao T (2014) Physiological and transcriptomic characterization of submergence and reoxygenation responses in soybean seedlings. Plant Cell Environ 37(10):2350–2365. https://doi.org/10.1111/pce.12277

Chapter 15

High-Throughput Targeted Transcriptional Profiling of Defense Genes Using RNA-Mediated Oligonucleotide Annealing, Selection, and Ligation with Next-Generation Sequencing in Arabidopsis

Sung-Il Kim, Yogendra Bordiya, Ji-Chul Nam, José Mayorga, and Hong-Gu Kang

Abstract

Tracking RNA transcription has been one of the most powerful tools to gain insight into the biological process. While a wide range of molecular methods such as northern blotting, RNA-seq, and quantitative RT-PCR are available, one of the barriers in transcript analysis is an inability to accommodate a sufficient number of samples to achieve high resolution in dynamic transcriptional changes. RASL-seq (RNA-mediated oligonucleotide Annealing, Selection, and Ligation with next-generation sequencing) is a sequencing-based transcription profiling tool that processes hundreds of samples assessing a set of over a hundred genes with a fraction of the cost of a conventional RNA-seq. We described a RASL-seq protocol for assessing 288 genes mostly including defense genes to capture their dynamic nature. We demonstrated that this transcriptional profiling method produced a highly reliable outcome comparable to a conventional RNA-seq and quantitative RT-PCR.

Key words *Arabidopsis*, Defense genes, High-throughput targeted transcriptional analysis, Transcript profiling, RASL-seq

1 Introduction

RNA transcription, the process by which RNA polymerase (RNAP) copies the genomic DNA into an RNA transcript, is the first stage in gene expression. As is the case in most regulatory points in the biological process, this initial step is critical to the regulation of cell activities. Consistent with this notion, modulating signaling pathways in response to internal/external stimuli involves substantial

---

Sung-Il Kim and Yogendra Bordiya contributed equally to this work.

Supplementary Information: The online version of this chapter (https://doi.org/10.1007/978-1-0716-1534-8_15) contains supplementary material, which is available to authorized users.

Shahid Mukhtar (ed.), *Modeling Transcriptional Regulation: Methods and Protocols*, Methods in Molecular Biology, vol. 2328, https://doi.org/10.1007/978-1-0716-1534-8_15, © Springer Science+Business Media, LLC, part of Springer Nature 2021

alterations in transcription [1]. For instance, transcriptional reprogramming is highly dynamic in plants undergoing defense responses, and its robust speed and magnitude are vital in fending off pathogens [2, 3]. Interestingly, the transcriptional induction of plant defense genes between resistant and susceptible defense responses in Arabidopsis is quantitative rather than qualitative [2]. Supporting the quantitative nature of defense responses, a few studies reported that susceptible mutants/lines display delayed the induction of defense genes rather than a lack thereof [4–6]. These findings argue that tracking the temporal transcriptional activity of defense genes is a critical element in determining how plant resistance would fare against pathogens. For instance, a recent report of a high-resolution time-course of the Arabidopsis transcriptome under biotic stress unraveled new insights that chromatin assembly and photosynthesis are dynamically regulated [7]. However, studies on the induction kinetics of defense genes in high resolution are very limited.

RNA quantitation has become routine since the introduction of northern blotting [8], along with a wide range of subsequent tools [9]. This quantitation became available for system-wide assessment for more than a decade when microarrays and RNA-seq were developed, which measures the transcription of essentially all the genes in an organism, giving the term "transcriptome" [10]. As next-generation DNA sequencing has become more affordable, RNA-seq is currently a dominant technique as a system-wide RNA quantitation tool [10]. However, in practice, RNA-seq or any modern tools thus far have seen limited usages in high-resolution transcriptome studies due to high cost, which appears to be one of the reasons why detailed transcription dynamics are rarely reported.

A high-throughput transcriptional analysis for a set of targeted RNA is currently available. NanoString nCounter (NanoString), for instance, utilizes a probe carrying multiple fluorophores that are tagged with targeted oligonucleotides [11, 12]. This tool analyzes hundreds of genes at a higher sensitivity via a highly sensitive CCD (charge-coupled device). However, because NanoString involves an advanced CCD camera and fluorophore-tagged oligonucleotides, the overall cost is currently high. RASL (RNA-mediated oligonucleotide Annealing, Selection, and Ligation with next-generation sequencing)-seq [13, 14], another high-throughput targeted approach, utilizes next-gen sequencing. While this tool was developed earlier than NanoString, it had a significant background issue, limiting its utility in analyzing low-expressing genes. Recently, a modification in RASL-seq overcame this shortcoming [15]. The original RASL-seq used a DNA ligase to anneal two DNA oligonucleotides that were bound to target RNA. In the improved RASL-seq, in contrast, the end of acceptor oligonucleotide probes carry two ribonucleotides, which an RNA ligase is used instead.

This modification substantially enhances the efficacy of oligonucleotide annealing at more than 100-fold and significantly reduces the background with enhanced sensitivity. With next-gen sequencing becoming cheaper, these developments argue that RASL-seq may be the best tool in high-resolution transcription analysis. However, since the improved RASL-seq was introduced, it has mainly been used in mammalian studies [16–21], except for a report revealing the role of the circadian clock in plant defense signaling, which used the original RASL-seq [22].

In this protocol, we described the RASL-seq procedure that analyzed the induction dynamics of defense genes in Arabidopsis challenged with *Pseudomonas syringae* pv. *tomato* DC3000 (*Pst*) by using probes targeting 288 genes, including 179 defense genes, 47 transposable elements, 35 hormone response genes, 17 mi/siRNA related genes, and ten housekeeping genes (Table 1). A pair of 20 base-long oligonucleotides was designed as a probe for a target RNA; the donor probe had a phosphate at the 5′ end, and the acceptor probe had two ribonucleotides at the 3′ end (Fig. 1a). The probes annealed with their target RNA were recovered by oligo-dT beads and subsequently ligated by Rnl2 (T4 dsRNA ligase 2). The ligated probes were amplified by PCR, which added a dual barcode and universal P5/P7 sequences (Fig. 1a). PCR products were pooled together and size-purified on an agarose gel for quantification and sequencing.

Expression of Arabidopsis *PR-1*, a defense marker gene, in response to bacterial infection was quantitated via RNA-seq, qRT-PCR, and RASL-seq (Fig. 2). The expression profile of *PR-1* in the RASL-seq showed a significant correlation coefficient with that of RNA-seq and qRT-PCR ($r > 0.98$), indicating a high reproducibility between these approaches (Fig. 2). A majority of the RASL-seq probes produced a highly correlative profile ($r > 0.8$) when compared with RNA-seq, while a notable portion did not (Table 2). Note that a similar outcome was observed in another RASL-seq study [15], suggesting that designing multiple probes for a single gene may be necessary. These weak-performing probes were removed from our subsequent studies.

2 Materials

2.1 Plant Material and RNA

1. Pathogen-treated plant tissue was stored in a −80 °C freezer.
2. RNA was extracted after the tissue was homogenized with mortar and pestle in liquid nitrogen using a standard Trizol-based or a silica column-based method.
3. RNA was quantified and evaluated in a standard UV spectrometer.

Table 1
List of the criteria and number of genes used in the RASL-seq

Criteria	Number of genes
Defense annotated genes	179
mi/siRNA related genes	17
TEs	47
Housekeeping genes	10
Hormone-related genes	35
Total	288

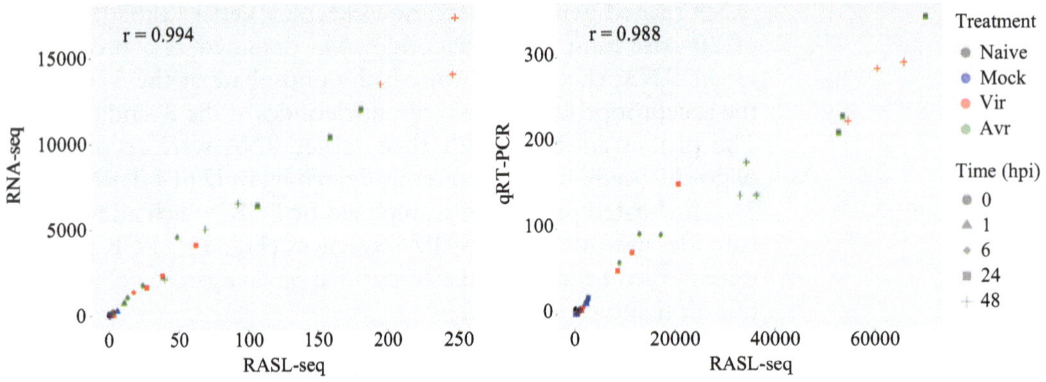

Fig. 1 RASL-seq shows outcomes comparable to qRT-PCR and RNA-seq in the induction profile of *PR-1*. Arabidopsis was infected with mock treatment (Mock), virulent *P. syringae* (Vir), and its avirulent counterpart (Avr) carrying *AvrRpt2* (10^6 cfu/mL) for 1, 6, 24, and 48 h. Arabidopsis, with no treatment, was also included (Naïve). Three biological replicates were presented. Scatter plots displayed a correlation of RASL-seq with RNA-seq (**a**) and qRT-PCR (**b**) analysis. The *x*- and *y*-axis shows a normalized expression value of the indicated approaches. A standard FPKM (Fragments Per Kilobase of transcript per Million mapped reads) was used for RNA-seq while relative values of qRT-PCR and RASL-seq compared with a housekeeping gene(s) were presented. Pearson's correlation coefficients (*r*) are shown between RASL-seq and RNA-seq or qRT-PCR. Note that normalization in Panel (**b**) was based on AT4G34270 (*Tip41 like*) only

2.2 Equipment

1. Centrifuge (Sovall, ST 40R).
2. Standard vortex mixer.
3. Cold room at 4 °C.
4. Magnet stand for PCR and 2 mL tube.
5. Rotating wheel.
6. Standard thermal cycler.
7. Agarose electrophoresis equipment.
8. UV illuminator.
9. Low binding PCR tubes and 2 mL tubes.

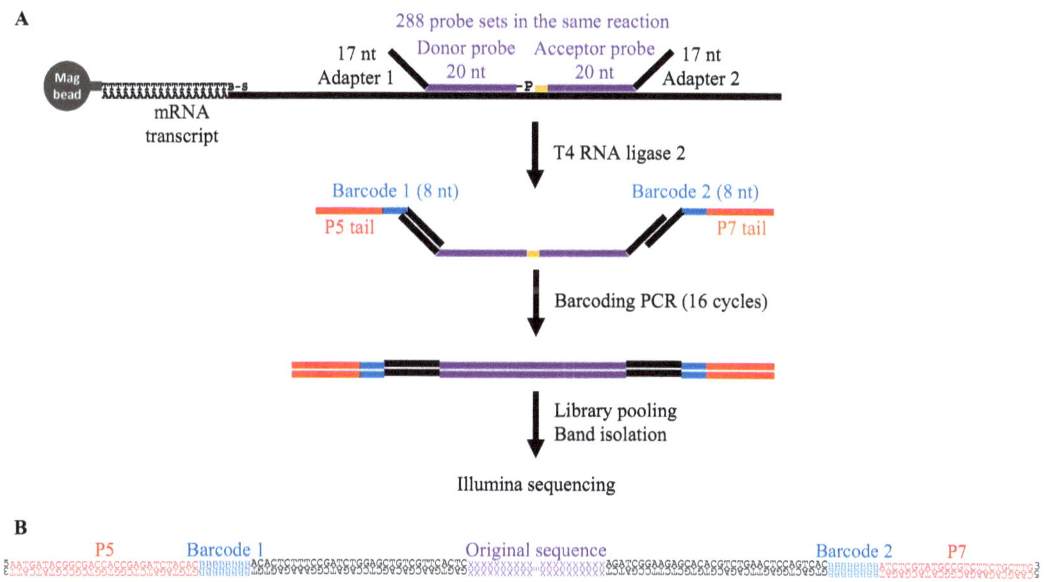

Fig. 2 A schematic diagram of the RASL-seq procedure (adapted from Larman et al., 2014). (**a**) mRNA is enriched using biotinylated oligo-dT from total RNA. An RNA-bound acceptor (with two ribonucleotides on the 3′ end) and donor probe (with a phosphate group on the 5′ end) are ligated by T4 RNA ligase 2 (Rnl2). A barcoding PCR is performed for multiplexing to accommodate a large number of samples. The amplicons are pooled and size-selected on an agarose gel, which is then subject to Illumina sequencing. (**b**) Sequence components of amplicons for the RASL-seq library are color-coded. Sequences for a target RNA, adapters, and barcodes are shown in purple, black, and blue, respectively. P5 and P7 sequences necessary for Illumina sequencing are shown in red

10. Standard band isolation kit.
11. Qubit® dsDNA HS Assay Kit (Invitrogen).
12. Bioanalyzer (Agilent).

2.3 Reagents

1. Deionized H_2O (dH_2O) at 18.2 Mohm.
2. Streptavidin magnetic beads (Thermo Scientific).
3. Biotin-oligo-dT (Promega).
4. T4 RNA ligase 2 (NEB).
5. Herculase II Fusion DNA polymerase (Agilent).
6. 2× binding and washing (B&W) buffer: 10 mM Tris–HCl/ 7.5, 1 mM EDTA, 2 M NaCl.
7. Sol A: DEPC-treated 0.1 M NaOH, DEPC-treated 0.05 M NaCl.
8. Sol B: DEPC-treated 0.1 M NaCl.
9. 10× saline-sodium citrate (SSC) buffer: 1.5 M NaCl, 150 mM Sodium Citrate.
10. Washing buffer: 20 mM Tris–HCl/pH 7.5, 0.1 M NaCl.

Table 2
List of RASL probes used in this study

RASL-seq probes					
#	Locus	Acceptor probe_sequence	Donor probe_sequence	Criteria	Corr_value
1	AT1G01470	ggagctgtcgttcactcaaactgatctcacagatc**gg**	[phos]aatcgaatgactgtaaggatagatcggaagagcacac	Defense ann	0.88
2	AT1G02450	ggagctgtcgttcactccttgtcttcgtttcgctc**uu**	[phos]ctgtttgatcttcttcttgtagatcggaagagcacac	Defense ann	0.98
3	AT1G02920	ggagctgtcgttcactcacattcaaaccaaaaaaaa**aa**	[phos]aaacagaggagacacacacaagatcggaagagcacac	Defense ann	0.67
4	AT1G02930	ggagctgtcgttcactcattatcgaagattacatt**ca**	[phos]aatcaaacactcggcagcagagatcggaagagcacac	Defense ann	0.66
5	AT1G05010	ggagctgtcgttcactcatatttgattgccttatt**ug**	[phos]tctgcagtttattcgtattgagatcggaagagcacac	Defense ann	0.63
6	AT1G10585	ggagctgtcgttcactcgaaaaaacacactagcgt**ua**	[phos]gcgtagacttatcatttgggagatcggaagagcacac	Defense ann	0.78
7	AT1G13340	ggagctgtcgttcactctctccaatagcttcacag**gg**	[phos]tcttctctgaatgacatcacagatcggaagagcacac	Defense ann	0.98
8	AT1G14870	ggagctgtcgttcactccaagcatattaacaacac**aa**	[phos]acagaggattaaacctcgttagatcggaagagcacac	Defense ann	0.70
9	AT1G19100	ggagctgtcgttcactctttttacacattttgttt**ug**	[phos]tctttccgagacagcccataagatcggaagagcacac	Defense ann	0.09
10	AT1G19250	ggagctgtcgttcactcgatgctgaacgtggaaat**gc**	[phos]aatgacgtttgtagaatctgagatcggaagagcacac	Defense ann	0.94
11	AT1G19570	ggagctgtcgttcactcatacttttcctcggtctt**ag**	[phos]ttttctcaaaagagtcgagcagatcggaagagcacac	Defense ann	0.97
12	AT1G19670	ggagctgtcgttcactcaatcgaatctccgctttt**uc**	[phos]accccacaaactatacttgaagatcggaagagcacac	Defense ann	0.99
13	AT1G21240	ggagctgtcgttcactcaagagaagaatcaaaaat**gg**	[phos]aaccgtgcaagtgatcgaaaagatcggaagagcacac	Defense ann	0.92
14	AT1G21250	ggagctgtcgttcactcgctactgtctccaagtcg**gg**	[phos]ctttctttatagcaactatgagatcggaagagcacac	Defense ann	0.97
15	AT1G21270	ggagctgtcgttcactcaatgtccaatgttgttac**au**	[phos]ttctgatgctgtcatagccaagatcggaagagcacac	Defense ann	0.96
16	AT1G21310	ggagctgtcgttcactcccggtggtggaggagaag**aa**	[phos]tagaaatagttagcggttgaagatcggaagagcacac	Defense ann	0.95
17	AT1G26390	ggagctgtcgttcactcagctttaacatccatcaa**uc**	[phos]tcttcaaattccccaagaaaagatcggaagagcacac	Defense ann	0.67
18	AT1G28480	ggagctgtcgttcactcccggagatatgagtagcc**au**	[phos]aaccctatctaaccctccaaagatcggaagagcacac	Defense ann	0.97
19	AT1G32640	ggagctgtcgttcactcagctgttcttgcgtatag**au**	[phos]cctaaaacccatcttcaccgagatcggaagagcacac	Defense ann	0.96

20	AT1G33960	ggagctgtcgttcactctgctcttctgaatgccct**uu**	[phos]tgattctagctcctcttgttagatcggaagagcacac	Defense ann	0.99
21	AT1G35710	ggagctgtcgttcactcgcggaccagtttgatgaa**uc**	[phos]cgtctttagaagttttgctgagatcggaagagcacac	Defense ann	0.97
22	AT1G42990	ggagctgtcgttcactcgaagttatgcctcaaaat**ca**	[phos]acattttaaaggcagaagcaagatcggaagagcacac	Defense ann	0.36
23	AT1G43160	ggagctgtcgttcactcaaggcaataatcagactg**aa**	[phos]gttgtattgggagagaaaaaagatcggaagagcacac	Defense ann	0.51
24	AT1G43910	ggagctgtcgttcactcacatcaacatgtagtttt**ag**	[phos]ctacattttgtttcatctggagatcggaagagcacac	Defense ann	0.94
25	AT1G44350	ggagctgtcgttcactcgtcttcgatcatattctt**ug**	[phos]caccatttccagcttcttcaagatcggaagagcacac	Defense ann	0.98
26	AT1G45145	ggagctgtcgttcactcagaccaccatgcttcatc**ag**	[phos]cttctcattgatctcatcttagatcggaagagcacac	Defense ann	0.98
27	AT1G51760	ggagctgtcgttcactcttgcatctcgacgtaatt**cu**	[phos]caatccctaacatatcgcctagatcggaagagcacac	Defense ann	0.97
28	AT1G51820	ggagctgtcgttcactccccttagagcctgaatct**cu**	[phos]gcaatgtgtacgtaagagtaagatcggaagagcacac	Defense ann	0.99
29	AT1G54320	ggagctgtcgttcactcctcaccgaagaaaccatt**uu**	[phos]gcaagcaaggattacatagtagatcggaagagcacac	Defense ann	0.69
30	AT1G55210	ggagctgtcgttcactctccgttatatttccctgt**cu**	[phos]tgaaagcaaagttcatcgccagatcggaagagcacac	Defense ann	0.98
31	AT1G56280	ggagctgtcgttcactctatgtgcattagacttca**uc**	[phos]gaaaatggctgatgacaagaagatcggaagagcacac	Defense ann	0.94
32	AT1G62300	ggagctgtcgttcactcttccctcgtaggttgtaa**uc**	[phos]agaattgatctgtcttccgcagatcggaagagcacac	Defense ann	0.97
33	AT1G64280	ggagctgtcgttcactcagaagtcgaatctgtcag**gg**	[phos]acgaatttcctaattccaaaagatcggaagagcacac	Defense ann	0.78
34	AT1G71100	ggagctgtcgttcactcagaatccatcgaaaatca**aa**	[phos]atcaaaccataaaccctaatagatcggaagagcacac	Defense ann	0.62
35	AT1G72520	ggagctgtcgttcactctcgttggcgtatgggtag**uc**	[phos]ttcaacaagtaatttaagccagatcggaagagcacac	Defense ann	0.98
36	AT1G72900	ggagctgtcgttcactctccaatgaagagtgacat**uu**	[phos]ttatacaccccaagagaaccagatcggaagagcacac	Defense ann	0.85
37	AT1G73805	ggagctgtcgttcactcaattccgctggagtcgtt**au**	[phos]atacccttcgtttactatctagatcggaagagcacac	Defense ann	0.98
38	AT1G74710	ggagctgtcgttcactctgtcactaaacattttct**gg**	[phos]acaaaagctcgtacctgagaagatcggaagagcacac	Defense ann	0.92
39	AT1G75040	ggagctgtcgttcactcagggcagaaagtgatttc**gu**	[phos]agttagctccggtacaagtgagatcggaagagcacac	Defense ann	0.97
40	AT1G76490	ggagctgtcgttcactcttctgcacacctttagaa**ac**	[phos]catattcatccccatagcatagatcggaagagcacac	Defense ann	0.91
41	AT1G78300	ggagctgtcgttcactcttcaaattgatgagaaaa**ga**	[phos]aagcaagtttcgattacacaagatcggaagagcacac	Defense ann	0.48
42	AT1G80590	ggagctgtcgttcactctattaatgttcaatcctg**ga**	[phos]aagggttctaatccaaagcaagatcggaagagcacac	Defense ann	0.94

(continued)

Table 2
(continued)

RASL-seq probes					
#	Locus	Acceptor probe_sequence	Donor probe_sequence	Criteria	Corr_value
43	AT1G80840	ggagctgtcgttcactcgagcacaagcacatttgaa**ag**	[phos]taagctcttggagatggattagatcggaagagcacac	Defense ann	0.97
44	AT2G04400	ggagctgtcgttcactctaagtatgagaaatgttc**cu**	[phos]gccaaaaagtccagctattcagatcggaagagcacac	Defense ann	0.96
45	AT2G04430	ggagctgtcgttcactcacttctcccgcgaccttt**uu**	[phos]aagcagatgttagctattaaagatcggaagagcacac	Defense ann	0.96
46	AT2G04450	ggagctgtcgttcactctccacactcctgtacctt**ug**	[phos]aaatgaccatcaatctcctgagatcggaagagcacac	Defense ann	0.99
47	AT2G06050	ggagctgtcgttcactcgcatcaccttgttgaaca**gc**	[phos]ttgcatacctagttccttatagatcggaagagcacac	Defense ann	1.00
48	AT2G13810	ggagctgtcgttcactcaacccgcaaacttagaga**au**	[phos]gatgaaacttcaatcgcgacagatcggaagagcacac	Defense ann	0.96
49	AT2G14610	ggagctgtcgttcactcgatcacatcattacttca**uu**	[phos]agtatggcttctcgttcacaagatcggaagagcacac	Defense ann	0.99
50	AT2G17420	ggagctgtcgttcactcaacaaatcttaaaagatg**aa**	[phos]tttggtggaagctgtgacaaagatcggaagagcacac	Defense ann	0.67
51	AT2G17720	ggagctgtcgttcactcagcatgtataaacggaaa**au**	[phos]tcatggttcgtattgagttgagatcggaagagcacac	Defense ann	0.78
52	AT2G18660	ggagctgtcgttcactcgtgtgtatacgacacgaa**ug**	[phos]ttaccggcatcagtattagcagatcggaagagcacac	Defense ann	0.95
53	AT2G19190	ggagctgtcgttcactcgtcatatctctctcttag**ac**	[phos]gctgatccacgattcctctaagatcggaagagcacac	Defense ann	0.97
54	AT2G20760	ggagctgtcgttcactcacattatacatacattgc**uu**	[phos]gcctggaaagagacgaaacaagatcggaagagcacac	Defense ann	0.50
55	AT2G21900	ggagctgtcgttcactcggattgttcgtatctctt**uc**	[phos]gatcttcttcttcacgttgcagatcggaagagcacac	Defense ann	0.97
56	AT2G23810	ggagctgtcgttcactcatcaatatcaaggtttag**cc**	[phos]caaaaggcagaacatactgaagatcggaagagcacac	Defense ann	0.87
57	AT2G24360	ggagctgtcgttcactcaaaatggtggtggtgggg**uu**	[phos]acttacttctcctatcttgaagatcggaagagcacac	Defense ann	0.81
58	AT2G24850	ggagctgtcgttcactcaagattcgggtttcttgg**ga**	[phos]caaaagagacaaggaatatcagatcggaagagcacac	Defense ann	0.97
59	AT2G25520	ggagctgtcgttcactcggactttcaatttgacat**uu**	[phos]aacaatctcaatggagggaaagatcggaagagcacac	Defense ann	0.83
60	AT2G26400	ggagctgtcgttcactcatgccaaaagacacatga**ac**	[phos]caaaatgtcttcggtttccaagatcggaagagcacac	Defense ann	0.99
61	AT2G27690	ggagctgtcgttcactcggatttcatctccatgat**ag**	[phos]ccatctctttcccgatacaaagatcggaagagcacac	Defense ann	1.00

62	AT2G29460	ggagctgtcgttcactctctactgatcgaagagta**uc**	[phos]attgttgtctcttaggctgaagatcggaagagcacac	Defense ann	0.99
63	AT2G30550	ggagctgtcgttcactctgacgccaaaacggtgga**au**	[phos]ttgtaaatgctctttcaaaaagatcggaagagcacac	Defense ann	0.98
64	AT2G30750	ggagctgtcgttcactcggcaaacatcgagaccaa**aa**	[phos]gcttcagttagatcaggttgagatcggaagagcacac	Defense ann	0.95
65	AT2G30770	ggagctgtcgttcactctgtccctgcggctatgtt**au**	[phos]atccctttactttgacatctagatcggaagagcacac	Defense ann	0.91
66	AT2G32190	ggagctgtcgttcactcagaaaaaacgatttattt**au**	[phos]aagaaggtaaaaattacacaagatcggaagagcacac	Defense ann	0.90
67	AT2G35980	ggagctgtcgttcactctgaggtacttaaaggaag**cc**	[phos]ttagatcatcgcaatcaaccagatcggaagagcacac	Defense ann	1.00
68	AT2G37040	ggagctgtcgttcactccggatcaatgattttacc**uu**	[phos]cacaaatcgccgtgaaaaccagatcggaagagcacac	Defense ann	0.98
69	AT2G38470	ggagctgtcgttcactcttgtgaccaattagcaga**ac**	[phos]cattcatgttttgtctggttagatcggaagagcacac	Defense ann	0.62
70	AT2G39030	ggagctgtcgttcactctcgccagtgagcctacaa**ag**	[phos]tctccattccttaaaaacctagatcggaagagcacac	Defense ann	0.99
71	AT2G39518	ggagctgtcgttcactcataaaacagcaagacaga**ug**	[phos]aaagcaaagagaagaagactagatcggaagagcacac	Defense ann	0.81
72	AT2G39530	ggagctgtcgttcactctgcgaagagaagaagact**gg**	[phos]ccgatgcatatcctttactaagatcggaagagcacac	Defense ann	0.99
73	AT2G40750	ggagctgtcgttcactctgatgatcatcaaacatc**au**	[phos]cgtctcttgccaaaccaatgagatcggaagagcacac	Defense ann	0.79
74	AT2G43530	ggagctgtcgttcactccgataacacaacgacgga**ua**	[phos]aatcctagccatacaatagcagatcggaagagcacac	Defense ann	0.98
75	AT2G43550	ggagctgtcgttcactctttacagtagtcgcagaa**gc**	[phos]acttgatatgtttgtgtttgagatcggaagagcacac	Defense ann	0.95
76	AT2G44240	ggagctgtcgttcactctgaaacatagatgcgtaa**ua**	[phos]tcaaaagcttgaacacacagagatcggaagagcacac	Defense ann	1.00
77	AT2G45220	ggagctgtcgttcactcacttttgcttccggtaat**aa**	[phos]tcgtcttccctataccatcaagatcggaagagcacac	Defense ann	0.99
78	AT2G45760	ggagctgtcgttcactcatttgtcaccatactcat**cu**	[phos]ctcaacctgtaactcaagaaagatcggaagagcacac	Defense ann	0.99
79	AT2G46400	ggagctgtcgttcactcttgtcataaagtacaatc**ca**	[phos]acttcatagggaaatcataaagatcggaagagcacac	Defense ann	0.81
80	AT3G01080	ggagctgtcgttcactcaacctcaaaagaaccgga**ga**	[phos]ttggcctgtgttattattgtagatcggaagagcacac	Defense ann	0.66
81	AT3G02520	ggagctgtcgttcactcatgggttaacacaaattc**ug**	[phos]ttgcgttaagcaaaaatcagagatcggaagagcacac	Defense ann	0.00
82	AT3G03470	ggagctgtcgttcactctgaactgttgcttctcgg**au**	[phos]agatcaacttcttcaccttcagatcggaagagcacac	Defense ann	0.98
83	AT3G05500	ggagctgtcgttcactcaaaaagaacgagaaccag**ag**	[phos]atcaatcgctaaccaagatcagatcggaagagcacac	Defense ann	0.85
84	AT3G07390	ggagctgtcgttcactctgtggtgaaacaaagtaa**au**	[phos]ggccctaagactaaaacagtagatcggaagagcacac	Defense ann	0.74

(continued)

Table 2
(continued)

RASL-seq probes					
#	Locus	Acceptor probe_sequence	Donor probe_sequence	Criteria	Corr_value
85	AT3G12580	ggagctgtcgttcactccgtgtagagtattatgcc**ca**	[phos]gtcgtctttcataggtcagaagatcggaagagcacac	Defense ann	0.64
86	AT3G12740	ggagctgtcgttcactctgggagaaaaagagaaaa**gg**	[phos]taatgagcatgataggaagcagatcggaagagcacac	Defense ann	0.78
87	AT3G13610	ggagctgtcgttcactccaataactgactctggtt**uu**	[phos]gggttcacaaagataggaacagatcggaagagcacac	Defense ann	0.99
88	AT3G13950	ggagctgtcgttcactccccacgcaattaattaac**uu**	[phos]tacaacaaaccaacccacacagatcggaagagcacac	Defense ann	0.61
89	AT3G17410	ggagctgtcgttcactcaagtaagattttcagtga**ag**	[phos]tggcagggggtttatgagagagatcggaagagcacac	Defense ann	0.83
90	AT3G17810	ggagctgtcgttcactcctcagcttttctctgctc**aa**	[phos]ctgcttctttctgtctcttaagatcggaagagcacac	Defense ann	0.88
91	AT3G18250	ggagctgtcgttcactctcggcacaagagaataac**ag**	[phos]agcggagaggatacaacaacagatcggaagagcacac	Defense ann	0.98
92	AT3G20510	ggagctgtcgttcactcatagctcaaaacagtttg**ug**	[phos]gttacaaaaacacgatgacaagatcggaagagcacac	Defense ann	0.78
93	AT3G21230	ggagctgtcgttcactctctgagggtagcttagct**uu**	[phos]aggtttcctagatttgttgaagatcggaagagcacac	Defense ann	0.79
94	AT3G22600	ggagctgtcgttcactctaaatagggacaaataaa**ga**	[phos]atgcaatctgagtggcacaaagatcggaagagcacac	Defense ann	0.98
95	AT3G24503	ggagctgtcgttcactctgaacaaaaaataaaaag**ug**	[phos]gttgcacaaaaagaaaaagaagatcggaagagcacac	Defense ann	0.67
96	AT3G25882	ggagctgtcgttcactcccgttacaatccaacgag**uu**	[phos]tctcaacccaagattctgacagatcggaagagcacac	Defense ann	0.99
97	AT3G26830	ggagctgtcgttcactcaccaaaccatatattcag**ug**	[phos]gtgaagaacttgaaagaaggagatcggaagagcacac	Defense ann	0.99
98	AT3G28510	ggagctgtcgttcactctgtgtttcaatctccaag**ua**	[phos]gttcttagccaaaaccttgaagatcggaagagcacac	Defense ann	0.98
99	AT3G28540	ggagctgtcgttcactctcaatatggttgtccatt**cu**	[phos]tcctctacgtatcaaagctgagatcggaagagcacac	Defense ann	0.90
100	AT3G28930	ggagctgtcgttcactcggaaaccatgaccggagc**ac**	[phos]attcgagaattaaattaacaagatcggaagagcacac	Defense ann	0.93
101	AT3G29240	ggagctgtcgttcactctcataccataaaccatta**gu**	[phos]gaaatgcagtacaaaaacaaagatcggaagagcacac	Defense ann	0.90
102	AT3G44300	ggagctgtcgttcactcggagtgacatgaactgac**ga**	[phos]atcacaaccgattacttgttagatcggaagagcacac	Defense ann	1.00
103	AT3G44720	ggagctgtcgttcactctcgatgcctcaaaatcca**cg**	[phos]taaaacatgtactcgaagttagatcggaagagcacac	Defense ann	0.95

104	AT3G46080	ggagctgtcgttcactccgttcttcccaactccaa**cu**	[phos]tccaattgactaaactctccagatcggaagagcacac	Defense ann	0.97
105	AT3G46090	ggagctgtcgttcactcctccatcgaatctaagtc**ca**	[phos]aatccaaacaagccactctcagatcggaagagcacac	Defense ann	0.98
106	AT3G47540	ggagctgtcgttcactcctacaatagtctctatag**ua**	[phos]cccaatccttgccttgaccgagatcggaagagcacac	Defense ann	0.98
107	AT3G48090	ggagctgtcgttcactctcatatagtctcgcagag**ga**	[phos]gaatgcgatttgtgatttttagatcggaagagcacac	Defense ann	0.87
108	AT3G48890	ggagctgtcgttcactcaatgtattttgaagcttc**ca**	[phos]taaccagaattagatgtcatagatcggaagagcacac	Defense ann	0.63
109	AT3G50480	ggagctgtcgttcactcaaagggtaaaaccctaga**aa**	[phos]catctagatgatgggcttagagatcggaagagcacac	Defense ann	0.91
110	AT3G50770	ggagctgtcgttcactctttcaccacaattcaata**au**	[phos]agcacttctaacataatcccagatcggaagagcacac	Defense ann	0.60
111	AT3G52430	ggagctgtcgttcactcaaacctccttcttcgtca**cc**	[phos]aatgtattcgcataactctcagatcggaagagcacac	Defense ann	0.97
112	AT3G52870	ggagctgtcgttcactcattcacaatcatccatgt**ga**	[phos]aaagatgacaacaaaaacctagatcggaagagcacac	Defense ann	0.87
113	AT3G53180	ggagctgtcgttcactcattagtatcggtgaatga**gu**	[phos]tgcttatatgcatccggattagatcggaagagcacac	Defense ann	0.95
114	AT3G56400	ggagctgtcgttcactctgagttgttaagtcatgg**cc**	[phos]ttcaacgagttggttcataaagatcggaagagcacac	Defense ann	0.98
115	AT3G57260	ggagctgtcgttcactcccgagtcgagatttgcgt**cg**	[phos]aataggttttggtatgagtaagatcggaagagcacac	Defense ann	1.00
116	AT3G57280	ggagctgtcgttcactcaacaggtcacaaacaaaa**cc**	[phos]aagtgtatcaattcgtaaagagatcggaagagcacac	Defense ann	0.80
117	AT3G59920	ggagctgtcgttcactcaagagcatttataagtct**uu**	[phos]aaccgaggaatccaacatcaagatcggaagagcacac	Defense ann	0.24
118	AT3G60450	ggagctgtcgttcactcactaaacaggtgaatggc**uu**	[phos]atgacaggttcataactgacagatcggaagagcacac	Defense ann	0.90
119	AT3G62290	ggagctgtcgttcactccatcgcatttggaagatc**uu**	[phos]gcttgttagcaaaaacgagaagatcggaagagcacac	Defense ann	0.86
120	AT3G63380	ggagctgtcgttcactctcaaggtttcttgagaga**ug**	[phos]ggcttcttgaacccttaaatagatcggaagagcacac	Defense ann	0.87
121	AT4G00330	ggagctgtcgttcactcaatcaacaattctcgtac**aa**	[phos]gaatgcatcgtgtatgacaaagatcggaagagcacac	Defense ann	0.82
122	AT4G01370	ggagctgtcgttcactctttcaccgagtatacaac**cg**	[phos]acagaccaaatatcaattgcagatcggaagagcacac	Defense ann	0.82
123	AT4G04490	ggagctgtcgttcactcccgaggatacaagactgt**aa**	[phos]tcgagacctcatccactgaaagatcggaagagcacac	Defense ann	0.65
124	AT4G08555	ggagctgtcgttcactcaaaagaacaaaatgaacc**aa**	[phos]tcataaaaagttaagaaaaaagatcggaagagcacac	Defense ann	0.44
125	AT4G08870	ggagctgtcgttcactcccaccttccatgatacga**gc**	[phos]aaaagaagatgcatgagagtagatcggaagagcacac	Defense ann	0.96
126	AT4G11170	ggagctgtcgttcactcttcttgctccaaccttaa**uc**	[phos]catctcataacagatttcctagatcggaagagcacac	Defense ann	0.95

(continued)

Table 2
(continued)

RASL-seq probes					
#	Locus	Acceptor probe_sequence	Donor probe_sequence	Criteria	Corr_value
127	AT4G14365	ggagctgtcgttcactctgtcgtgattctatatttcg	[phos]acttctttgacttcaccacaagatcggaagagcacac	Defense ann	0.87
128	AT4G15470	ggagctgtcgttcactctttggtatcaatgtttgcuu	[phos]ttgctcttaagcaatcagaaagatcggaagagcacac	Defense ann	0.22
129	AT4G16890	ggagctgtcgttcactcaaactccacatcgttgaaau	[phos]tcaatttagatggaagatgaagatcggaagagcacac	Defense ann	0.65
130	AT4G18440	ggagctgtcgttcactccccaacagcacctgcaaauu	[phos]tccccttaatcttagtttctagatcggaagagcacac	Defense ann	0.96
131	AT4G21830	ggagctgtcgttcactctcgcctttgaaaacatggcc	[phos]taaatgtccatcacacactgagatcggaagagcacac	Defense ann	0.93
132	AT4G21840	ggagctgtcgttcactcgtccatcacacactgcacau	[phos]gttatctccattcttcttccagatcggaagagcacac	Defense ann	0.97
133	AT4G23140	ggagctgtcgttcactcctttgcaggatcttcttgaa	[phos]cacataaaagaccgatatggagatcggaagagcacac	Defense ann	0.93
134	AT4G23150	ggagctgtcgttcactcaaaatccgagaatcctaaca	[phos]agatttttgtgccgaagattagatcggaagagcacac	Defense ann	0.96
135	AT4G31800	ggagctgtcgttcactcttgtaacctttttgtccgtau	[phos]ttcctccattgaaatccatcagatcggaagagcacac	Defense ann	0.96
136	AT4G35180	ggagctgtcgttcactcatggatgcaggctttttcuu	[phos]cttggttatgtacaccatctagatcggaagagcacac	Defense ann	0.85
137	AT4G36270	ggagctgtcgttcactcattaacatttttactaaacu	[phos]tgttgcatctccctcctcttagatcggaagagcacac	Defense ann	0.54
138	AT4G36280	ggagctgtcgttcactccgctacaaatatgcacatgc	[phos]tattaaacaaatacaagaacagatcggaagagcacac	Defense ann	−0.18
139	AT4G36290	ggagctgtcgttcactctgacaggctattaagcaaau	[phos]acaagaacaccatgaaagttagatcggaagagcacac	Defense ann	0.32
140	AT4G37150	ggagctgtcgttcactcttgatctccatcacttctuu	[phos]aaccggagagttctcaatcaagatcggaagagcacac	Defense ann	0.65
141	AT4G37370	ggagctgtcgttcactccgccaacttcttaacccgug	[phos]tctcgtaatctgaaaccaacagatcggaagagcacac	Defense ann	0.97
142	AT4G37640	ggagctgtcgttcactcacaaaaatggaggaaatcag	[phos]atatttgtccttaatagcacagatcggaagagcacac	Defense ann	0.85
143	AT4G39030	ggagctgtcgttcactcggatcgcaaaagagtaagca	[phos]ttgtagccttctttgttcagagatcggaagagcacac	Defense ann	0.98
144	AT4G39950	ggagctgtcgttcactcttgacgtagtttagttttgg	[phos]gatgtcggattcttgaacgaagatcggaagagcacac	Defense ann	0.98
145	AT5G01900	ggagctgtcgttcactcatagtgtcatgatgataagu	[phos]cgtgagatgtccagaaaggaagatcggaagagcacac	Defense ann	0.98

146	AT5G03290	ggagctgtcgttcactccatcactttttccagtgt**ua**	[phos]tgaatcttcttgccttacggagatcggaagagcacac	Defense ann	0.60
147	AT5G03610	ggagctgtcgttcactcatcgtttctttcccgcgt**aa**	[phos]tctttccaaaagtatgggatagatcggaagagcacac	Defense ann	0.99
148	AT5G05730	ggagctgtcgttcactcaatgcatcctctagcctg**aa**	[phos]taacagaacgaaaacagcatagatcggaagagcacac	Defense ann	0.78
149	AT5G07100	ggagctgtcgttcactccactgacttgggatcttg**aa**	[phos]atgctctttcaacgtgtttcagatcggaagagcacac	Defense ann	0.88
150	AT5G08300	ggagctgtcgttcactctggaaaaatcatatcaat**au**	[phos]agcaaaacaaaccccaaaaaagatcggaagagcacac	Defense ann	0.64
151	AT5G08790	ggagctgtcgttcactcataaacttctcatgaaag**aa**	[phos]caagaatgtgcctgctaatgagatcggaagagcacac	Defense ann	0.70
152	AT5G17380	ggagctgtcgttcactcaaatggaccggtttatga**uc**	[phos]cggtttacaatagtgaaaaaagatcggaagagcacac	Defense ann	0.55
153	AT5G17990	ggagctgtcgttcactcatgtctctcctttagctc**uc**	[phos]agaagaactagaaaggcactagatcggaagagcacac	Defense ann	0.89
154	AT5G19590	ggagctgtcgttcactcgaagggtatttagcagtt**au**	[phos]cccagcaacctcaaagacaaagatcggaagagcacac	Defense ann	0.95
155	AT5G21020	ggagctgtcgttcactctattaaacatatccaact**ca**	[phos]tacaaggcctaagaaaagccagatcggaagagcacac	Defense ann	0.91
156	AT5G22570	ggagctgtcgttcactccattactggttatctcac**gg**	[phos]tactgatctatagcttgctcagatcggaagagcacac	Defense ann	0.97
157	AT5G24110	ggagctgtcgttcactctttcatcagatctttgga**cu**	[phos]tgtttagtggcttcacatccagatcggaagagcacac	Defense ann	0.94
158	AT5G24200	ggagctgtcgttcactcatgatctgtctgaaaatc**cg**	[phos]gtgatatgattgtgttcaccagatcggaagagcacac	Defense ann	0.98
159	AT5G24530	ggagctgtcgttcactctggagaccgcaaacagta**gu**	[phos]gtcttgaagaagaatggttaagatcggaagagcacac	Defense ann	0.97
160	AT5G25250	ggagctgtcgttcactcacacaagaataagccaag**uc**	[phos]ctccaacaggattaaaataaagatcggaagagcacac	Defense ann	0.88
161	AT5G25260	ggagctgtcgttcactccaaacacttacgaataaa**cc**	[phos]aagttttccaacgggattaaagatcggaagagcacac	Defense ann	0.92
162	AT5G26340	ggagctgtcgttcactccgacgtatgtgcagatca**uu**	[phos]accacgactagaattgcgaaagatcggaagagcacac	Defense ann	0.98
163	AT5G27760	ggagctgtcgttcactcttagtttctgtacgaatc**uc**	[phos]tcttacaatttcgcatcccgagatcggaagagcacac	Defense ann	0.82
164	AT5G35735	ggagctgtcgttcactcaagaatccaaaatcgagt**ag**	[phos]aagtagaacaaacaagaacaagatcggaagagcacac	Defense ann	0.77
165	AT5G39050	ggagctgtcgttcactcagctctacaccaccactc**cc**	[phos]atctctactctccgcaaaagagatcggaagagcacac	Defense ann	0.76
166	AT5G39510	ggagctgtcgttcactccttcttgctctttccaat**gu**	[phos]tatcgtctactccatgaagcagatcggaagagcacac	Defense ann	0.84
167	AT5G39670	ggagctgtcgttcactccattctcctgcagttctc**aa**	[phos]ggttagatccttgctttaagagatcggaagagcacac	Defense ann	0.98
168	AT5G40780	ggagctgtcgttcactcaaaatacaaaccctaaaa**gc**	[phos]tacatttagattcagaccagagatcggaagagcacac	Defense ann	0.70

(continued)

Table 2
(continued)

RASL-seq probes					
#	**Locus**	**Acceptor probe_sequence**	**Donor probe_sequence**	**Criteria**	**Corr_value**
169	AT5G42380	ggagctgtcgttcactcgttttgacaaattcagaa**ua**	[phos]tcaaaacaagacacgtttatagatcggaagagcacac	Defense ann	0.68
170	AT5G44568	ggagctgtcgttcactctcctgcactatgatgact**ua**	[phos]tggcttatgggcctttatctagatcggaagagcacac	Defense ann	0.92
171	AT5G46350	ggagctgtcgttcactcgttcttcgagggtagatc**aa**	[phos]aacctacttgtgatgatgagagatcggaagagcacac	Defense ann	0.91
172	AT5G47200	ggagctgtcgttcactccgtcgaaggcattaaaga**gu**	[phos]gctctacattatttgcataaagatcggaagagcacac	Defense ann	0.63
173	AT5G52750	ggagctgtcgttcactcatagcttccttagcttca**uc**	[phos]acgataaccggtacatcaacagatcggaagagcacac	Defense ann	0.96
174	AT5G53560	ggagctgtcgttcactctcatcaaagtcgaaaaca**cu**	[phos]gtcatcataaactcaacacaagatcggaagagcacac	Defense ann	0.49
175	AT5G54500	ggagctgtcgttcactcttaagataattaaacccg**aa**	[phos]aaatcacaaaagcccccaaaagatcggaagagcacac	Defense ann	0.70
176	AT5G55450	ggagctgtcgttcactcttttctaatcccttattc**uu**	[phos]tctcactcaactctgttgtgagatcggaagagcacac	Defense ann	0.97
177	AT5G59420	ggagctgtcgttcactctcgaaatggaacaaagat**ac**	[phos]aatgagacagaaacacaaaaagatcggaagagcacac	Defense ann	0.22
178	AT5G59820	ggagctgtcgttcactccgtcggcaaaataggcta**au**	[phos]agattattcactaaatgctaagatcggaagagcacac	Defense ann	0.91
179	AT5G61210	ggagctgtcgttcactcaaaaacgataccaaagtt**gc**	[phos]ttctgtttgatcaataagttagatcggaagagcacac	Defense ann	0.80
180	AT1G06160	ggagctgtcgttcactcaatcctcaagaaccacaa**gu**	[phos]gttgtattactttcttgcgtagatcggaagagcacac	Hormone	0.69
181	AT1G17380	ggagctgtcgttcactctacttccataatctcttt**ag**	[phos]ctttgtctacggggaactcaagatcggaagagcacac	Hormone	0.99
182	AT1G17990	ggagctgtcgttcactctgatacatacagatttgg**ug**	[phos]tattgactggtcaaagcggtagatcggaagagcacac	Hormone	0.00
183	AT1G19180	ggagctgtcgttcactcgcccggcgtagaatatag**uc**	[phos]aatggtgcagtttgagactcagatcggaagagcacac	Hormone	0.98
184	AT1G19220	ggagctgtcgttcactcctgcatgaaagttgaagc**ug**	[phos]ggaagctgttattaccatgtagatcggaagagcacac	Hormone	0.63
185	AT1G30135	ggagctgtcgttcactcagtgggaaaaagacgaag**uu**	[phos]ccaagtcacaattttgctgtagatcggaagagcacac	Hormone	0.81
186	AT1G50640	ggagctgtcgttcactcacaacagatcatcgtgtt**cg**	[phos]gtcttcttcttcttggttttagatcggaagagcacac	Hormone	0.94
187	AT1G54040	ggagctgtcgttcactctaataaaacagccagcca**ua**	[phos]aataaattggctccttattgagatcggaagagcacac	Hormone	0.51

188	AT1G66340	ggagctgtcgttcactccaaagctcatgcatttct**cu**	[phos]tttgtggatttgtcagtgttagatcggaagagcacac	Hormone	0.80
189	AT1G70700	ggagctgtcgttcactcccaaagcattacaaacaa**ug**	[phos]atcaaatacagagacgccctagatcggaagagcacac	Hormone	0.95
190	AT1G72260	ggagctgtcgttcactcagaccattccgttcaaag**ca**	[phos]aatgaaaatggtcgagagaaagatcggaagagcacac	Hormone	0.00
191	AT1G74950	ggagctgtcgttcactcaagctcagatctccaaaa**cu**	[phos]acccttctccttcaggtaacagatcggaagagcacac	Hormone	0.96
192	AT2G23170	ggagctgtcgttcactctcttgaccacataccgaa**gu**	[phos]tatgaaatcagccagttcttagatcggaagagcacac	Hormone	0.84
193	AT2G24570	ggagctgtcgttcactctcaccggtgatcgcagaa**ga**	[phos]catgaaagaagagttagaagagatcggaagagcacac	Hormone	0.57
194	AT2G34600	ggagctgtcgttcactcgttccaagtcgcattttg**uu**	[phos]tgcatctccatctctttgaaagatcggaagagcacac	Hormone	0.90
195	AT2G39940	ggagctgtcgttcactctttctagctatggtttcc**aa**	[phos]gtctttgggactgattttggagatcggaagagcacac	Hormone	0.28
196	AT2G43710	ggagctgtcgttcactcttgtttttgtctttgtcc**uu**	[phos]ttagagctgcacttctctgtagatcggaagagcacac	Hormone	0.85
197	AT2G46370	ggagctgtcgttcactcactcttctccaatcttga**cu**	[phos]tgagttaaaccaaccggtttagatcggaagagcacac	Hormone	0.96
198	AT3G04720	ggagctgtcgttcactccaaaatcattacataata**uc**	[phos]aaacgcgatcaatggccgaaagatcggaagagcacac	Hormone	0.98
199	AT3G12500	ggagctgtcgttcactccactccaatccaccgtta**au**	[phos]gatgttcgtaatcactccatagatcggaagagcacac	Hormone	0.94
200	AT3G17860	ggagctgtcgttcactcggttgcagagctgagaga**ag**	[phos]aactaatgcattcagacattagatcggaagagcacac	Hormone	0.83
201	AT3G23030	ggagctgtcgttcactcaagataaagatggtgact**gg**	[phos]atgttggttggtgatgttccagatcggaagagcacac	Hormone	0.15
202	AT3G23150	ggagctgtcgttcactcgccacttcataaccgtcc**au**	[phos]ctctgccatttgaagatcaaagatcggaagagcacac	Hormone	0.65
203	AT3G23240	ggagctgtcgttcactcttaaggtccctaatacaa**au**	[phos]cctaatctttcaccaagtccagatcggaagagcacac	Hormone	0.75
204	AT3G45140	ggagctgtcgttcactcttttctggcgactcatag**aa**	[phos]ctcttttaaggcttcatctgagatcggaagagcacac	Hormone	0.97
205	AT3G62980	ggagctgtcgttcactccttccacattcagctttg**gc**	[phos]atcttctgtcctagtaacttagatcggaagagcacac	Hormone	0.58
206	AT4G14560	ggagctgtcgttcactctttgtagccttctctctc**gg**	[phos]aatattcacctactgtgaacagatcggaagagcacac	Hormone	0.76
207	AT4G23810	ggagctgtcgttcactcttgaattgaaaatgtaat**cu**	[phos]tggcgatgatgactctcgctagatcggaagagcacac	Hormone	0.92
208	AT4G31550	ggagctgtcgttcactcgcctgaagaagaaatatt**cu**	[phos]cctgcatcgcggattggttaagatcggaagagcacac	Hormone	0.93
209	AT4G38850	ggagctgtcgttcactctttcttgggtgctaagca**aa**	[phos]ttattcgaagggaatcatcgagatcggaagagcacac	Hormone	0.27
210	AT5G03280	ggagctgtcgttcactcccatgctaacaatcttct**cc**	[phos]acaggactcattggttcaatagatcggaagagcacac	Hormone	0.77

(continued)

Table 2
(continued)

RASL-seq probes					
#	Locus	Acceptor probe_sequence	Donor probe_sequence	Criteria	Corr_value
211	AT5G13320	ggagctgtcgttcactctttgcacagaggatctag**au**	[phos]tgatcccaaggtagtctccagatcggaagagcacac	Hormone	0.99
212	AT5G20900	ggagctgtcgttcactccactatcatacaacacat**ua**	[phos]cacacagcagcgtacatgatagatcggaagagcacac	Hormone	0.91
213	AT5G24770	ggagctgtcgttcactctgatctccgatattgcca**ac**	[phos]gatgttgtatcctttcttcaagatcggaagagcacac	Hormone	1.00
214	AT5G24780	ggagctgtcgttcactctggtgccaaaacggctac**aa**	[phos]agataaacgaaacgacatagagatcggaagagcacac	Hormone	0.98
215	AT5G44420	ggagctgtcgttcactcaaggttaatgcactgatt**cu**	[phos]tgcatgcattactgtttccgagatcggaagagcacac	Hormone	1.00
216	AT1G13320	ggagctgtcgttcactcacattgtcaatagattgg**ag**	[phos]agcttgatttgcgaaataccagatcggaagagcacac	Housekeeping	0.19
217	AT1G13440	ggagctgtcgttcactccctcagtgtatcccaaaa**uu**	[phos]cccttcattttgccttcagaagatcggaagagcacac	Housekeeping	0.87
218	AT1G58050	ggagctgtcgttcactcaatcaacaggaagttttg**cu**	[phos]aaatgatgtcctagtggtgtagatcggaagagcacac	Housekeeping	0.00
219	AT2G28390	ggagctgtcgttcactctgcaagtggatcaaatgc**ug**	[phos]catagagttcaaaatctggtagatcggaagagcacac	Housekeeping	−0.05
220	AT3G18780	ggagctgtcgttcactcaaaccccagctttttaag**cc**	[phos]tttgatcttgagagcttagaagatcggaagagcacac	Housekeeping	0.68
221	AT4G27960	ggagctgtcgttcactcaaaggatcatctgggttt**gg**	[phos]atccgttaacaaagaacagaagatcggaagagcacac	Housekeeping	0.87
222	AT4G34270	ggagctgtcgttcactcccaaatcaatctgatctt**ca**	[phos]cagttctcccactgaagagtagatcggaagagcacac	Housekeeping	0.29
223	AT5G46630	ggagctgtcgttcactctaaaatttcaggtgagag**au**	[phos]tttgtggatagccaaagtccagatcggaagagcacac	Housekeeping	0.45
224	AT5G60390	ggagctgtcgttcactctaagagagtcgatcataa**cg**	[phos]aaagtctcatcatttggcacagatcggaagagcacac	Housekeeping	0.86
225	AT1G01040	ggagctgtcgttcactcttgaaccgtgacgaaggt**ac**	[phos]aagtggagtttatgtttaacagatcggaagagcacac	mi/siRNA	0.25
226	AT1G31280	ggagctgtcgttcactcactacgaaaaacccaatt**ag**	[phos]attcacgcacaaactcctttagatcggaagagcacac	mi/siRNA	0.67
227	AT1G31290	ggagctgtcgttcactcgacgccgtctctacccac**gu**	[phos]ctccgacgccgtctctacccagatcggaagagcacac	mi/siRNA	0.75
228	AT1G48410	ggagctgtcgttcactcgtgtgcatcttgcatacg**ug**	[phos]taacataagttattggtcagagatcggaagagcacac	mi/siRNA	0.03
229	AT1G63020	ggagctgtcgttcactctatgagaagtcttcgaat**gu**	[phos]cccgtctattcttacaacctagatcggaagagcacac	mi/siRNA	0.49

230	AT1G69440	ggagctgtcgttcactcagtaagctcgtttatgaa**cu**	[phos]tgggaatggtagatttctgaagatcggaagagcacac	mi/siRNA	0.59
231	AT2G27040	ggagctgtcgttcactcttaattaaacaaccgaat**ug**	[phos]ttaagggggacaaaagcataagatcggaagagcacac	mi/siRNA	0.36
232	AT2G27880	ggagctgtcgttcactcaaaagttggccagatcct**uc**	[phos]aaaatattaacagccaaaccagatcggaagagcacac	mi/siRNA	0.00
233	AT2G32940	ggagctgtcgttcactcgagatgaagatgtctcca**uc**	[phos]cttccaggggaaaataatctagatcggaagagcacac	mi/siRNA	0.06
234	AT2G40030	ggagctgtcgttcactcgaagcagcccttcactat**uc**	[phos]caaaatcaccagatgaagaaagatcggaagagcacac	mi/siRNA	0.30
235	AT3G03300	ggagctgtcgttcactcaatagtccaggggacaga**ca**	[phos]aggatatttgtcgtatagatagatcggaagagcacac	mi/siRNA	0.67
236	AT3G43920	ggagctgtcgttcactcatacatcacagcctcacg**au**	[phos]tgggtatccttagggttatgagatcggaagagcacac	mi/siRNA	0.30
237	AT5G20320	ggagctgtcgttcactccaagtagacctttgtgga**gg**	[phos]aacttatagtcaagctggttagatcggaagagcacac	mi/siRNA	0.64
238	AT5G21030	ggagctgtcgttcactccaaagatagaaatcgttg**uu**	[phos]gtgttggtgacagatgttgcagatcggaagagcacac	mi/siRNA	0.21
239	AT5G21150	ggagctgtcgttcactcagcaatgcactgagtcac**aa**	[phos]ttccaagatcaacaagattcagatcggaagagcacac	mi/siRNA	0.00
240	AT5G43810	ggagctgtcgttcactctaaatctaaagccgaact**ac**	[phos]aagcaaacaaccaaactactagatcggaagagcacac	mi/siRNA	0.77
241	AT1G11260	ggagctgtcgttcactcggcgaaaacaaggaataa**cc**	[phos]caaacttcaaatgacaaagcagatcggaagagcacac	TE	0.94
242	AT1G11265	ggagctgtcgttcactcccaatcgctatcgctata**uc**	[phos]caacaagcttgtaatcactaagatcggaagagcacac	TE	0.03
243	AT1G57630	ggagctgtcgttcactcgcaatcttagatcctttg**au**	[phos]cgcctgtttgagtacaggacagatcggaagagcacac	TE	0.99
244	AT1G59860	ggagctgtcgttcactctctactttaacctcttct**uu**	[phos]cttcatccccggtaaatccgagatcggaagagcacac	TE	0.74
245	AT1TE40120	ggagctgtcgttcactcgttaagggagttgcaatt**au**	[phos]gctagaagaaaattcatttcagatcggaagagcacac	TE	−0.21
246	AT1TE40170	ggagctgtcgttcactcaattagttaagggagttg**ca**	[phos]attatgctagaagaaaattcagatcggaagagcacac	TE	−0.20
247	AT1TE70490	ggagctgtcgttcactcaaccatacagagatttat**ga**	[phos]agacgacataccttgctaggagatcggaagagcacac	TE	0.30
248	AT1TE72060	ggagctgtcgttcactcgaataggtgagctccaaa**ca**	[phos]tcagaataaatgtattcaagagatcggaagagcacac	TE	0.00
249	AT2G04140	ggagctgtcgttcactctatgagtatattcgtcga**ga**	[phos]tctttgagacggcgtaatccagatcggaagagcacac	TE	0.00
250	AT2G04240	ggagctgtcgttcactcgagcttgttgatctctga**au**	[phos]ccccttggaatttcgacaaaagatcggaagagcacac	TE	0.84
251	AT2G14560	ggagctgtcgttcactcgaagttgagcttaggtaa**aa**	[phos]catagcaacaagaagtggtaagatcggaagagcacac	TE	0.85
252	AT2G44180	ggagctgtcgttcactccctctccatctctctctt**cu**	[phos]cttcagatgttgttctccacagatcggaagagcacac	TE	0.60

(continued)

Table 2
(continued)

RASL-seq probes					
#	**Locus**	**Acceptor probe_sequence**	**Donor probe_sequence**	**Criteria**	**Corr_value**
253	AT2TE25335	ggagctgtcgttcactcctctcaaaactcttaaag**gu**	[phos]ataccacactatggcaactcagatcggaagagcacac	TE	0.88
254	AT2TE25440	ggagctgtcgttcactctttgatgctaaactgtgg**uu**	[phos]ccatatatgactagatatgaagatcggaagagcacac	TE	−0.29
255	AT2TE45420	ggagctgtcgttcactctcctactttaaggaggtg**au**	[phos]gatttttatgatgcttatagagatcggaagagcacac	TE	0.00
256	AT2TE82905	ggagctgtcgttcactctatcgtggaaatatgatt**uu**	[phos]tatgtgaagatacaacgtcaagatcggaagagcacac	TE	0.00
257	AT3G42883	ggagctgtcgttcactccaaatttctggataaagt**ac**	[phos]aacgccattcttctgctcatagatcggaagagcacac	TE	0.00
258	AT3G43190	ggagctgtcgttcactcaggttaactcttacgtat**uc**	[phos]ccaaactccaggtcttggtcagatcggaagagcacac	TE	0.00
259	AT3G44480	ggagctgtcgttcactcgaaagtcagcaagaggag**ac**	[phos]cacgacatgatcgacaagggagatcggaagagcacac	TE	0.61
260	AT3G50490	ggagctgtcgttcactctttgaaatgctgcctcat**ua**	[phos]aaaaccttaccaagaaaaccagatcggaagagcacac	TE	0.00
261	AT3G50500	ggagctgtcgttcactcgctaactccagatataag**cu**	[phos]gctgaaagaagaaccgagccagatcggaagagcacac	TE	0.69
262	AT3G60190	ggagctgtcgttcactctctcatctcttgctttct**ug**	[phos]tataactctagcctcttcgcagatcggaagagcacac	TE	0.95
263	AT3G61330	ggagctgtcgttcactctactctttcggtcatcta**cg**	[phos]tctccaccccaatcgctatcagatcggaagagcacac	TE	0.00
264	AT3G61430	ggagctgtcgttcactccaccggagataccagcgg**ua**	[phos]cagtagactaaggcaaatatagatcggaagagcacac	TE	0.95
265	AT3TE64915	ggagctgtcgttcactcgcgtaatccgatacgata**ga**	[phos]ctaacttcttgaaagtagcaagatcggaagagcacac	TE	−0.04
266	AT3TE90530	ggagctgtcgttcactcaatcgacttatcaaacat**ac**	[phos]tccatacattaagtcaggccagatcggaagagcacac	TE	0.07
267	AT4G04220	ggagctgtcgttcactctcgttaaccgtgatacag**ac**	[phos]gacggtatggaactattgaaagatcggaagagcacac	TE	0.74
268	AT4G11650	ggagctgtcgttcactcacctgaggagtcaaagtt**ac**	[phos]aattggtcctaccccaaatcagatcggaagagcacac	TE	0.97
269	AT4G12400	ggagctgtcgttcactcgaatctccttcatcgtct**ua**	[phos]acaaaatcatcctgctccaaagatcggaagagcacac	TE	0.99
270	AT4G12426	ggagctgtcgttcactctctttgagacggcgtaat**cc**	[phos]gatacgatagactaacttctagatcggaagagcacac	TE	0.00
271	AT4G16860	ggagctgtcgttcactccaggattacccttggact**uu**	[phos]gcacgatcaatttctctaccagatcggaagagcacac	TE	0.90

272	AT4G18430	ggagctgtcgttcactcgtttctggtgaagcgaga**ca**	[phos]aaagattggattttccgacgagatcggaagagcacac	TE	0.00
273	AT4TE09225	ggagctgtcgttcactctctcgtttcttctctttc**uc**	[phos]tctagaactctctttgtgtcagatcggaagagcacac	TE	−0.79
274	AT4TE30225	ggagctgtcgttcactcctctgataccatgttgga**uu**	[phos]attgagagagagaagaaaaaagatcggaagagcacac	TE	0.00
275	AT4TE42880	ggagctgtcgttcactctaagagactcaacaaatt**cg**	[phos]tcctaggtataataaaccgtagatcggaagagcacac	TE	0.19
276	AT4TE46360	ggagctgtcgttcactcttagaaatcctctacttt**ug**	[phos]attatgtatctctatggaaaagatcggaagagcacac	TE	0.00
277	AT4TE56270	ggagctgtcgttcactcgcctctaaaccagctaaa**cc**	[phos]aaacctatacagactccaatagatcggaagagcacac	TE	0.85
278	AT5G13220	ggagctgtcgttcactcgtaggtaacgtaatctcc**ua**	[phos]aaaacgtttgtcgtgaatagagatcggaagagcacac	TE	0.98
279	AT5G13330	ggagctgtcgttcactcgattcactaactcctctt**gg**	[phos]tgcatgagaaatggttgtggagatcggaagagcacac	TE	0.87
280	AT5G33315	ggagctgtcgttcactcgaaatgctgcctcattaa**aa**	[phos]accttaccaggaaaactcaaagatcggaagagcacac	TE	0.00
281	AT5G35080	ggagctgtcgttcactcaatcttcaggaaggaact**ca**	[phos]atcttgtactttggttcacgagatcggaagagcacac	TE	0.81
282	AT5G39680	ggagctgtcgttcactcaaccacagtagctacaaa**cu**	[phos]cattaggtctcgattcaccaagatcggaagagcacac	TE	0.01
283	AT5G46470	ggagctgtcgttcactccaaatatccttcaaatcc**ac**	[phos]aaggaactcccgttctccagagatcggaagagcacac	TE	0.70
284	AT5TE15250	ggagctgtcgttcactcatgacaagtggtcgactg**aa**	[phos]tcctagagatattatttcatagatcggaagagcacac	TE	0.00
285	AT5TE15460	ggagctgtcgttcactctaggaatcaaagcttgta**ug**	[phos]agtgtatccatttgaaaatcagatcggaagagcacac	TE	0.43
286	AT5TE57335	ggagctgtcgttcactcatgtgtagtcattcttat**aa**	[phos]ccagtccaaaaatcacccaaagatcggaagagcacac	TE	0.54
287	AT5TE67700	ggagctgtcgttcactcgcatgccataaaatgaca**uc**	[phos]tttaagaacaacgtcgagatagatcggaagagcacac	TE	0.00
288	AT5TE67885	ggagctgtcgttcactcattggacaaagaatgatt**ug**	[phos]agttttactggggttttataagatcggaagagcacac	TE	0.07

RNA residues in the acceptor probe are underlined in bold

11. Probe mix: All the probes in dH_2O were pooled to a final concentration of 10 nM (*see* **Note 1**).
12. P5/P7 barcode primer: These primers carry a complementary sequence to the 3′ end of each probe oligo following the barcoding sequence and P5/P7 sequences for the Illumina flow cell (Fig. 1). 100μM stock for long-term storage and 3 μM diluted solution for immediate use were prepared.

3 Methods

3.1 Designing RASL Probes

1. The NCBI primer BLAST tool (https://www.ncbi.nlm.nih.gov/tools/primer-blast/) was used for the probe design. The design parameters are as following: minimum, optimum, and maximum Tm was set at 60, 68, and 85, respectively. The GC percentage allowed was set at 30 and 70 minimum and maximum, respectively. The maximum length of the primer designed using this tool was 36 bases. Four bases were manually added to the 3′ end of the primer sequence to create a total length of 40. The 40 base sequence was split in half, which resulted in the donor (with 5′ phosphate) and the acceptor (with two 3′ ribose nucleotides) probe.
2. 17 base adapter sequence for the sequencing library was added to each probe to make a total length of 37 bases (Fig. 1a). The probe sequence information is listed in Table 2. Barcode primers for dual multiplexing are listed in Table 3. These primers are schematically represented in Fig. 1.

3.2 Preparation of Equilibrated Biotinylated Oligo-dT Streptavidin-Coated Beads

1. Transfer 3 μL of MagnaBind streptavidin bead slurry into a PCR tube on a magnet stand (*see* **Note 2**).
2. Wait until all beads were bound to the magnetic side (*see* **Note 3**).
3. Remove supernatant and wash the beads by gentle pipetting using 6 μL of Sol A twice.
4. Wash the beads by gentle pipetting using 6 μL of Sol B.
5. Resuspend the beads in 9 μL of 1× B&W buffer, and mix with 2 μL of biotinylated oligo-dT (10 μM) by pipetting.
6. Incubate at 4 °C for 1 h with a slow rotation, and place the tube on a magnet stand for 2 min.
7. Wash beads with 3 μL of 1× B&W by gentle pipetting twice, and wash with 3 μL of 3× SSC to remove unbound poly dT-biotin.
8. Move the tube to a regular rack and resuspend the beads in 10 μL of 3× SSC.

Table 3
List of barcode, custom Read1 and RT-PCR primers. The portion of primers used for barcoding is underlined in bold

Barcoding PCR primers		
#	**Name**	**Sequence**
1	P5_barcode_1_F	aatgatacggcgaccaccgagatctacac**gactgact**acactctttccgatctggagctgtcgttcactc
2	P5_barcode_2_F	aatgatacggcgaccaccgagatctacac**gcatgcat**acactctttccgatctggagctgtcgttcactc
3	P5_barcode_3_F	aatgatacggcgaccaccgagatctacac**atcgatcg**acactctttccgatctggagctgtcgttcactc
4	P5_barcode_4_F	aatgatacggcgaccaccgagatctacac**ctagctag**acactctttccgatctggagctgtcgttcactc
5	P5_barcode_5_F	aatgatacggcgaccaccgagatctacac**gtacgtac**acactctttccgatctggagctgtcgttcactc
6	P5_barcode_6_F	aatgatacggcgaccaccgagatctacac**gtcagtca**acactctttccgatctggagctgtcgttcactc
7	P5_barcode_7_F	aatgatacggcgaccaccgagatctacac**acgtacgt**acactctttccgatctggagctgtcgttcactc
8	P5_barcode_8_F	aatgatacggcgaccaccgagatctacac**atgcatgc**acactctttccgatctggagctgtcgttcactc
9	P5_barcode_9_F	aatgatacggcgaccaccgagatctacac**ctgactga**acactctttccgatctggagctgtcgttcactc
10	P5_barcode_10_F	aatgatacggcgaccaccgagatctacac**agtcagct**acactctttccgatctggagctgtcgttcactc
11	P5_barcode_11_F	aatgatacggcgaccaccgagatctacac**cagtcgac**acactctttccgatctggagctgtcgttcactc
12	P5_barcode_12_F	aatgatacggcgaccaccgagatctacac**acgtagca**acactctttccgatctggagctgtcgttcactc
13	P5_barcode_13_F	aatgatacggcgaccaccgagatctacac**gatcgata**acactctttccgatctggagctgtcgttcactc
14	P5_barcode_14_F	aatgatacggcgaccaccgagatctacac**cgtatcga**acactctttccgatctggagctgtcgttcactc
15	P5_barcode_15_F	aatgatacggcgaccaccgagatctacac**catgtcag**acactctttccgatctggagctgtcgttcactc
16	P5_barcode_16_F	aatgatacggcgaccaccgagatctacac**cgtacatg**acactctttccgatctggagctgtcgttcactc
17	P5_barcode_17_F	aatgatacggcgaccaccgagatctacac**actgagtc**acactctttccgatctggagctgtcgttcactc
18	P5_barcode_18_F	aatgatacggcgaccaccgagatctacac**cgatcgtg**acactctttccgatctggagctgtcgttcactc
19	P5_barcode_19_F	aatgatacggcgaccaccgagatctacac**tgactgcg**acactctttccgatctggagctgtcgttcactc
20	P5_barcode_20_F	aatgatacggcgaccaccgagatctacac**tgcatgag**acactctttccgatctggagctgtcgttcactc
21	P5_barcode_21_F	aatgatacggcgaccaccgagatctacac**cgtacggc**acactctttccgatctggagctgtcgttcactc
22	P5_barcode_22_F	aatgatacggcgaccaccgagatctacac**tcagttga**acactctttccgatctggagctgtcgttcactc
23	P5_barcode_23_F	aatgatacggcgaccaccgagatctacac**ctgaccag**acactctttccgatctggagctgtcgttcactc
24	P5_barcode_24_F	aatgatacggcgaccaccgagatctacac**tcaggacg**acactctttccgatctggagctgtcgttcactc
25	P5_barcode_25_F	aatgatacggcgaccaccgagatctacac**tcgatggt**acactctttccgatctggagctgtcgttcactc
26	P5_barcode_26_F	aatgatacggcgaccaccgagatctacac**gtaccatc**acactctttccgatctggagctgtcgttcactc
27	P5_barcode_27_F	aatgatacggcgaccaccgagatctacac**tgcatcct**acactctttccgatctggagctgtcgttcactc
28	P5_barcode_28_F	aatgatacggcgaccaccgagatctacac**gtcacgtt**acactctttccgatctggagctgtcgttcactc
29	P5_barcode_29_F	aatgatacggcgaccaccgagatctacac**ctgagctt**acactctttccgatctggagctgtcgttcactc

(continued)

Table 3
(continued)

Barcoding PCR primers		
#	**Name**	**Sequence**
30	P5_barcode_30_F	aatgatacggcgaccaccgagatctacac**ctgagagc**acactctttccgatctggagctgtcgttcactc
31	P7_barcode_1_R	caagcagaagacggcatacgagat**gtacgagt**gtgactggagttcagacgtgtgctcttccgatct
32	P7_barcode_2_R	caagcagaagacggcatacgagat**tagctcat**gtgactggagttcagacgtgtgctcttccgatct
33	P7_barcode_3_R	caagcagaagacggcatacgagat**gctagtat**gtgactggagttcagacgtgtgctcttccgatct
34	P7_barcode_4_R	caagcagaagacggcatacgagat**catgctgt**gtgactggagttcagacgtgtgctcttccgatct
35	P7_barcode_5_R	caagcagaagacggcatacgagat**acgtgatg**gtgactggagttcagacgtgtgctcttccgatct
36	P7_barcode_6_R	caagcagaagacggcatacgagat**cgatacta**gtgactggagttcagacgtgtgctcttccgatct
37	P7_barcode_7_R	caagcagaagacggcatacgagat**agctcaga**gtgactggagttcagacgtgtgctcttccgatct
38	P7_barcode_8_R	caagcagaagacggcatacgagat**gactgtgc**gtgactggagttcagacgtgtgctcttccgatct
39	P7_barcode_9_R	caagcagaagacggcatacgagat**tcagtctc**gtgactggagttcagacgtgtgctcttccgatct
40	P7_barcode_10_R	caagcagaagacggcatacgagat**catgcaca**gtgactggagttcagacgtgtgctcttccgatct
41	P7_barcode_11_R	caagcagaagacggcatacgagat**ctagcgct**gtgactggagttcagacgtgtgctcttccgatct
42	P7_barcode_12_R	caagcagaagacggcatacgagat**tacgtgta**gtgactggagttcagacgtgtgctcttccgatct
43	P7_barcode_13_R	caagcagaagacggcatacgagat**tgcatatc**gtgactggagttcagacgtgtgctcttccgatct
44	P7_barcode_14_R	caagcagaagacggcatacgagat**ctgacact**gtgactggagttcagacgtgtgctcttccgatct
45	P7_barcode_15_R	caagcagaagacggcatacgagat**agctacac**gtgactggagttcagacgtgtgctcttccgatct
46	P7_barcode_16_R	caagcagaagacggcatacgagat**tacgtagg**gtgactggagttcagacgtgtgctcttccgatct
47	P7_barcode_17_R	caagcagaagacggcatacgagat**gtcaactg**gtgactggagttcagacgtgtgctcttccgatct
48	P7_barcode_18_R	caagcagaagacggcatacgagat**gcattagc**gtgactggagttcagacgtgtgctcttccgatct
49	P7_barcode_19_R	caagcagaagacggcatacgagat**gctagccg**gtgactggagttcagacgtgtgctcttccgatct
50	P7_barcode_20_R	caagcagaagacggcatacgagat**atgccgta**gtgactggagttcagacgtgtgctcttccgatct
51	P7_barcode_21_R	caagcagaagacggcatacgagat**gatcggag**gtgactggagttcagacgtgtgctcttccgatct
52	P7_barcode_22_R	caagcagaagacggcatacgagat**tagcctcg**gtgactggagttcagacgtgtgctcttccgatct
53	P7_barcode_23_R	caagcagaagacggcatacgagat**atcgcaat**gtgactggagttcagacgtgtgctcttccgatct
54	P7_barcode_24_R	caagcagaagacggcatacgagat**gtacagga**gtgactggagttcagacgtgtgctcttccgatct
55	P7_barcode_25_R	caagcagaagacggcatacgagat**gattgctc**gtgactggagttcagacgtgtgctcttccgatct
56	P7_barcode_26_R	caagcagaagacggcatacgagat**gtacgcca**gtgactggagttcagacgtgtgctcttccgatct
57	P7_barcode_27_R	caagcagaagacggcatacgagat**tacggcgt**gtgactggagttcagacgtgtgctcttccgatct
58	P7_barcode_28_R	caagcagaagacggcatacgagat**tacgcttc**gtgactggagttcagacgtgtgctcttccgatct
59	P7_barcode_29_R	caagcagaagacggcatacgagat**gatcttca**gtgactggagttcagacgtgtgctcttccgatct

(continued)

Table 3
(continued)

#	Name	Sequence	
Barcoding PCR primers			
60	P7_barcode_30_R	caagcagaagacggcatacgagat**gcatatgg**gtgactggagttcagacgtgtgctcttccgatct	
61	Custom Read1 seq primer	acactctttccgatctggagctgtcgttcactc	
qRT-PCR			
1	PR1 (AT2G14610)	ccaccattgttacacctcacttt	aaaacttagcctggggtagcgg
2	Tip41 (AT4G34270)	gcgattttggctgagagttgat	ggataccctttcgcagatagagac

3.3 Preparing RASL-seq Library

1. Add 20 μL of total RNA (1 μg) to 10 μL of biotinylated oligo-dT streptavidin-coated beads (prepared at Subheading 3.2, **step 8**).
2. Incubate at 4 °C for 1 h with slow rotation and place the tube on a magnet stand for 2 min.
3. Discard the supernatant and wash with 20 μL of 1× SSC twice by pipetting.
4. Add 20 μL of 10 nM probe mix and 10 μL of 3× SSC.
5. Incubate at 70 °C for 10 min, followed by 45 °C for 1 h. Then keep the tubes at 30 °C (*see* **Note 4**).
6. Place the tube on a magnet stand and wait for 2 min until the beads are bound to the magnetic side.
7. Wash the beads with 50 μL of washing buffer twice and with 20 μL of 1× Rnl2 ligase buffer by pipetting the beads (*see* **Note 5**).
8. Resuspend the beads in 10 μL of ligation master mix containing 5 U of T4 Rnl2 by pipetting. Incubate at 37 °C for 1 h. The single reaction mix is prepared by adding the following components:
   (a) 1 μL of 10× T4 Rnl2 buffer.
   (b) 0.5 μL of T4 Rnl2 enzyme.
   (c) 8.5 μL of dH_2O.
9. Place the tube on a magnet stand for 2 min and discard the supernatant.
10. Resuspend the beads in 10 μL of dH_2O. At this point, the samples can be stored at −20 °C for an extended time.
11. Prepare a PCR master mix by adding the following components to a PCR tube:

(a) 5 μl of the ligated probe (prepared at **step 10**).

(b) 0.9 μL of 3 μM P5_barcode primer.

(c) 0.9 μL of 3 μM P7_barcode primer.

(d) 1 μL of 2.5 mM dNTP.

(e) 2 μL of 5× Herculase II buffer.

(f) 0.2 μL of Herculase II DNA polymerase.

12. Incubate the tube at 95 °C for 2 min followed by 16 cycles of 95 °C for 15 s, 54 °C for 20 s, and 72 °C for 25 s (*see* **Note 6**).
13. Briefly centrifuge the tube and place it on a magnet stand for 2 min.
14. Mix all the PCR reactions into a single tube and run it on 2% agarose gel.
15. Cut out a band at the expected size (176 bp) from the gel and extract DNA using a standard band isolation kit.
16. To quantify the library, measure the concentration using Qubit dsDNA HS Assay Kit (*see* **Note 7**).
17. To evaluate the library, check the quality of the sequencing library using Bioanalyzer (*see* **Note 8**).
18. Load the library into Illumina sequencer with Custom Read1 seq primer (*see* **Note 9**).

3.4 Quantitation from Sequencing Data

1. A reference genome file (Supplement Data File 1) was created by pooling all the probe sequences; this was used for the read alignment and counting below.
2. A standard bowtie2 alignment and an intersectBed counting were used to count raw reads for each gene.
3. For normalization of raw counts, ten housekeeping genes including AT4G34270 (*Tip41 like*), AT3G18780 (*Actin 2*), AT4G27960 (*UBC9*), AT5G46630 (*CACS*), AT1G13440 (*GAPDH*), AT4G05320 (*UBQ10*), AT1G58050 (*Helicase*), AT1G13320 (*PDF2*), AT2G28390 (*SAND family*) and AT5G60390 (*EF-1a*) were chosen from an early study [23] and tested. Among these housekeeping genes, three of them (AT1G13320, AT2G28390, and AT5G60390) were found to be the most stable and used in the subsequent studies.
4. The ratio between the three chosen housekeeping genes was calculated, based on their total reads. This ratio was used to adjust reads from the three housekeeping genes to have a comparable number among them. These corrected reads were then averaged and used as the normalization factor for each sample.
5. Raw counts of each gene in each sample were divided by the normalization factor to give a normalization value. This normalization value was used for a transcriptional profile.

4 Notes

1. For multiple uses, prepare 10 mL of 10 nM probe mix and make 0.5 mL aliquots in a 1.5 mL tube. Store the probe pool at −20 °C and thaw right before use.
2. Equilibrated beads should be thoroughly resuspended by vortexing right before use.
3. The settling of the streptavidin bead towards the magnet side takes some time, depending on the amount of beads. To expedite the resuspension, slowly pipette the solution in the middle of the settled beads.
4. Do not cool down the reaction below 25 °C. It may increase the background noise signal due to non-specific probe interactions. After the annealing step, centrifuge briefly and move the tubes immediately to a magnet stand.
5. The washing step is critical to remove un-annealed probes and RNA, which is important in reducing the background ligation. The beads should be completely resuspended during the washing step to remove un-annealed probes.
6. PCR at a higher cycle number tends to lead to a band of around 250 bp. Use a minimal number of cycles to produce a visible and expected band at 176 bp.
7. Qubit or qPCR is recommended for library quantification. Spectrophotometry-based measurement frequently provided a misleading number. An accurate library concentration is critical for optimum Illumina sequencer clustering.
8. Microfluidic systems can be used to determine library qualification. You should have only a single band at 176 bp long which is expected size of RASL-seq library. If you increase barcoding primer amount for PCR step, it may create non-specific 75 bp size band which can interfere the Illumina sequencing.
9. Illumina "universal read 1 sequencing primer" is not compatible with this library. Use "Custom Read 1 seq primer" for the sequencing.

Acknowledgments

We thank Angela H. Kang for critical comments on the manuscript and Dr. Benjamin Larman for sharing information on his RASL-seq. This work is supported by National Science Foundation Grant (IOS-1553613) to HGK.

References

1. Lee TI, Young RA (2013) Transcriptional regulation and its misregulation in disease. Cell 152:1237–1251
2. Tao Y, Xie Z, Chen W et al (2003) Quantitative nature of Arabidopsis responses during compatible and incompatible interactions with the bacterial pathogen *Pseudomonas syringae*. Plant Cell 15:317–330
3. Tsuda K, Sato M, Stoddard T et al (2009) Network properties of robust immunity in plants. PLoS Genet 5:e1000772
4. Bordiya Y, Zheng Y, Nam JC et al (2016) Pathogen infection and MORC proteins affect chromatin accessibility of transposable elements and expression of their proximal genes in Arabidopsis. Mol Plant-Microbe Interact 29:674–687
5. Pan Y, Liu Z, Rocheleau H et al (2018) Transcriptome dynamics associated with resistance and susceptibility against fusarium head blight in four wheat genotypes. BMC Genomics 19:642
6. Wang Y, An C, Zhang X et al (2013) The Arabidopsis elongator complex subunit2 epigenetically regulates plant immune responses. Plant Cell 25:762–776
7. Rodrigues DF, Costa VM, Silvestre R et al (2019) Methods for the analysis of transcriptome dynamics. Toxicol Res (Camb) 8:597–612
8. Alwine JC, Kemp DJ, Stark GR (1977) Method for detection of specific RNAs in agarose gels by transfer to diazobenzyloxymethyl-paper and hybridization with DNA probes. Proc Natl Acad Sci U S A 74:5350–5354
9. Suzuki T, Higgins PJ, Crawford DR (2000) Control selection for RNA quantitation. BioTechniques 29:332–337
10. Stark R, Grzelak M, Hadfield J (2019) RNA sequencing: the teenage years. Nat Rev Genet 20:631–656
11. Geiss GK, Bumgarner RE, Birditt B et al (2008) Direct multiplexed measurement of gene expression with color-coded probe pairs. Nat Biotechnol 26:317–325
12. Reis PP, Waldron L, Goswami RS et al (2011) mRNA transcript quantification in archival samples using multiplexed, color-coded probes. BMC Biotechnol 11:46
13. Li H, Qiu J, Fu XD (2012) RASL-seq for massively parallel and quantitative analysis of gene expression. Curr Protoc Mol Biol Chapter 4(Unit 4.13):11–19
14. Yeakley JM, Fan JB, Doucet D et al (2002) Profiling alternative splicing on fiber-optic arrays. Nat Biotechnol 20:353–358
15. Larman HB, Scott ER, Wogan M et al (2014) Sensitive, multiplex and direct quantification of RNA sequences using a modified RASL assay. Nucleic Acids Res 42:9146–9157
16. Qiu J, Zhou B, Thol F et al (2016) Distinct splicing signatures affect converged pathways in myelodysplastic syndrome patients carrying mutations in different splicing regulators. RNA 22:1535–1549
17. Scekic-Zahirovic J, Sendscheid O, El Oussini H et al (2016) Toxic gain of function from mutant FUS protein is crucial to trigger cell autonomous motor neuron loss. EMBO J 35:1077–1097
18. Shao C, Yang B, Wu T et al (2014) Mechanisms for U2AF to define 3′ splice sites and regulate alternative splicing in the human genome. Nat Struct Mol Biol 21:997–1005
19. Simon JM, Paranjape SR, Wolter JM et al (2019) High-throughput screening and classification of chemicals and their effects on neuronal gene expression using RASL-seq. Sci Rep 9:4529
20. Ying Y, Wang XJ, Vuong CK et al (2017) Splicing activation by rbfox requires self-aggregation through its tyrosine-rich domain. Cell 170:312–323.e310
21. Zhou Z, Qiu J, Liu W et al (2012) The Akt-SRPK-SR axis constitutes a major pathway in transducing EGF signaling to regulate alternative splicing in the nucleus. Mol Cell 47:422–433
22. Wang W, Barnaby JY, Tada Y et al (2011) Timing of plant immune responses by a central circadian regulator. Nature 470:110–114

Chapter 16

Rapid Validation of Transcriptional Enhancers Using a Transient Reporter Assay

Yuan Lin and Jiming Jiang

Abstract

Enhancers are one of the main classes of *cis*-regulatory elements (CREs) in the regulation of plant gene expression. Plant enhancers can be predicted based on genomic signatures associated with open chromatin. However, predicted enhancers need to be validated experimentally. We developed an experimental system for rapid enhancer validation. Predicted enhancer candidates are cloned into a vector containing a minimal 35S promoter and a luciferase reporter gene. The construct is then agroinfiltrated into *Nicotiana benthamiana* leaves followed by bioluminescence signal detection and analysis. Positive bioluminescence signals indicate the enhancer function of each candidate, and the relative signal strength from different enhancers can be quantitatively measured and compared. In summary, we have developed an efficient and rapid plant enhancer validation assay based on a bioluminescent luciferase reporter and agroinfiltration-based *N. benthamiana* leaf transient expression. This assay can be used for the initial screening of candidate enhancers that are active in leaf tissue. The system can potentially be used to examine the activity of candidate enhancers under different environmental conditions.

Key words Enhancer, Luciferase, Transient assay, Agroinfiltration, *Nicotiana benthamiana*

1 Introduction

Enhancers are one of the most important classes of *cis*-regulatory elements (CREs) of gene expression and play a key role in plant growth and development. The first reported enhancer, a 72-bp sequence derived from the SV40 virus, can cause a 200-fold increase in expression of a nearby gene [1]. Since then, enhancers have been widely found in all higher eukaryotes. Nevertheless, enhancers are difficult to identify due to their lack of positional constraints. Enhancers can be located a few kb to several megabases (Mb) away from their cognate genes. The "enhancer trapping" methodology was developed to capture functional enhancers in several plant species [2–6]. However, this method has been restricted by several major limitations [7]. It requires the production of a large number of transgenic lines, for example, 31,443

Shahid Mukhtar (ed.), *Modeling Transcriptional Regulation: Methods and Protocols*, Methods in Molecular Biology, vol. 2328, https://doi.org/10.1007/978-1-0716-1534-8_16, © Springer Science+Business Media, LLC, part of Springer Nature 2021

independent transgenic lines were developed for "enhancer trapping" in rice [2]. This has limited its application in most plant species.

Excitingly, several recent genome-wide studies showed that plant enhancers can be predicted based on their distinct features associated with open chromatin [8, 9]. The genomic regions associated with open chromatin are hypersensitive to DNase I (Deoxyribonuclease I) digestion and are known as DNase I hypersensitive sites (DHSs) [10, 11]. Nearly 70–80% DHSs located in intergenic regions in maize (*Zea mays*) and *Arabidopsis thaliana* showed enhancer function [8, 9]. Therefore, the development of DHS-based prediction methodology has opened a new venue for enhancer identification in plants.

The *Agrobacterium*-mediated transient assay provides a rapid and high-throughput method to transfer foreign DNA into plant cells and to survey reporter gene expression [12]. This transient assay has been successfully applied in many plant species [12–15]. Among various transient assays, *Nicotiana benthamiana* leaf-based agroinfiltration is by far the most widely used assay due to its simple operation and efficient transformation [16, 17]. In addition, the bioluminescent luciferase (LUC)-based reporter, which has high sensitivity and low background, can be implemented for high-throughput live imaging [18–20]. Here we report a rapid and efficient enhancer validation system based on the LUC reporter and *N. benthamiana*-based leaf agroinfiltration (Fig. 1) [21].

2 Materials

2.1 Vector Construction

1. Target enhancer sequences are synthesized individually, tagged with 5′-TGCA*CTGCAG* -*Pst*I and 3′-*ACTAGT* CC-*SpeI* digestion tag.
2. Vector used for enhancer activity detection: pCAMBIA1381Z-noskan-LUC [22].
3. Sanger sequencing primer pair: Forward primer 5′ CAGGAAA CAGCTATGAC 3′; reverse primer 5′ TCTCTTCATAGCCT TATGCAG 3′.
4. Digestion buffer: 1× CutSmart Buffer (NEB), 1 unit/μl *Pst*I-HF® (NEB, R3140), 1 unit/μl *Spe*I-HF® (NEB, R3133).

2.2 Agroinfiltration Preparation

1. Agroinfiltration buffer: 10 mM $MgCl_2$, pH 7.0, 200 μM acetosyringone.
2. Plant growth chamber condition: 150 $\mu mol/m^2$ s light, humidity 60%, fan speed 45%.
3. *Agrobacterium* culture: liquid Luria-Bertani (LB) broth medium (100 μg/ml gentamicin and kanamycin).

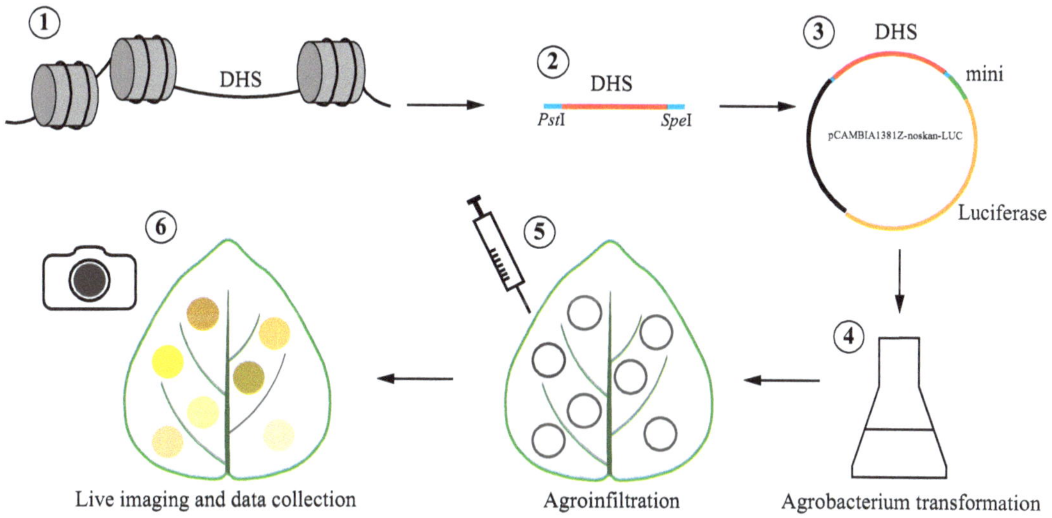

Fig. 1 Schematic workflow of the transient reporter assay. 1 Enhancer candidates are predicted based on open chromatin signatures such as DHSs. 2 Synthesis of the predicted enhancer sequence together with tags of restriction sites (blue bar). 3 The candidate enhancer is inserted into the pCAMBIA1381Z-noskan-LUC vector containing a downstream mini 35S promoter and the bioluminescent firefly luciferase reporter gene. 4 Transfer of the vector into *Agrobacterium* strain GV3101. 5 Agroinfiltrate multiple enhancer constructs into *N. benthamiana* leaves, including positive and negative controls. 6 Detect the bioluminescent live data using the NightSHADE LB 985 plant imaging system

3 Methods

3.1 Enhancer DNA Fragments Preparation

1. Analyze the published dicots leaf open chromatin DHS data, and select a few interested DHSs as enhancer candidates for vector construction (*see* **Note 1**) (Fig. 1).
2. Synthesize the predicted enhancer fragments individually, tagged with digestion tags.
3. Dissolve the synthesized enhancer DNA fragment in ddH_2O as 50 ng/μl.

3.2 Vector Construction and Transformation

1. Digest the synthesized enhancer DNA fragment and the pCAMBIA1381Z-noskan-LUC plasmid in digestion buffer. Then the enhancer DNA fragment is ligated into the lined pCAMBIA1381Z-noskan-LUC plasmid (*see* **Note 2**) (Fig. 1).
2. Send this plasmid for Sanger sequencing using sequencing primer pairs to confirm the insertion.
3. Transfer the enhancer-mini-LUC vector into *Agrobacterium* GV3101 for downstream functional validation (Fig. 1).

3.3 Plant Growth

1. Sow *Nicotiana benthamiana* seeds on PRO-MIX soil mix and cover lightly, grow the plants in the growth chamber at 26 °C under 12 h light/12 h dark cycle for 14 days before transplanting.
2. Transplant young plants into 3-in. pots, one plant per pot, and water on the first day, then fertilize twice a week (*see* **Note 3**).
3. When the plants reach to 6 extended leaves at around 20 days after transplanting, take the second extended leaf for further research (*see* **Note 4**).
4. Water the plants one day before agroinfiltration (*see* **Note 5**).

3.4 Agrobacterium-Mediated Leaf Transformation

1. Inoculate a single colony of *Agrobacterium*, containing enhancer-mini-LUC plasmid, in 15 ml tube with 5 ml freshly prepared LB culture. Culture overnight at 28 °C, 250 rpm.
2. Subculture 100 μl in 50 ml flask with 10 ml LB with the same antibiotics for around 12 h at 28 °C, 250 rpm.
3. Measure the OD_{600}, and harvest the agrobacteria when it reaches to 0.5–0.8 (*see* **Note 6**).
4. Centrifuge at 5000 × *g* for 10 min, resuspend the pellet in 10 mM $MgCl_2$.
5. Centrifuge again at 5000 × *g* for 10 min, discard the liquid.
6. Resuspend the pellet in agroinfiltration buffer, and adjust the OD_{600} to around 0.6 (*see* **Note 7**).
7. Leave at room temperature for 2 h before infiltration.
8. Transfer 100 μl (good for three times agroinfiltration) into a new tube, add 10 μl luciferin stock solution (10 mM) (*see* **Note 8**).
9. Poke leaf with a 27 G needle, one poking for each sample (*see* **Note 9**).
10. Use 1 ml luer-slip blunt end syringe to perform the infiltration from the underside of each leaf, cover the poking site with a syringe, and slightly press it with a finger on the other side. A spreading dark circled "wetting" area will be observed for successful agroinfiltration. Limit the diameter of each infiltration circle area to 1–1.5 cm (*see* **Note 10**) (Fig. 1).
11. Using lab wipes gently dry the infiltration spot to avoid cross-contamination, and mark the margin of each spot (*see* **Note 11**).

3.5 Photographing Photon-Counting Experiments and Data Normalization

The NightSHADE LB 985 (Berthold Technologies USA) in vivo plant imaging system was used to detect leaf bioluminescent signals (*see* **Note 12**) (Fig. 1).

3.5.1 For Enhancer Characterization

1. After agroinfiltration, keep the plants in darkness for 12 h.
2. For leaf data collection, bioluminescent signals are collected under 40 s scanning using the camera system. Data are analyzed by IndiGO™ software (*see* **Note 13**).

3.5.2 For Time-Lapse Tracking

1. Place plants into the chamber of the camera system after agroinfiltration, set the temperature to 27 °C (*see* **Notes 14** and **15**).
2. Leave it overnight (12 h) under dark before subjected to a 12 h light/12 h dark cycle, then collect the bioluminescent signals every 5 h. Water the plants every 2–3 days. Data are analyzed by IndiGO™ software.

4 Notes

1. This method is a preferred platform for leaf-specific enhancers in dicot plant species [21]. Here we choose the *Arabidopsis* leaf DHS data as an example.
2. The original pCAMBIA1381Z-LUC vector [22] is modified following two steps: First, place the minimal 35S promoter at upstream of the luciferase gene. Second, the gene conferring Hygromycin resistance (*Hph*) (including the original 35S promoter and terminator) is replaced by a reversed Nopaline synthase (NOS) promoter-Kanamycin-NOS terminator. This replacement is aimed to avoid false-positive signals or strong background produced by the bidirectional 35S enhancer from the close by *Hph* gene [21].
3. Aerate the soil every week to prevent soil compaction or mossy surface, always keep the soil semi-wet and neither overwet nor dry.
4. The selected leaves should be uniform bright green, fully extended and thinner than old leaves, the plant should not start flowing or become yellowish.
5. Leaves at the same developmental stage, from uniformly grown plants, and under standard light regimes, should be used to maximize reproducibility.
6. Make sure there is no floating or precipitating floccus in the bacteria culture. Restart if any floccus present.
7. It is necessary to adjust all samples including controls to the same concentration.
8. Luciferin is light-sensitive, make sure the tube is placed in the darkness.
9. Select the middle region from each leaf side, and be away from leaf vein.

10. A young leaf is always not easy to infiltrate, try to press gently to avoid breaking the young leave tissue. Each leaf contains at least two negative controls (mini empty vector and no DHS-mini vector) [21] and 35S positive control.
11. Infiltrate the 35S as a positive control at a fixed position from each leaf for future normalization.
12. Up to four live plants after agroinfiltration can be placed inside the chamber of the NightSHADE instrument at the same time.
13. The 35S signal usually is extraordinarily strong compared to most enhancers, we recommend taking one extra picture by covering the 35S with a small piece of black paper [21].
14. Plant transpiration increases the chamber humidity, especially under the daytime lighting period. We set the chamber temperature to 27 °C, which is higher than the room temperature 25 °C. This can help reduce the chamber condensation caused by the plant. Absorbent tower paper or desiccant could also be used if condensation is found in the chamber.
15. The NightSHADE LB 985 does not have an autofocus lens, and lighting will regulate leaf vertical movements, which will cause mis-focusing along with weakened bioluminescent signal. Here, we simply use a twisted paper clip to stabilize leaf vertical movements.

Acknowledgments

We thank Guilherme Braz for the discussion and manuscript editing and Dr. Huazhong Shi for providing the pCAMBIA1381Z-LUC vector. This work was supported by National Science Foundation grant MCB-1412948 to J.J.

References

1. Banerji J, Rusconi S, Schaffner W (1981) Expression of a β-globin gene is enhanced by remote SV40 DNA sequences. Cell 27:299–308
2. Wu C, Li X, Yuan W et al (2003) Development of enhancer trap lines for functional analysis of the rice genome. Plant J 35:418–427
3. Weber B, Zicola J, Oka R et al (2016) Plant enhancers: a call for discovery. Trends Plant Sci 11:974–987
4. Sundaresan V, Springer P, Volpe T et al (1995) Patterns of gene action in plant development revealed by enhancer trap and gene trap transposable elements. Genes Dev 9:1797–1810
5. Groover A, Fontana JR, Dupper G et al (2004) Gene and enhancer trap tagging of vascular-expressed genes in poplar trees. Plant Physiol 134:1742–1751
6. Pérez-Martín F, Fernando JY, Benito P et al (2017) A collection of enhancer trap insertional mutants for functional genomics in tomato. Plant Biotechnol J 11:1439–1452
7. Marand AP, Zhang T, Zhu B et al (2017) Towards genome-wide prediction and characterization of enhancers in plants. Biochim Biophys Acta Gene Regul Mech 1:131–139
8. Zhao H, Zhang W, Chen L et al (2018) Proliferation of regulatory DNA elements derived from transposable elements in the maize genome. Plant Physiol 176:2789–2803

9. Zhang W, Wu Y, Schnable JC et al (2012) High-resolution mapping of open chromatin in the rice genome. Genome Res 22:151–162
10. Zhu B, Zhang W, Zhang T et al (2015) Genome-wide prediction and validation of intergenic enhancers in Arabidopsis using open chromatin signatures. Plant Cell 27:2415–2426
11. Zhang W, Zhang T, Wu Y et al (2012) Genome-wide identification of regulatory DNA elements and protein-binding footprints using signatures of open chromatin in Arabidopsis. Plant Cell 24:2719–2731
12. Kapila J, De Rycke R, Van M et al (1997) An Agrobacterium-mediated transient gene expression system for intact leaves. Plant Sci 122:101–108
13. Van DH, R A L, Laurent F et al (2000) Agroinfiltration is a versatile tool that facilitates comparative analyses of Avr9/Cf-9-induced and Avr4/Cf-4-induced necrosis. Mol Plant-Microbe Interact 13:439–446
14. Wroblewski T, Tomczak A, Michelmore R (2005) Optimization of Agrobacterium-mediated transient assays of gene expression in lettuce, tomato and Arabidopsis. Plant Biotechnol J 3:259–273
15. Bhaskar PB, Venkateshwaran M, Wu L et al (2009) Agrobacterium-mediated transient gene expression and silencing: a rapid tool for functional gene assay in potato. PLoS One 4: e5812
16. Gerasymenko IM, Sheludko YV (2017) Synthetic cold-inducible promoter enhances recombinant protein accumulation during Agrobacterium-mediated transient expression in Nicotiana excelsior at chilling temperatures. Biotechnol Lett 39:1059–1067
17. Banu SA, Huda K, Tuteja N (2014) Isolation and functional characterization of the promoter of a DEAD-box helicase Psp68 using Agrobacterium-mediated transient assay. Plant Signal Behav 9:e28992
18. Thorne N, Inglese J, Auld DS (2010) Illuminating insights into firefly luciferase and other bioluminescent reporters used in chemical biology. Cell Press 17:646–657
19. Xie Q, Soutto M, Xu X et al (2011) Bioluminescence resonance energy transfer (BRET) imaging in plant seedlings and mammalian cells. Methods Mol Biol 680:3–28
20. Van Leeuwen W, Hagendoorn M, Tom R et al (2000) The use of the luciferase reporter system for in planta gene expression studies. Plant Mol Biol Report 18:143–144
21. Lin Y, Meng F, Fang C et al (2019) Rapid validation of transcriptional enhancers using agrobacterium-mediated transient assay. Plant Methods. https://doi.org/10.1186/s13007-019-0407-y
22. Jiang J, Wang B, Shen Y et al (2013) The Arabidopsis RNA binding protein with K homology motifs, SHINY1, interacts with the C-terminal domain phosphatase-like 1 (CPL1) to repress stress-inducible gene expression. PLoS Genet 7:e1003625

Chapter 17

Computational Identification of ceRNA and Reconstruction of ceRNA Regulatory Network Based on RNA-seq and Small RNA-seq Data in Plants

Xiangyuan Wan and Ziwen Li

Abstract

Competing endogenous RNAs (ceRNAs) are transcripts with the ability to competitively titrate microRNAs (miRNAs) against miRNA repressing target genes to post-transcriptionally regulate the expression of corresponding miRNAs. It is a newly discovered gene regulation pattern between longer RNA and miRNA molecules. Recent research has gradually revealed the functional significance of ceRNAs in regulating normal development and stress response processes in plants and animals, as well as in cancer genesis and metastasis. Therefore, ceRNA identification is an important and necessary step to deepen our understanding of the regulation mechanisms of various biological processes. Here, we provide a pipeline used to computationally identify plant ceRNAs and reconstruct ceRNA regulatory networks based on RNA-seq and small RNA-seq data.

Key words ceRNA, miRNA, Gene regulatory network, RNA-seq, Small RNA-seq

1 Introduction

MiRNA was firstly discovered by Lee et al. (1993) in *Caenorhabditis elegans* [1]. They are short, single-stranded and regulatory RNA molecules to post-transcriptionally and/or translationally control the expressions of target genes in both animals and plants. The amount of researches on the roles of miRNAs in normal development and stress response processes has grown vigorously in recent years. At the transcriptional level, the expression of miRNA is mainly controlled by *trans*-acting factors represented by transcription factors (TFs), or *cis*-regulatory elements in promoter regions of miRNAs. However, our understanding of the regulation mechanism of miRNA expression is relatively less known at the post-transcriptional regulation level. This prompts us to think about how the simple regulatory strategy (only transcriptional regulation) could support diverse molecular functions of miRNAs.

Shahid Mukhtar (ed.), *Modeling Transcriptional Regulation: Methods and Protocols*, Methods in Molecular Biology, vol. 2328, https://doi.org/10.1007/978-1-0716-1534-8_17,

Pandolfi et al. (2011) proposed a post-transcriptional regulation model controlling the miRNA expression, namely, the "ceRNA hypothesis" [2]. This hypothesis suggests that there exist some endogenously produced RNAs, which directly and competitively absorb miRNAs to form "ceRNA-miRNA" regulatory relationships in the organism. The absorbing of miRNA can lead to a reduced transcription level of the corresponding miRNA, and finally release or reduce the expression inhibition on the target gene of the sponged miRNA [2]. The first ceRNA in plants is discovered by Franco-Zorrilla et al. (2007) in the phosphorus metabolism of *Arabidopsis thaliana*. They found a long non-coding RNA (lncRNA), *INDUCED BY PHOSPHATE STAR VATION1* (*IPS1*), which can competitively bind *ath-miR399* and release the inhibitory effect on its target gene, *Phosphate Overaccumulator 2* (*PHO2*). Although *IPS1* and *ath-miR399* have the paired sequences similar to those between *ath-miR399* and *PHO2*, a protruding loop is formed at the cleavage site due to the base mismatches between *IPS1* and *ath-miR399*. Overexpression of *IPS1* leads to the enriched transcripts of *PHO2*, resulting in the decreased phosphorus content in shoot tips. The mutation of the mismatched base in *IPS1* reproduces the inhibition effect of *ath-miR399* on *PHO2* translation. Another typical ceRNA in plants is *MIKKI*, a retrotransposon-originated transcript in rice. *MIKKI* can decoy *osa-miR171* to de-repress the expression inhibition on SCARECROW-Like TF genes that function in rice root development [3]. Mutations in the binding sites of *osa-miR171* on *MIKKI* lead to developmental abnormality in root growth. Obviously, ceRNAs is a type of RNA molecules that could directly and negatively regulate the expressions of captured miRNAs, and further indirectly control the expressions of miRNA target genes.

Recent studies on ceRNA have revealed that transcripts originated from protein-coding genes [4], pseudogenes [5], transposable elements (TEs) [3], and simple sequence repeats (SSRs) [6], as well as circular RNAs (circRNAs) [7] can function as ceRNAs to regulate the expressions of sponged miRNAs. Therefore, the concept of ceRNA is mostly like a functional description of investigated RNA, rather than a special subclass in the RNA classification system (e.g., coding and non-coding RNAs, or linear and circular RNAs). More importantly, these results documented that almost all kinds of relatively longer transcripts in a transcriptome may have the ability to function as ceRNAs. This leads to the first key point in ceRNA identification that all transcripts in the investigated transcriptomes should be scanned. As official genome annotation files (GFF or GTF files) in most of genome-available species mainly contain coding genes, other type of transcripts (e.g., pseudogenes and circRNAs) should be identified according to professional identification processes before detecting ceRNA.

A systematical detection of miRNA should be also prior to ceRNA identification. As the regulated targets of ceRNAs, the identification and quantification of miRNAs are also important for ceRNA detection. Though the total number of annotated miRNA models in miRBase have reached to 38,589 in the recent version (Release 22.1, as of Oct. 2018), the amounts of annotated miRNAs vary in different species (e.g., 428 miRNAs in *A. thaliana* and 11 miRNAs in *Brassica oleracea* that also belong to Brassicaceae). Unlike genome data that are consistent between cells in different tissues, transcriptome represents transcriptional information of genes in a special tissue during a special time period. Thus, transcriptome data are different from each other for samples of different tissues at different developmental stages. The complexity and diversity may be the main reason to explain that miRNA annotation of investigated species is far from complete, even in the model species. Thus, to some extent, the prediction of novel miRNA is a necessary step in ceRNA identification for both relatively well-annotated genomes and newly sequenced species.

In addition, ceRNA identification mainly depends on two types of information, including miRNA recognition element (MRE) detection and expression correlation. Unlike miRNAs displaying stem-loop structures for their pre-miRNA sequences, there is no common primary or secondary structural feature for ceRNAs. So far, the mostly used characteristic to recognize ceRNAs by computational analysis at the sequence level is MREs localized in ceRNA sequences. On the other hand, the expression correlations between RNAs, as well as between RNAs and miRNAs, are also frequently used to infer ceRNAs. Furthermore, the sequence conservation in transcribed loci across species provides supporting information for identifying functional RNAs (including ceRNAs) at the evolutionary level.

Here, we systematically describe a pipeline (including required programs, parameters, databases, and commands) in ceRNA identification based on high-throughput sequencing data in plants.

2 Materials

2.1 Software and Programs

1. NGSQC (Next Generation Sequencing Quality Control) Toolkit (processing RNA-seq raw data) [8].

   Home page: http://www.nipgr.ac.in/ngsqctoolkit.html.

2. TopHat2 (mapping high-throughput sequencing reads to the reference genome) [9].

   Download source: http://ccb.jhu.edu/software/tophat/downloads.

   Manual page: http://ccb.jhu.edu/software/tophat/manual.shtml.

3. SAMtools (SAM means Sequence Alignment/Map format. This package can be used to process aligned files) [10].
   Download source: https://github.com/samtools/samtools/releases/download/1.10/samtools-1.10.tar.bz2.
   Manual page: http://www.htslib.org/doc/samtools.html.
4. Cufflinks (detecting transcribed loci) [11].
   Download source: https://github.com/cole-trapnell-lab/cufflinks.
   Manual page: http://cole-trapnell-lab.github.io/cufflinks/manual.
5. BLAST (checking whether detected transcripts are intact or not) [12].
   Download source: https://ftp.ncbi.nlm.nih.gov/blast/executables/blast+.
6. Pseudopipe (detecting pseudogene loci in the reference genome sequence) [13].
   Download source: http://pseudogene.org/DOWNLOADS/pipeline_codes/ppipe.tar.gz.
7. BEDTools (merging annotated gene regions from different sources).
   Download source: https://github.com/arq5x/bedtools2/releases.
   Manual page: https://bedtools.readthedocs.io/en/latest/content/bedtools-suite.html.
8. featureCounts (counting mapped reads for each transcript model based on GFF/GTF file) [14]. This program has an R application in Rsubread [15, 16].
   Home Page: http://subread.sourceforge.net.
9. edgeR (estimating expression levels for long transcripts and miRNAs) [17].
   Home Page: http://www.bioconductor.org/packages/release/bioc/html/edgeR.html.
10. Cutadapt (processing small RNA-seq raw data) [18].
    Download source: https://github.com/marcelm/cutadapt.
    Manual page: https://cutadapt.readthedocs.io/en/stable.
11. Infernal (filtering out small RNA-seq reads from small RNA loci other than miRNA) [19].
    Download source: https://github.com/EddyRivasLab/infernal.
    Manual page: http://eddylab.org/infernal/Userguide.pdf.
12. Bowtie (mapping small RNA-seq reads to the reference genome) [20].

Download source: https://github.com/BenLangmead/bowtie.

Manual page: http://bowtie-bio.sourceforge.net/manual.shtml.

13. miRDeep2 (detecting novel miRNA candidates) [21].

    Download source: https://github.com/rajewsky-lab/mirdeep2/releases.

    Manual page: https://www.mdc-berlin.de/content/mirdeep2-documentation.

14. psRobot (predicting miRNA target genes) [22].

    Home page: http://omicslab.genetics.ac.cn/psRobot.

15. TAPIR (target prediction for plant microRNAs and ceRNAs) [23].

    Download source: http://bioinformatics.psb.ugent.be/webtools/tapir/tapir-1.2.tar.gz.

16. R (supporting the running environment for bioconductor packages and performing statistic computations).

    Home page: https://www.r-project.org.

17. Perl (mainly support the running of miRDeep2).

    Home page: https://www.perl.org.

18. Cytoscape (visualizing the reconstructed network) [24].

    Download source: https://cytoscape.org/download.html.

    Manual page: http://manual.cytoscape.org/en/stable.

2.2 Databases and Online Services

1. miRBase (this database is a collection of published miRNA sequences; http://www.mirbase.org) [25].
2. Rfam database (this database is a collection of RNA families; http://rfam.xfam.org) [26].
3. TAPIR web services (the web services of TAPIR program; http://bioinformatics.psb.ugent.be/webtools/tapir) [23].
4. psRobot web services (the web services of psRobot program; http://omicslab.genetics.ac.cn/psRobot/) [22].
5. ViennaRNA web services (this web services can be used to analyze RNA secondary structures; http://rna.tbi.univie.ac.at) [27].

2.3 Input Files

1. The RNA-seq and small RNA-seq data generated in your lab or published previously.
2. The reference genome sequence of investigated species.

   It can be downloaded from National Center for Biotechnology Information (NCBI) genome database (https://ftp.ncbi.nlm.nih.gov/genomes), Ensembl Genomes (http://ensemblgenomes.org), or UCSC (University of California Santa Cruz) Genome Browser (http://hgdownload.soe.ucsc.edu/downloads.html). Specifically, Ensembl Plants (http://

plants.ensembl.org/index.html) and Phytozome (https://phytozome.jgi.doe.gov/pz/portal.html) are plant genome databases.

3. The genome annotation file(s) of investigated species (GFF or GTF format).
4. The non-coding RNA annotation information.
   Download "Rfam.cm.gz" and "Rfam.clanin" from Rfam database (ftp://ftp.ebi.ac.uk/pub/databases/Rfam) for small RNA annotation.
5. The annotated miRNA loci.
   Download "zma.gff3" from miRBase for maize or the file for the investigated species included in the database (ftp://mirbase.org/pub/mirbase/CURRENT/genomes).
6. The mature and precursor sequences of annotated miRNA.
   Download "mature.fa.gz" and "hairpin.fa.gz" from miRBase (ftp://mirbase.org/pub/mirbase/CURRENT).

3 Methods

In previous research, we predicted hundreds of ceRNAs and tens of potential novel miRNAs in maize anther transcriptomes according to the methods including the transcript identification based on RNA-seq data, the known and novel miRNA identification from small RNA-seq data, the ceRNA and target gene prediction, the estimation of expression correlation, and the reconstruction of ceRNA regulatory networks [28, 29]. The pipeline is shown in Fig. 1.

3.1 Identification of Transcribed Loci (See Note 1)

1. Filtering out low-quality reads (the default cut-off quality score is 20) from RNA-seq data using NGSQCToolkit program. If paired-end sequencing is performed, NGSQCToolkit can remove unpaired reads.
   (a) *IlluQC_PRLL.pl -c 8 -o ./output/sample/illuqc -pe ./input/sample.1.fq ./input/sample.2.fq A*
   (b) "-c" defined the number of threads, "-o" defined the output directory, "A" means the format of the input file is automatically detected by the program. Two output files of clean reads ("sample.1.fq_filtered" and "sample.2.fq_filtered") are generated in the assigned output directory.
2. Mapping read pairs to the reference genome using TopHat2.
   (a) *tophat2 -p 8 -m 2 -o . /output/sample/tophat2 reference_genome.fa ./output/sample/illuqc/sample.1.fq_filtered.gz ./output/sample/illuqc/sample.2.fq_filtered.gz*

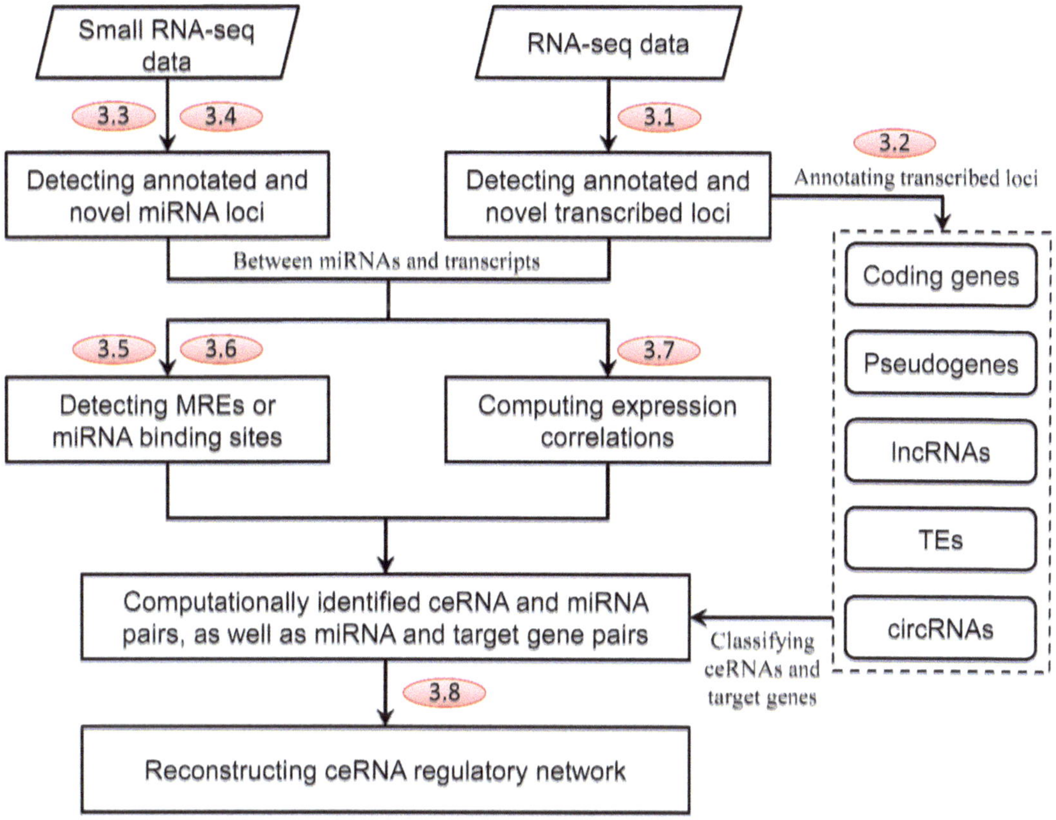

Fig. 1 A flowchart of ceRNA identification and ceRNA regulatory network reconstruction. Numbers in the red circles indicate detection steps in the pipeline

(b) "-p" defined the number of threads, "-o" defined the output directory. The FASTQ files of clean reads can be compressed as input. Output files include the aligned result ("accepted_hits.bam") and the mapping rate ("align_summary.txt").

3. Processing the aligned result using SAMtools before executing Cufflinks.
   (a) *samtools index ./output/sample/tophat2/accepted_hits.bam*
   (b) An output file of "accepted_hits.bam.bai" is generated.
4. Identifying transcribed loci using Cufflinks from aligned files.
   (a) *cufflinks -p 8 -o ./output/sample/cufflinks ./output/sample/tophat2/accepted_hits.bam*
   (b) "-p" defined the number of threads, "-o" defined the output directory. An output file of "transcripts.gtf" is generated in the assigned directory.

5. Combining transcribed loci detected from transcriptomes at different developmental stages or tissues into an integrated dataset using the cuffmerge program in Cufflikes package without genome annotation information.
   (a) *find ./output/sample* -name transcripts.gtf > transcripts_gtf_list.txt;*
   cuffmerge -p 8 -o merged.gtf transcripts_gtf_list.txt
   (b) "transcripts_gtf_list.txt" contains the absolute paths of "transcripts.gtf" files for different RNA-seq libraries (one path per line). "-p" defined the number of threads, "-o" defined the output filename. "merged.gtf" will be generated in the output directory and this file contains the identified transcribed loci among all transcriptomes from aligned samples listed in "transcripts_gtf_list.txt".
6. Calculating expression levels of detected transcribed loci in each sample.
   (a) *featureCounts -T 8 -t exon -g gene_id -a merged.gtf -o /out/sample/RNA_featureCounts.txt /output/sample/tophat2/accepted_hits.bam*
   (b) "-T" defined the number of threads, "-t" defined the minimal feature region for read count, "-g" defined the meta-feature region for read count.
   (c) Transforming the read count to the RPKM (reads per kilobase per million mapped reads) value for each transcribed locus by edgeR in R environment (*see* **Note 2**).
   (d) *featureCounts<-read.delim("/out/sample/RNA_featureCounts.txt",header=T, stringsAsFactor=F)RPKM_result<-rpkm(featureCounts, gene.length=featureCounts$Length, normalized.lib.sizes=T, log=FALSE, prior.count=0.25)*

3.2 Annotation of Transcribed Loci (See Note 3)

1. Protein-coding genes.
   The annotation information of protein-coding genes can be obtained from the official gene sets in genome-sequenced species.
2. LncRNAs.
   The information of lncRNA loci can be extracted from genome annotation results of some well-studied species or can be collected from published studies on lncRNAs.
3. Pseudogenes.
   Pseudogene loci in the genome sequence can be annotated by Pseudopipe with gene models of protein-coding genes as the "parent" sequences.
4. TEs.
   Identification and annotation of TEs in genome sequences is a specialized and systematic work that is difficult to be

completed by executing a single program in a simple step. However, information on TEs can be acquired from genome annotation results for most of the genome-available species.

5. CircRNAs.

   For the identification of circRNAs, please refer to a published review [30].

6. The annotation information of the above types of transcribed loci can be merged into an integrated database using BEDTools.

3.3 Investigation of Annotated miRNA Loci

1. Processing small RNA-seq data by the cutadapt program to remove adaptor sequences and filter out reads with poly (A) tracts or ambiguous bases and finally retain reads with lengths between 18 nt and 25 nt.
   (a) *cutadapt -a adaptor_sequence_1 -a adaptor_sequence_2 -a* AAAAAAAAAA *--max-n 0 --minimum-length 18 --maximum-length 25 -o ./output/sample/cutadapt/srna.clean.fa ./input/srna.raw.fa*
   (b) "-a" defined the adaptors or other sequences need to be removed from the short reads. "adaptor_sequence_1"and "adaptor_sequence_2" represent two adaptor sequences. "AAAAAAAAAA" defined the length of poly(A) sequence need to be truncated.
2. Annotating the processed reads by the cmscan program in Infernal package and discarding reads possessing high sequence similarities with rRNAs (ribosomal RNAs), tRNAs (transfer RNAs), snRNAs (small nuclear RNAs), and snoRNAs (small nucleolar RNAs).
   (a) *cmscan -E 50 --cpu 1 --rfam --nohmmonly --tblout sample.cmscan.tbl --fmt 2 --clanin Rfam.14.0.clanin Rfam.14.0.cm ./output/sample/cutadapt/srna.clean.fa*
   (b) Extracting IDs of reads annotated as rRNAs, tRNAs, snRNAs, and snoRNAs from "sample.cmscan.tbl" to generate a file of "sample.non-miRNA.list".
   (c) Remove short reads from "srna.clean.fa" based on "sample.non-miRNA.list" to generate a new file named "mirna.clean.fa", which only includes reads from miRNA loci.
3. Mapping reads considered from miRNA loci to the reference genome by Bowtie.
   (a) *bowtie -f -l 25 -n 0 -k 1 -S -p 8 reference_genome.fa ./output/sample/mirna.clean.fa > ./output/sample/bowtie/mirna.sam*
   (b) "-f" means the input file is in FASTA format, "-l" defined the configured seed length, "-n" defined the largest

number of mismatch bases, "-k" defined the number of reported alignment(s), "-S" defined that the output file is in SAM format, "-p" defined the number of threads.

4. Estimating the expression level of annotated miRNA by featureCounts and the CPM (read counts per million mapped reads) value.
   (a) *featureCounts -T 8 -t exon -g gene_id -a annotated_miRNA.gtf -o /out/sample/miRNA_featureCounts.txt ./output/sample/bowtie/mirna.sam*
   (b) Convert read counts to CPM values by *cpm(miRNA_featureCount_matrix)* in edgeR package under R environment (*see* **Note 2**).

3.4 Identification of Novel miRNA Loci (See Note 4)

1. Preparing small RNA sequencing reads considered from miRNA loci to ensure repetitive reads only appear once in the input file. "collapse_reads_md.pl" is a Perl script in miRDeep2. The input file used in this step is the output file of Subheading 3.3, **step 2**.
   (a) *collapse_reads_md.pl ./output/sample/cmscan/mirna.clean.fa zmay >./output/sample/mirdeep2/mirna.collapsed.fa*
   (b) "zmay" defined the species symbol.
2. Identifying novel miRNAs by miRDeep2.
   (a) *mapper.pl ./output/sample/mirdeep2/mirna.collapsed.fa -c -j -p reference_genome.fa -s ./output/sample/mirdeep2/read.fa -t ./output/sample/mirdeep2/result.arf*
   (b) "-c" means the input file is in FASTA format, "-j" means deleting irregular base(s) (standard letters include a, c, g, t, u, n, A, C, G, T, U, N), "-p" defined the genome template, "-s" defined the output file name of processed reads, "-t" design the output file name of mapping result.
   (c) Preparing mature and precursor sequences of annotated miRNA loci for miRDeep2 to predict novel miRNA loci. "mature_ref.fa" and "precursors_ref.fa" are the mature and precursor sequences of annotated maize miRNA loci, respectively. "mature_other.fa" is the mature sequences of annotated miRNA loci in plant species closely related to maize. The above three FASTA files can be generated by extracting corresponding sequences from "mature.fa.gz" and "hairpin.fa.gz" downloaded from the miRBase database.
   (d) *miRDeep2.pl ./output/sample/mirdeep2/read.fa reference_genome.fa ./output/sample/mirdeep2/result.arf ./input/mature_ref.fa ./input/mature_other.fa ./input / precursors_ref.fa*

(e) Hits with *P* values <0.05 or other threshold and scores >5 or other threshold can be considered as novel miRNA candidates. The threshold values can be modified according to your need.

3.5 Identification of miRNA Target Gene

1. Combining mature sequences of annotated and predicted novel miRNAs into a single FASTA file "sample.mature.fa".
2. Extracting genome sequences of all types of detected transcribed loci into a single FASTA file "sample.transcript.fa".
3. Predicting miRNA targets by psRobot (*see* **Note 5**). Hits with penalty scores ≤2.5 or another threshold value and without any mismatched base pair in the 2nd to 7th nucleotides of mature miRNA can be retained.
   (a) *psRobot_tar -s sample.mature.fa -t sample.transcript.fa -o target.psrobot.txt -p 8*
   (b) To improve the computation speed, "sample.mature.fa" and "sample.transcript.fa" can be divided into several components for parallel calculations.
4. Predicting miRNA targets by TAPIR under the RNAhybrid engine. Hits fulfilling the following three criteria are retained: scores ≤4, ratios of minimum free energy (MFE) ≥ 0.7, and no mismatched base pair in the 2nd to 7th nucleotides of mature miRNAs. The threshold values can be modified according to your need.
   (a) *tapir_hybrid sample.mature.fa sample.transcript.fa | hybrid_parser > target.tapir.txt*
   (b) Similarly, we can divide the "sample.mature.fa" and "sample.transcript.fa" files and use the parallel computation strategy.
5. Combining detected targets from psRobot and TAPIR.

3.6 Identification of ceRNA

1. Identifying ceRNA candidates by TAPIR under the target decoy search method (*see* **Note 5**).
   (a) *tapir_hybrid sample.miRNA.fa sample.transcript.fa | hybrid_parser --mimic 1 > ceRNA.txt*
   (b) Hits with following three characteristics are considered as ceRNAs: MFE ratios ≥0.7 or another threshold value, no more than one mismatched base pair in the 2nd to 7th nucleotides of mature miRNAs, and a mismatch loop (3 nt) behind the 10th complementary base pair in the mature miRNA for each ceRNA-miRNA candidate pair.
   (c) Similarly, we can divide the "sample.mature.fa" and "sample.transcript.fa" files and use the parallel computation strategy.

3.7 Expression Correlation Among ceRNA, miRNA, and Target Genes

As ceRNAs are capable of increasing the expressions of miRNA target genes by decoying corresponding miRNAs, information of the expression correlation between ceRNA and miRNA or between ceRNA and its regulated RNA could provide additional evidence supporting a predicted ceRNA, while this method can be effectively used in studies with RNA-seq and small RNA-seq data from samples at several developmental stages or under different treatment conditions. Specifically, expression correlation can be computed by three strategies that are described below.

1. Negative correlations between ceRNA and miRNA expressions.

   Expression correlations between ceRNAs and miRNAs can be directly estimated by the Pearson correlation coefficients ($r_{ceRNA\ -\ miRNA}$) in R using the command of *cor.test(expressions_ceRNA, expressions_miRNA)*. A significant negative $r_{ceRNA\ -\ miRNA}$ is an indicator of a potential ceRNA.

$$r_{ceRNA-miRNA} = \frac{\sum\left(E_{ceRNA} - \overline{E}_{ceRNA}\right)\left(E_{miRNA} - \overline{E}_{miRNA}\right)}{n \times \sigma_{ceRNA}\sigma_{miRNA}}$$

2. Co-expressions between ceRNAs and other RNAs.

   Gene co-expression analysis can be also performed by direct estimation of the Pearson correlation coefficient between two RNAs ($r_{ceRNA\ -\ regulated\ RNA}$) in R, while it can be also executed by an R package named WGCNA (Weighted Gene Co-expression Network Analysis; http://www.genetics.ucla.edu/labs/horvath/CoexpressionNetwork/Rpackages/WGCNA) [31] to detect several groups of co-expressed RNAs. Obviously positive relationships between ceRNAs and other RNAs could increase the credibility of computationally identified ceRNAs.

$$r_{ceRNA-regulated\ RNA} = \frac{\sum\left(E_{ceRNA} - \overline{E}_{ceRNA}\right)\left(E_{regulated\ RNA} - \overline{E}_{regulated\ RNA}\right)}{n \times \sigma_{ceRNA}\sigma_{regulated\ RNA}}$$

3. "Sensitivity correlation" of a ceRNA and its regulated RNA.

   A significant co-expression pattern between a ceRNA and the regulated RNA candidate may result from the other RNA that can regulate the expressions of both the predicted ceRNA and the regulated RNA candidate. In this situation, the predicted regulatory function of the ceRNA on the regulated RNA is very fragile. To exclude expression influence from transcripts other than the corresponding miRNAs for the predicted ceRNA and the target genes, estimating a sensitivity correlation ($S_{ceRNA\ -\ regulated\ RNA}$) based on the partial correlation coefficient between expressions of the ceRNA and the regulated candidate ($r_{ceRNA\ -\ regulated\ RNA(miRNA)}$) has been documented as an effective method for ceRNA identification [32].

$$r_{ceRNA-regulated\ RNA(miRNA)} = \frac{r_{ceRNA-regulated\ RNA} - r_{ceRNA-miRNA} r_{miRNA-regulated\ RNA}}{\sqrt{1 - r^2_{ceRNA-miRNA}}\sqrt{1 - r^2_{miRNA-regulated\ RNA}}}$$

$$S_{ceRNA-regulated\ RNA} = r_{ceRNA-regulated\ RNA} - r_{ceRNA-regulated\ RNA(miRNA)}$$

3.8 Reconstruction and Display ceRNA Regulatory Network

1. Reconstructing ceRNA regulatory networks by combining ceRNA-miRNA pairs and miRNA-target gene pairs.
2. Visualizing the ceRNA regulatory network in Cytoscape.

4 Notes

1. If gene annotation results (GFF/GTF files) downloaded from the public database are sufficient to meet the requirement of the planned research, the identification of transcribed loci (Subheading 3.1) becomes an optional step in ceRNA detection.
2. The expression levels of long transcripts and miRNAs are estimated in two different ways. In detail, expression levels of long transcripts are estimated by RPKM or FPKM (fragments per kilobase per million mapped reads) values, while miRNA expression levels are calculated by CPM values.
3. The annotation of transcripts (Subheading 3.2) is not a necessary step in ceRNA detection or it can be performed after ceRNA identification (Subheadings 3.5–3.7) only based on transcripts considered as ceRNA candidates.
4. When a large number of annotated miRNA loci have been published and accumulated in the public database in investigated species, the identification of novel miRNA (Subheading 3.4) can be not executed. However, for species with limited numbers of annotated miRNAs or for rarely studied tissues, we suggest that the detection of potential novel miRNAs using small RNA-seq data or based on genome sequences is a required step.
5. The method used to identify MREs or miRNA binding sites in animals has some differences from that described herein for plants. However, the workflow described above can be used to identify ceRNAs and reconstruct ceRNA regulatory networks in animals.

Acknowledgments

The National Natural Science Foundation of China (31771875), The National Key Research and Development Program of China (2018YFD0100806, 2017YFD0102001, 2017YFD0101201), the

Fundamental Research Funds for the Central Universities of China (06500136, 2302019FRF-TP-19-013A1), and the Beijing Science & Technology Plan Program (Z191100004019005) supported this work.

References

1. Lee RC, Feinbaum RL, Ambros V (1993) The *C. elegans* heterochronic gene lin-4 encodes small RNAs with antisense complementarity to lin-14. Cell 75:843–854
2. Salmena L, Poliseno L, Tay Y, Kats L, Pandolfi PP (2011) A ceRNA hypothesis: the Rosetta stone of a hidden RNA language? Cell 146:353–358
3. Cho J, Paszkowski J (2017) Regulation of rice root development by a retrotransposon acting as a microRNA sponge. Elife 6:30038
4. Tay Y, Kats L, Salmena L, Weiss D, Tan SM, Ala U et al (2011) Coding-independent regulation of the tumor suppressor PTEN by competing endogenous mRNAs. Cell 147:344–357
5. Poliseno L, Salmena L, Zhang J, Carver B, Haveman WJ, Pandolfi PP (2010) A coding-independent function of gene and pseudogene mRNAs regulates tumour biology. Nature 465:1033–1038
6. Witkos TM, Krzyzosiak WJ, Fiszer A, Koscianska E (2018) A potential role of extended simple sequence repeats in competing endogenous RNA crosstalk. RNA Biol 15:1399–1409
7. Zheng Q, Bao C, Guo W, Li S, Chen J, Chen B et al (2016) Circular RNA profiling reveals an abundant circHIPK3 that regulates cell growth by sponging multiple miRNAs. Nat Commun 7:11215
8. Patel RK, Jain M (2012) NGS QC Toolkit: A toolkit for quality control of next generation sequencing data. PLoS One 7:e30619
9. Trapnell C, Pachter L, Salzberg SL (2009) TopHat: discovering splice junctions with RNA-Seq. Bioinformatics 25:1105–1111
10. Li H, Handsaker B, Wysoker A, Fennell T, Ruan J, Homer N et al (2009) Genome project data processing S: the sequence alignment/map format and SAMtools. Bioinformatics 25:2078–2079
11. Trapnell C, Williams BA, Pertea G, Mortazavi A, Kwan G, van Baren MJ et al (2010) Transcript assembly and quantification by RNA-Seq reveals unannotated transcripts and isoform switching during cell differentiation. Nat Biotechnol 28:511–515
12. Camacho C, Coulouris G, Avagyan V, Ma N, Papadopoulos J, Bealer K et al (2009) BLAST+: architecture and applications. BMC Bioinformatics 10:421
13. Zhang Z, Carriero N, Zheng D, Karro J, Harrison PM, Gerstein M (2006) PseudoPipe: an automated pseudogene identification pipeline. Bioinformatics 22:1437–1439
14. Liao Y, Smyth GK, Shi W (2014) featureCounts: an efficient general purpose program for assigning sequence reads to genomic features. Bioinformatics 30:923–930
15. Liao Y, Smyth GK, Shi W (2019) The R package Rsubread is easier, faster, cheaper and better for alignment and quantification of RNA sequencing reads. Nucleic Acids Res 47:e47
16. Liao Y, Smyth GK, Shi W (2013) The subread aligner: fast, accurate and scalable read mapping by seed-and-vote. Nucleic Acids Res 41:e108
17. Robinson MD, McCarthy DJ, Smyth GK (2010) edgeR: a Bioconductor package for differential expression analysis of digital gene expression data. Bioinformatics 26:139–140
18. Martin M (2011) Cutadapt removes adapter sequences from high-throughput sequencing reads Embnet. Journal 17:10–12
19. Nawrocki EP, Eddy SR (2013) Infernal 1.1: 100-fold faster RNA homology searches. Bioinformatics 29:2933–2935
20. Langmead B, Trapnell C, Pop M, Salzberg SL (2009) Ultrafast and memory-efficient alignment of short DNA sequences to the human genome. Genome Biol 10:R25
21. Friedlander MR, Mackowiak SD, Li N, Chen W, Rajewsky N (2012) miRDeep2 accurately identifies known and hundreds of novel microRNA genes in seven animal clades. Nucleic Acids Res 40:37–52
22. Wu HJ, Ma YK, Chen T, Wang M, Wang XJ (2012) PsRobot: a web-based plant small RNA meta-analysis toolbox. Nucleic Acids Res 40: W22–W28
23. Bonnet E, He Y, Billiau K, Van de Peer Y (2010) TAPIR, a web server for the prediction of plant microRNA targets, including target mimics. Bioinformatics 26:1566–1568

24. Shannon P, Markiel A, Ozier O, Baliga NS, Wang JT, Ramage D et al (2003) Cytoscape: a software environment for integrated models of biomolecular interaction networks. Genome Res 13:2498–2504
25. Kozomara A, Birgaoanu M, Griffiths-Jones S (2019) miRBase: from microRNA sequences to function. Nucleic Acids Res 47:D155–D162
26. Kalvari I, Argasinska J, Quinones-Olvera N, Nawrocki EP, Rivas E, Eddy SR et al (2018) Rfam 13.0: shifting to a genome-centric resource for non-coding RNA families. Nucleic Acids Res 46:D335–D342
27. Lorenz R, Bernhart SH, Honer Zu Siederdissen C, Tafer H, Flamm C (2011) ViennaRNA Package 2.0. Algorithms Mol Biol 6:26
28. Li Z, An X, Zhu T, Yan T, Wu S, Tian Y et al (2019) Discovering and constructing ceRNA-miRNA-target gene regulatory networks during anther development in maize. Int J Mol Sci 20:3480
29. Wan X, Li Z (2019) Plant comparative transcriptomics reveals functional mechanisms and gene regulatory networks involved in anther development and male sterility. In: Blumenberg M (ed) Transcriptome analysis. IntechOpen, London
30. Szabo L, Salzman J (2016) Detecting circular RNAs: bioinformatic and experimental challenges. Nat Rev Genet 17:679–692
31. Langfelder P, Horvath S (2008) WGCNA: an R package for weighted correlation network analysis. BMC Bioinformatics 9:559
32. Paci P, Colombo T, Farina L (2014) Computational analysis identifies a sponge interaction network between long non-coding RNAs and messenger RNAs in human breast cancer. BMC Syst Biol 8:83

Chapter 18

In Silico Prediction for ncRNAs in Prokaryotes

Amanda Carvalho Garcia

Abstract

The identification and characterization of non-coding RNAs (ncRNAs) in prokaryotes is an important step in the study of the interaction of these molecules with mRNAs—or target proteins, in the post-transcriptional regulation process. Here, we describe one of the main in silico prediction methods in prokaryotes, using the TargetRNA2 tool to predict target mRNAs.

Key words ncRNAs, Prediction in silico, Prokaryote

1 Introduction

Bioinformatics is dedicated to the development of computer programs for the treatment of biological data and the identification of gene sequences, the prediction of the three-dimensional configuration of proteins, the identification of enzymatic inhibitors, and the promotion of protein grouping, to establish phylogenetic trees and to analyze expression genic [1]. It provides tools for the development of Genomics, Transcriptomics, Proteomics, and Metabolomics [2].

Regarding the prediction of ncRNAs, several computational approaches are used for the identification of genes in intergenic regions of prokaryotic genomes [3, 4]. Many of them are based on the search for transcriptional signals, conserved promoter sequences, rho-independent terminator, transcription factor site, such as sRNAPredict [5], putative prediction algorithm analysis using TransTermHP [6] or TRANSFARC [7] database. sRNAPredict3 and SIPHT are recent computational versions of ncRNA prediction in bacteria [8]. sRNAscanner and sRNAfinder [4] were developed to overcome the limitation of the prediction of transcription signals available in all genomic sequences and proved efficient.

NocoRNAc (non-coding RNA characterization) is a computational tool developed to study the interactions between ncRNA and mRNA in conjunction with the prediction of ncRNAs in bacterial

Shahid Mukhtar (ed.), *Modeling Transcriptional Regulation: Methods and Protocols*, Methods in Molecular Biology, vol. 2328, https://doi.org/10.1007/978-1-0716-1534-8_18, © Springer Science+Business Media, LLC, part of Springer Nature 2021

genomes. This program uses transcription termination signals predicted through the TransTermHP tool, promoters are identified by the SIDD model (Stress-Induced Duplex Destabilization) and checks possible regions to be destabilized [9, 10]. Thus, in regions of the genome flanked by promoter sequence and Rho-independent terminator sequence, **ncRNA sequences are identified**. In a different approach, the Cufflinks tool locates regions of the genome with considerable levels of transcription and free of ORFs [11].

The comparative genomic analyzes that we applied to predict new ncRNAs, and conserved sequences identified for the first time in the intergenic region (IGR), resulting from the comparison of multiple alignments, were classified into ncRNAs. Programs such as QRNA [12] ERPIN [13], ISI [14], INFERNAL [15], MSARI [16]; RNAModif [17] and RNAz [18] compare thermodynamic stability by predicting conserved stable RNA structures, predicting ncRNAs in bacteria [19].

With regard to target prediction, model development is very important, as it integrates bioinformatics for prediction and experimental validation for confirming target mRNA. The classification of ncRNAs provides information about the complementarity of bases (perfect or imperfect) with their target mRNAs and the eventual binding to proteins, changing their activity [20–24].

The TargetRNA2 target mRNA prediction tool is one of the most widely used today [25]. It uses several resources including the conservation of ncRNA in other bacteria, the secondary structure of both ncRNA and each target candidate, and the hybridization energy between interactions. It has the ability to integrate RNA-seq material data when available. Another computational approach used to predict mRNA targets, the IntaRNA tool [26], is also considered to be quite efficient and fast in predicting interactions between ncRNA mRNA. It uses free hybridization energy and has integration with the CopraRNA tool, which predicts ncRNAs by comparing the query sequence with sequences available in the program [26, 27].

2 Materials

2.1 Reference Prokaryotic Genomes

We obtained the sequence data and annotation of the prokaryotic genome to be analyzed, from the database of the National Center for Biotechnology Information Search (NCBI), in *fasta* format.

2.2 RNA-Seq Data

We obtained RNA-seq sequencing data from the Gene Expression Omnibus (GEO) database, or readings derived from the RNA-seq experiments.

2.3 Annotation of ncRNA Sequences

We annotated the sequences of ncRNAs using the online database Rfam versions 14.2 (https://rfam.xfam.org/). This bank contains a collection of ncRNA families represented by manually cured alignment sequences, secondary consensus structures, and notes taken from taxonomy and ontology sources [15]. The database is a broad and diverse source of ncRNAs, including 3024 families of ncRNAs, with information on different types of ncRNAs across the three domains of life and viruses. We used Infernal 1.1.2 to perform multiple sequence alignments using the covariance model and, in addition to the annotation of ncRNAs, Rfam 14.2 classifies ncRNAs and provides bibliographic references for each family, links to the PDB (from English Protein Data Bank), miRBase, ENA and Gene Ontology (GO).

2.4 Bioinformatics Tools

2.5 Artemis

Genome annotation tool developed by the Sanger Institute (http://www.sanger.ac.uk/science/tools/artemis/), which allows the visualization of nucleotide sequences and annotations in EMBL and GENBANK or FASTA formats [28]. This program we written in Java is available for several platforms, such as UNIX, Linux, and Windows. It is possible to view the alignments of the RNA-seq readings in a file in BAM format. In addition, we calculated RPKM values for each candidate ncRNA [29].

2.6 BLAST

The algorithm used by the BLAST program (from the English Basic Local Alignment Search Tool; https://blast.ncbi.nlm.nih.gov/) compares nucleotide or protein sequences with database sequences and calculates the statistical significance of the alignments. This algorithm allows different sequence comparisons: Blastp compares sequences of amino acids (query) against a database of protein sequences (subject); Blastn compares nucleotide sequences against a database of nucleotide sequences.

Blastx compares translated nucleotide sequences for all possible reading phases against a protein sequence database; tBlastn compares amino acid sequences against a database of translated nucleotide sequences and tBlastx compares the six reading phases of a nucleotide sequence against a translated nucleotide database for possible reading phases [30].

2.7 Infernal 1.1.2 (INFERence of RNA Alignment)

Computational tool that uses a covariance model to generate probabilistic profiles of sequences and secondary structures of RNA families [15]. The covariance model is a special case of a probabilistic model that values a combination of the consensus sequence and the secondary consensus structure of a given RNA so that, in many cases, it is able to identify homologous RNAs that have a conserved but low secondary structure primary sequence conservation

[15]. Infernal 1.1.2 (http://eddylab.org/infernal/) consists of five programs: cmbuild, cmcalibrate, cmsearch, cmscan, and cmalig.

2.8 TargetRNA2

TargetRNA2 is a tool for identifying targets of small regulatory RNAs in bacteria (http://cs.wellesley.edu/~btjaden/TargetRNA2/). Regulatory ncRNAs exert a post-transcriptional response on their target mRNAs, through base-pair interactions. This RNA: RNA interaction results in the formation of a regulatory circuit or regulatory network, in which some ncRNAs may be involved in different cellular regulatory mechanisms. Prediction tool for target mRNAs that base-base with ncRNAs [20, 25]. Calculates the hybridization score and the statistical significance of the mRNA–ncRNA interactions [22, 31]. Uses a variety of resources to identify target mRNAs ncRNA conservation—comparison of the sequence available on the GenBank according to the deposited genome (replicon) and indication of regions with greater conservation of sequences prone to undergo mRNA–ncRNA interactions:

(a) Accessibility—the structure and stability of the target ncRNAs-mRNAs interaction in the region of ncRNAs
(b) Hybridization energy—Regions of mRNAs with a **high** index of hybridization energy in one or more regions of interactions with potential targets [25]

2.9 SAMTools

It is a package for manipulating mappings in SAM format [32]. In this work, we used the following commands:

(a) Conversion of SAM to BAM mapping.

```
samtools view -bS file.SAM> file.BAM
```

(b) Alignment of coordinates.

```
samtools sort file.BAM filebam.sorted
```

(c) Classify the index to speed up access to the file.

```
samtools index file.sorted.bam
```

We run all programs on the Linux® platform.

2.10 Transcriptome Analysis

The normalization of the sequenced samples on the SoliD platform uses the RPKM method (reads per kilobase of transcript per million mapped reads) which estimates the expression value of a gene where it measures the density of the readings in a gene region of interest, normalizing the count of the readings in their exonic regions compared to the original size of the gene or exon [33]. This method considers the size of the library, the size of the genes, and the number of readings per gene according to the formula:

$$R = \frac{10^9 C}{NL}$$

C = number of readings per gene

L = size of the gene (kb)

N = library size (total number of readings per biological replicate)

To determine the coverage value, we used the following mathematical expression:

$$x = \frac{C \times 35}{L}$$

C = number of readings per gene

L = size of the gene (Kb)

35 = size of the RNA-seq reading after the trimming process

3 Methods

Although there are several computational tools for prediction of sRNAs in prokaryotes, as shown in (Fig. 1), here we will use the infernal computational approach 1.1.2.

1. Install the Linux operating system.
2. Download the infernal tool 1.1.2 using the following command:

```
cd
ls
cd (folder in the tool)
ls
cd infernal-1.1.2
ls
~\infernal-1.1.2 $ sudo apt-get install infernal
cmpress Rfam.cm
```

3. Download the sequence (fasta format) of the prokaryotic genome to be analyzed.
4. Use the following command line.

```
cmscan -o sequence.out -tblout sequence.tbl -T 24 -notrunc
Rfam.cm sequence
```

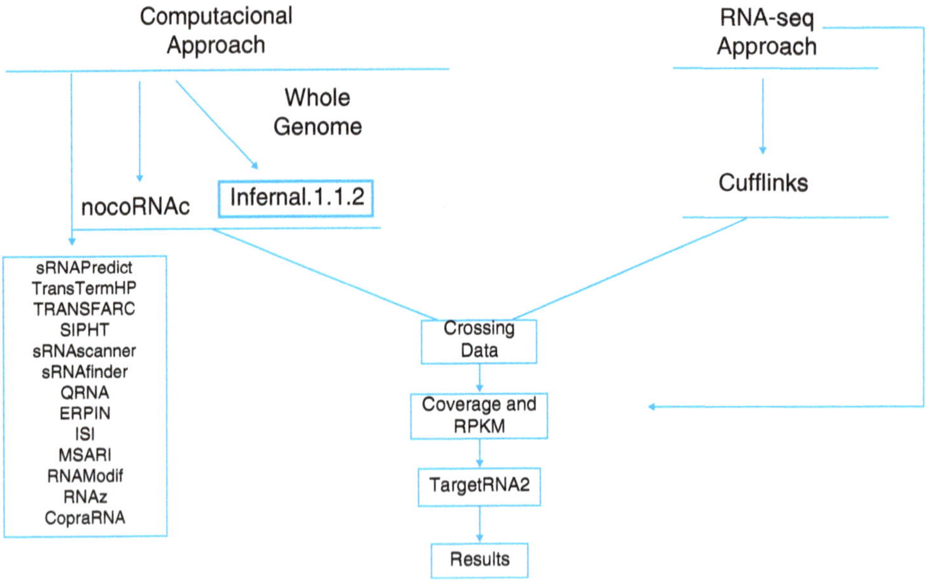

Fig. 1 Protocol flowchart

5. After processing the analysis, generate two directories (Fig. 2) in separate files.

```
Output directed to file (sequence.out).
Tabular output of hits (sequence.tbl).
Query sequence file is the reference genome sequence.
Target CM database is the Rfam database.
```

6. Visualize the searched sequences lined up in a section of our cDNA library by Artemis.

   Use of files: arquivo.fasta, arquivo.sorted.bam e arquivo.sorted.bam.bai.

7. Follow the commands

```
~$./art &
File → Open (ou control O) → arquivo.fasta
File → Read BAM /VCF → arquivo.sorted.bam
```

8. With that, generate an image showing the position where each read stopped (Fig. 3).
9. Use TargetRNA2 to identify targets of small regulatory RNAs in bacteria. This RNA: RNA interaction results in the

```
>> mir-1255
rank     E-value   score  bias  mdl mdl  from    mdl   to        seq  from      seq to         acc  trunc   gc
 ----    -------  -----  ----- -- -----  -----          --------  --------       --- ---   ---
  (6) !   2e-12    69.4   0.0  cm          1     66 []     44397     44462 + . . 1.00    no 0.38

                        v           v                          v          v          NC
                   <<<<<-<-<<---<<--<<-<<<<<-<<<<______>>>>->>>>>->>-->>--->>->>>>>:    CS
   mir-1255      1 UCuuAuGGAuGAGCAAAgaAaGUgGUUUCuUgAGAUaGAAuCuACuuucGGUGAAgaUgCUgaGAA    66
                   UC:UAU  GAUGAGCAAA : AAGUG U: ::UGAGAU:: :U UACU+ :G UGAAGAUG UG: GAA
           7 44397 UCUUAUAGAUGAGCAAAGAAAGUGAUUUUUUGAGAUGGGGUUUACUCCUGCUGUGAAGAUGCUGUGAA    44462
                   ******************************************************************    PP
```

Fig. 2 Visualization by WordPad of the local alignment between the predicted mir-1255 sequence and the target sequence of the reference genome. The statistical value of the expected-value (e-value) of 2e-12 with a score of 69.4, indicates a real alignment and not by chance

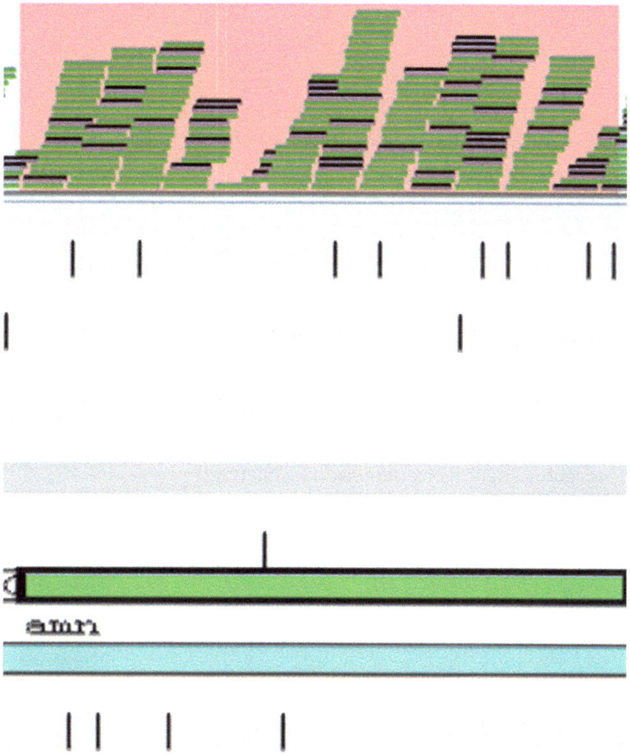

Fig. 3 Visualization of amn position and expression profile in the reference genome. The sequence was located using the Artemis program. The readings are delimited by the pink rectangle

formation of a regulatory circuit or regulatory network, in which some ncRNAs may be involved in different cellular regulatory mechanisms (Fig. 4).

```
amn
     sRNA 20    3' ACUACCUCUUAGUACUAC-A 5'  1
                   |||||||||||||||||| |
     amn  -12   5' UGAUGGAGAAUCAUGAUGUU 3'  8

Energy: -19.09
p-value: 0.000
Gene product: AMP nucleosidase
```

Fig. 4 According to TargetRNA2, a pairing energy value of −19.09 pairing position and energy value make amn an unlikely target for sRNAs

Acknowledgment

Colaborador J.C Hamashia.

References

1. Araújo NDD, Farias RPD, Pereira PB, Figueirêdo FM, Morais AMBD, Saldanha lC, Gabriel JE (2008) The effects and applications of bioinformatics on the biomedical area. Estud Biol 30(70/72):143–148
2. Binneck E (2004) The omics: integrating bioinformation. Magazine Biotechnol 32: 28–37
3. Altuvia S (2007) Identification of bacterial small non-coding RNAs; experimental approaches. Curr Opin Microbiol 10:257–261
4. Sridhar J, Sambaturu N, Sabarinathan R (2010) sRNAscanner: a computational tool for intergenic small RNA detection in bacterial genomes. PLoS One 5(8):e11970
5. Livny J, Waldor MK (2007) Identification of small RNAs in diverse bacterial species. Curr Opin Microbiol 2:1096–1101
6. Kingsford CL, Ayanbule K, Salzberg SL (2007) Rapid, accurate, computational discovery of rho-independent transcription terminators illuminates their relationship to DNA uptake. Genome Biol 8(2):R22
7. Wingender E, Chen X, Hehl R, Karas H, Liebich I, Matys V, Meinhardt T, Pruss M, Reuter I, Schacherer F (2000) TRANSFAC: an integrated system for gene expression regulation. Nucleic Acids Res 28(1):316–319
8. Livny J, Teonadi H, Livny M, Waldor MK (2008) High-throughput, Kingdom wide prediction and annotation of bacterial non-coding RNAs. PLoS One 3(9):e3197
9. Herbig A, Nieselt K (2011) nocoRNAc: characterization of non-coding RNAs in prokaryotes. BMC Bioinformatics 12:40
10. Sridhar J, Gunasekaran P (2013) Computational small RNA prediction in bacteria. Bioinform Biol Insights 7:83–95
11. Trapnell C, Williams BA, Pertea G, Mortazavi A, Kwan G, Van Baren MJ, Salzberg SL, Wold BJ, Pachter L (2010) Transcript assembly and quantification by RNA-Seq reveals unannotated transcripts and isoform switching during cell differentiation. Nat Biotechnol 28(5):511–515
12. Rivas E, Klein RJ, Jones TA, Eddy SR (2001) Computational identification of noncoding RNAs in E. coli by comparative genomics. Curr Biol 11(17):1369
13. Gautheret D, Lambert A (2001) Direct RNA motif definition and identification from multiple sequence alignments using secondary structure profiles. J Mol Biol 313(5):1003–1011
14. Pichon C, Felden B (2003) Intergenic sequence inspector: searching and identifying bacterial RNAs. Bioinformatics 19:1707–1709
15. Nawrocki EP, Kolbe DL, Eddy SR (2009) Infernal 1.0: inference of RNA alignments. Bioinformatics 25(10):1335–1337
16. Coventry A, Kleitman DJ, Berger B (2004) MSARI: multiple sequence alignments for statistical detection of RNA secondary structure. PNAS 101(33):12102–12107
17. Macke TJ, Ecker DJ, Gutell RR, Gautheret D, Case DA, Sampath R (2001) RNAMotif, an RNA secondary structure definition and search algorithm. Nucleic Acids Res 29 (22):4724–4735
18. Pedersen JS, Bejerano G, Siepel A, Rsenbloom K, lndblad-toh K, Lander ES, Kent J, Miller W, Haussler D (2006) Identification and classification of conserved RNA secondary structures in the human genome. PLoS Comput Biol 4:e33
19. Gruber AR, Findeiβ S, Washietl S, Hofacker IL, Stadler PF (2010) RNAz 2.0: improved

noncoding RNA detection. Pac Symp Biocomput 69–79. PMID: 19908359

20. Tjaden B, Goodwin SS, Opdyke JA, Guillie RM, Fu DX, Gottesman S, Storz G (2006) Target prediction for small, noncoding RNAS in bacteria. Nucleic Acids Res 34 (9):2791–2802
21. Pichon C, Felden B (2007) Proteins that interact with bacterial small rna regulators. FEMS Microbiol Rev 31:614–625
22. Tjaden B (2008) TargetRNA: a tool for predicting targets of small RNA action in bacteria. Nucleic Acids Res 36:109–113
23. Cao S, Xu X, Chen SJ (2014) Predicting structure and stability for RNA complexes with intermolecular loop-loop base-pairing. RNA 20(6):835–845
24. Li W, Ying XLQ, Chen L (2012) Predicting sRNAs and their targets in bacteria. Genomics Proteomics Bioinformatics 10(5):276–284
25. Kery MB, Feldman M, Livny J, Tjaden B (2014) TargetRNA2: identifying targets of small regulatory RNAS in bacteria. Nucleic Acids Res 42:124–129
26. Wright PR, Georg J, Mann M, Sorescu DA, Richter AS, Lott S, Kleinkauf R, Hess WR, Backofen R (2014) CopraRNA and IntaRNA: predicting small RNA targets, networks and interaction domains. Nucleic Acids Res 42:119–123
27. Busch A, Richter AS, Backofen R (2008) IntaRNA: efficient prediction of bacterial sRNA targets incorporating target site accessibility and seed regions. Bioinformatics 24 (24):2849–2856
28. Rutherford K, Parkhill J, Crook J, Horsnell T, Rice P, Rajandream MA, Barrell B (2000) Artemis: sequence visualization and annotation. Bioinformatics 16(10):944–945
29. Carver T, Böhme U, Otto TD, Parkhill J, Berriman M (2010) BamView: viewing mapped read alignment data in the context of the reference sequence. Bioinformatics 26(5):676–677
30. Altschul SF, Gish W, Miller W, Myers EW, Lipmanl DJ (1990) Basic local alignment search tool. J Mol Biol 215:403–410
31. Backofen R, Hess WR (2010) Computational prediction of sRNAs and their targets in bacteria. RNA Biol 7(1):33–42
32. Li H, Handsaker B, Wysoker A, Fennell T, Ruan J, Homer N, Marth G, Abecasis G, Durbin R (2009) The sequence alignment/map format and SAMtools. Bioinformatics 25 (16):2078–2079
33. Mortazavi A, Williams BA, Mccue K, Schaeffer L, Wold B (2008) Mapping and quantifying mammalian transcriptomes by RNA-Seq. Nat Methods 5(7):621–628

Chapter 19

Mathematical Linear Programming to Model MicroRNAs-Mediated Gene Regulation Using Gurobi Optimizer

Vijaykumar Yogesh Muley

Abstract

Genes are transcribed into various RNA molecules, and a portion of them called messenger RNA (mRNA) is then translated into proteins in the process known as gene expression. Gene expression is a high-energy demanding process, and aberrant expression changes often manifest into pathophysiology. Therefore, gene expression is tightly regulated by several factors at different levels. MicroRNAs (miRNAs) are one of the powerful post-transcriptional regulators involved in key biological processes and diseases. They inhibit the translation of their mRNA targets or degrade them in a sequence-specific manner, and hence control the rate of protein synthesis. In recent years, in response to experimental limitations, several computational methods have been proposed to predict miRNA target genes based on sequence complementarity and structural features. However, these predictions yield a large number of false positives. Integration of gene and miRNA expression data drastically alleviates this problem. Here, I describe a mathematical linear modeling approach to identify miRNA targets at the genome scale using gene and miRNA expression data. Mathematical modeling is faster and more scalable to genome-level compared to conventional statistical modeling approaches.

Key words Gene expression, Gene regulation, Gurobi, Linear modeling, Linear programming, Mathematical optimization, MicroRNA, miRNA, Post-transcriptional gene regulation, RNA interference

1 Introduction

Proteins are functional and structural parts of all vital processes that take place in living organisms. The genome (DNA sequence) of an organism contains a genetic code that determines how proteins are built. A gene is a region of DNA that contains the genetic code to make an RNA molecule, which subsequently can be used to synthesize a particular protein in a process called *gene expression* [1]. Genes code for various types of RNA molecules. Most RNA types are direct functional products, which act as catalysts or participate in cellular functions. Only a subset of RNA molecules acts as a

Shahid Mukhtar (ed.), *Modeling Transcriptional Regulation: Methods and Protocols*, Methods in Molecular Biology, vol. 2328, https://doi.org/10.1007/978-1-0716-1534-8_19, © Springer Science+Business Media, LLC, part of Springer Nature 2021

message or template to build proteins and is hence referred to as *messenger RNA* (mRNA). The human genome consists of thousands of genes, but only a fraction of them are used in each cell type at each particular time point. The cellular machinery that copies DNA to RNA in a process called *transcription* regulates which genes should be expressed, whereas proteins that are made from the transcribed genes are further controlled by the machinery that copies mRNA to protein in a process called *translation*. Gene expression is also tightly regulated at post-transcriptional and post-translational levels to maintain the necessary pool of RNA molecules and proteins as per the functional requirements of the cell. This very high precision execution of gene expression is essential due to its high energy requirement. Moreover, the cellular functions of genes are affected when they are expressed above or below their predetermined thresholds in the cell.

In the early 1980s and 1990s, molecular biologists obtained unexpected but consistent results on the post-transcriptional inhibition of gene expression [2–5]. It was evident that there are specific RNA molecules binding to mRNA in a sequence-complementary manner, and do not allow protein synthesis from it. The phenomenon was referred to as *post-transcriptional gene silencing*, and was enigmatic until the discovery of RNA interference (RNAi) by Fire and Mello in 1998, for which they received the Nobel prize [6]. They demonstrated that the silencing of mRNA depends on its mature sequence that is homologous to double-stranded RNA molecules available in the cell. These RNA molecules are referred to small interfering RNA (siRNA). When mRNA is bound into a complex of double-stranded RNA, the targeted mRNA degrades. Mello coined the term RNA interference for the phenomenon later on, which has shown great promise not only in understanding basic biology but also using it as a tool to silence a specific gene and understand how it affects cellular physiology.

It was evident from the earlier study that RNAi is a protective mechanism used by cells against RNA viruses and also maintains genome stability by keeping mobile genes silent [7]. There is another class of endogenous small RNA molecules referred to as microRNAs (miRNAs), which are processed from larger hairpin-like precursors RNAs by a machinery similar to siRNAs [8–10]. This miRNA-mediated gene regulation (silencing) involves complementary nucleotide base-pairing with their target mRNA, followed by the recruitment of specific proteins forming an RNA induced silencing complex (RISC). mRNA in a RISC complex becomes unavailable for protein synthesis and leads to its translational inhibition or eventual degradation [11]. The expression of 60% protein-coding human genes is estimated to be controlled post-transcriptionally [12]. Further studies revealed the presence of miRNAs in phylogenetically diverse organisms and play an important regulatory role in a growing number of diverse biological processes.

Using sequence homology, it is possible to identify target genes of miRNAs. The obtained results can often be improved by integrating transcriptomic data. Monga and Kumar have comprehensively reviewed tools and resources available for miRNA and their target predictions—a good resource to start with [13]. However, it is often unclear how miRNA influences the expression of its target genes in various physiological conditions [14]. This chapter describes a protocol demonstrating a linear modeling approach to identify miRNAs influencing the expression of their target gene using expression data by employing the Gurobi software, one of the leading software for mathematical programming. The assumption here is an inversely linear relationship between expression of the gene and its miRNA regulators [11, 14], which can be modeled using linear programming.

1.1 A Linear Model Function

Mathematical modeling originally began with the invention of linear programming (LP) in 1947 by George B. Dantzig. It has three major components: The formulation of a real-world problem in detailed mathematical terms as equation is called the model (models). The model needs to be solved in a reasonable amount of time and in an efficient way, which needs algorithms for optimization of the model, and use of suitable software and hardware to run the algorithms. Hence, mathematical modeling is also called mathematical optimization.

Suppose expression profiling of genes and miRNAs had been carried out in several physiological samples or time-points. The expression profiling data can be organized into a matrix as shown in Table 1. In the table, columns represent variables (i.e., genes and miRNAs) and rows represent systems (or samples) in which the variables were measured. In this case, rows are skin cutaneous melanoma tissue samples of human. The numbers in the table are expression levels of a gene of interest and miRNAs in the corresponding samples. This data can be used to build a linear model.

The simplest linear model finds the relationship between one independent variable, which is also called the predictor, and the dependent (or response) variable as a straight line according to the values of two constant parameters (as shown in Table 1). This model can be written in the familiar mathematical form:

$$\tilde{y}_j = \beta_0 + \beta_t \times x_{t,j} \tag{1}$$

In the equation, y is a response variable (predicted expression of a gene), whose output depends on the predictor variable x ($x_{t,j}$ is the expression of miRNA t in jth sample), β_0 is the value of y when $x = 0$ (y-intercept), and β_t is the degree to which y changes per unit

Table 1
An example of a gene expression data matrix for linear modeling

Samples	Target gene (Dependent variable)	miRNAs (Independent variables)				
	KRAS	miR_181a_1	let_7a_1	miR_487b	miR_143	miR_193b
TCGA.FS. A1ZR.06	9.146	13.417	14.489	2.925	17.504	6.288
TCGA.EE. A3JB.06	10.086	10.939	14.359	2.649	17.505	5.737
TCGA.D3. A3CF.06	10.272	12.051	13.189	−1.127	15.325	6.827
TCGA.EE. A2GI.06	8.737	13.41	14.244	−0.753	16.15	5.817
TCGA.D3. A1Q3.06	10.146	13.239	13.069	−0.28	13.959	5.694
TCGA.EE. A2MC.06	10.173	12.527	13.682	2.294	16.931	6.243
TCGA.BF. AAP0.06	8.726	12.273	13.709	−0.659	14.652	6.394
TCGA.WE. A8K1.06	9.645	12.346	13.919	2.15	15.747	6.467
TCGA.D3. A3MU.06	8.502	12.211	14.237	−0.666	17.667	5.692
TCGA.FS. A1Z4.06	9.556	12.656	13.607	1.436	15.776	5.944

Notes: Columns represent variables (a gene of interest and miRNAs)
Rows represent systems (or samples) in which variables were measured
The numbers in the table represent the expression levels of the variables

of change in x (gradient of the line). The goal of the linear model is to use the j independent measurements to determine a mathematical function that describes the relationship between the response (gene expression) and the predictor variable (miRNA expression). Since there are j measurements to estimate the parameter coefficients (β values) which can take up any positive or negative number, the model needs to explore a combination of all possible numbers and find out the coefficients which will provide the best solution of Eq. 1. Hence, this becomes a mathematical optimization problem and needs an objective function that will keep the track of the best solution over a combination of numbers used as coefficients in Eq. 1. In this case, the objective is to minimize the difference between the measured (real) and the predicted gene expression. It can be mathematically formulated as shown below:

$$\min \sum_{j=1}^{S} \left| y_j - \widetilde{y}_j \right| = \sum_{j=1}^{S} e_j \tag{2}$$

where S is the total number of samples or measurements, y_j and $\widetilde{y}_j$ is the measured and the predicted expression of a gene in the jth sample, respectively. e_j is the difference between the measured and the predicted gene expression in the jth sample, conventionally referred to as an error.

The linear optimizer is required to solve the objective function. This chapter demonstrates the utility of the optimizer Gurobi [15], which uses highly efficient mathematical programming to find the solution.

1.2 A Linear Model for Multiple Independent Variables

Before going into details on how to solve Eq. 2 (i.e., the objective function) using Gurobi, let's generalize the simple linear model for multiple predictor variables (i.e., as many miRNAs as possible). It is just a simple addition of a new predictor variable to Eq. 1 as follows:

$$\widetilde{y}_j = \beta_0 + \beta_1 \times x_{1,j} + \beta_2 \times x_{2,j} + \quad \ldots \quad \beta_n \times x_{n,j} \tag{3}$$

where n is the total number of miRNAs used as predictor variables and all β are optimization parameters. The equation can be summarized in mathematical form as follows:

$$\widetilde{y}_j = \beta_0 + \sum_{t=1}^{n} \beta_t \times x_{t,j} \tag{4}$$

The objective function to optimize this linear model is the same as having shown in Eq. 2.

1.3 Gurobi Optimizer for Linear Modeling

Gurobi has been used in several industries for mathematical programming to solve complex problems [15]. Gurobi is a highly configured tool for mathematical optimization problems. It captures the key features of an optimization problem effectively, and efficiently solves them in a reasonable amount of time. Gurobi uses leading-edge mathematical and computer science technology in solving optimization problems and has perhaps the best performance. Users do not need to worry about the mathematical background of how to solve the optimization problems. This is in-built in the Gurobi optimizer. The mathematics and computer technology behind Gurobi optimization is rather complex and details are beyond the scope of this chapter. However, users are encouraged to explore documentation and tutorials available at Gurobi website. The users only need to efficiently formulate the mathematical model that captures the main characteristics of the optimization problem and the required data for the model. Gurobi optimizes the model automatically behind the scenes. For instance, Gurobi cannot handle absolute values or terms as shown in Eq. 2. Therefore, it is essential to transform Eq. 2 into two inequalities for each sample j as shown below:

$$y_j - \widetilde{y}_j - e_j \leq 0 \tag{5}$$

$$-y_j + \widetilde{y}_j - e_j \leq 0 \tag{6}$$

With this, the optimization problem of modeling gene expression has been formulated in the mathematical form which Gurobi can solve. It has to be noted that the above equations represent the linear model to find the best line fit between response and predictor variables. The subsequent sections provide a detailed workflow to solve this optimization problem practically.

2 Material

1. The workflow can be performed on a standard UNIX or MacOS laptop with 4–8 Gb of RAM. Gurobi automatically uses available computational processing and users do not need to worry about it. It should not be difficult to adopt this workflow on Windows OS.
2. It is expected that users are familiar with at least one programming language to convert a gene and miRNA expression matrix into equations in the specific format that Gurobi can handle.
3. Users lacking basic programming skills are encouraged to collaborate with a good computer programmer.
4. Expression profiling data for a gene(s) of interest and miRNAs should be obtained from the same source.

3 Methods

3.1 Gurobi Optimizer Installation

1. Go to https://www.gurobi.com/
   (a) Click on *Downloads and Licenses.*
   (b) Click on *Gurobi Optimizer-Download Software*, accept the terms and conditions, then download the version appropriate for your operating system.
   (c) Install the software by following the instructions given on the Gurobi webpage for the choice of your operating system.
   (d) Click on *Academic license*, accept the terms and conditions, and generate an academic license.
2. Open command prompt or terminal and type command *grbgetkey* followed by license key code and hit enter to activate the license.

(a) Gurobi will ask where to save the license file. It is recommended to save the file to its default location by hitting the enter key.

(b) The license key code can be obtained from the *Academic license* menu of Gurobi webpage after its creation.

(c) The command looks like "*grbgetkey XXXXXXXX-XXXX-XXXX-XXXX-XXXXXXXXXXXX*," where Xs represent key code.

3. To test if Gurobi works fine, type *gurobi_cl* command in the terminal and hit the enter key. *gurobi_cl* is a Gurobi executable file, which seeks for the input model file, solves optimization problems therein and writes the output solution file. Some typical errors or warnings can be expected (*see* **Note 1**) but usually Gurobi installation is straightforward.

3.2 Gene Expression Data

1. The minimal requirement for linear modeling is a matrix containing expression levels of a gene of interest and for the set of miRNAs under the same physiological conditions in several replicates (or samples).
2. It's possible to select all miRNAs encoded by an organism for modeling. But many miRNAs could have linear expression relationship with a gene of interest even though they do not regulate it. It can lead to a high number of false predictions.
3. On the other hand, since miRNAs regulate gene expression in a sequence-dependent manner, it makes sense to include only miRNAs which are computationally predicted to target a gene of interest based on sequence and structural analysis. Many databases provide potential target genes of miRNAs, which can be used to select potential miRNA regulators for a gene of interest [13]. This approach can overcome the problem of false predictions, though it will miss out on miRNAs that regulate the gene expression but cannot be established by available computational prediction techniques. I recommend modeling with all miRNAs and then filtering using sequence information if necessary.
4. Linear models assume that the predictor variables (expression of miRNAs) are independent. Therefore, predictor variables should not be co-linear (i.e., having strong correlated expression). However, it is not often obvious how each of the variables depends on others especially in biological systems. The Gurobi algorithm deals with multicollinearity quite well but not infallibly. Gurobi underscores variables having multicollinearity, especially on gene expression datasets with small number of samples (personal experience). There are several ways to find multicollinearity in the data and avoid incorporating them in the model (*see* **Note 2**).

5. For brevity, conceptualization and demonstration purposes, expression measurements of the human KRAS gene and five miRNAs were used for linear modeling. The selected miRNAs are known to negatively regulate KRAS in multiple experimental settings [16]. Briefly, the expression matrices for genes and miRNAs were obtained from The Cancer Genome Atlas database, which were measured in skin cutaneous melanoma tissue. Samples, genes, or miRNAs were excluded if they contained more than 10% missing expression values. This criterion retained 349 miRNAs and 223 samples for further analysis. The missing expression values were filled using the nearest neighbor averaging method available in the impute Bioconductor package for R [17]. Then, expression values for KRAS and five selected miRNAs from randomly selected ten samples were used to demonstrate linear modeling (Table 1).

3.3 Preparing the Gurobi Input Model File

1. This workflow uses the Gurobi command-line version for scalability. Also, the input model file format captures an optimization model in a way that is easier for the user to understand and often more intuitive to produce. Experienced programmers may explore the implementation of Gurobi optimization problems in Python or R.
2. Gurobi reads a model from a file, optimizes it and writes the solution to a file. The model input file can be written in various formats, but the LP format is simple and easy to implement.
3. An example LP format file is shown in Fig. 1a, which contains a structured list of sections, where each section captures one logical piece of the whole optimization model. Sections begin with particular keywords and must generally be in the fixed order.
4. The first section in an example LP file is the objective section.
   (a) The goal is to minimize the errors between observed and predicted gene expression, hence it begins with the term **Minimize** on its own line.
   (b) The next line then specifies equation containing the sum of errors from ten samples, which need to be minimized.
5. The second section is the mandatory constraint section and begins with the term **Subject To**, followed by a linear combination of decision variables and parameters that need to be estimated.
   (a) Each sample is represented by two equations written in the LP format syntax, which are equivalent to Eqs. 5 (first block of 10 equations) and 6 (second block of 10 equations). The aim here is to restrict the predicted expression of the gene from too great deviations in either direction

A) Gurobi model file format

```
Minimize

e01 + e02 + e03 + e04 + e05 + e06 + e07 + e08 + e09 + e10

Subject To

e01 + b + 13.417 miR_181a_1 + 14.489 let_7a_1 + 2.925 miR_487b + 17.504 miR_143 + 6.288 miR_193b >= 9.14600
e02 + b + 10.939 miR_181a_1 + 14.359 let_7a_1 + 2.649 miR_487b + 17.505 miR_143 + 5.737 miR_193b >= 10.0860
e03 + b + 12.051 miR_181a_1 + 13.189 let_7a_1 - 1.127 miR_487b + 15.325 miR_143 + 6.827 miR_193b >= 10.2720
e04 + b + 13.410 miR_181a_1 + 14.244 let_7a_1 - 0.753 miR_487b + 16.150 miR_143 + 5.817 miR_193b >= 8.73700
e05 + b + 13.239 miR_181a_1 + 13.069 let_7a_1 - 0.280 miR_487b + 13.959 miR_143 + 5.694 miR_193b >= 10.1460
e06 + b + 12.527 miR_181a_1 + 13.682 let_7a_1 + 2.294 miR_487b + 16.931 miR_143 + 6.243 miR_193b >= 10.1730
e07 + b + 12.273 miR_181a_1 + 13.709 let_7a_1 - 0.659 miR_487b + 14.652 miR_143 + 6.394 miR_193b >= 8.72600
e08 + b + 12.346 miR_181a_1 + 13.919 let_7a_1 + 2.150 miR_487b + 15.747 miR_143 + 6.467 miR_193b >= 9.64500
e09 + b + 12.211 miR_181a_1 + 14.237 let_7a_1 - 0.666 miR_487b + 17.667 miR_143 + 5.692 miR_193b >= 8.50200
e10 + b + 12.656 miR_181a_1 + 13.607 let_7a_1 + 1.436 miR_487b + 15.776 miR_143 + 5.944 miR_193b >= 9.55600

e01 - b - 13.417 miR_181a_1 - 14.489 let_7a_1 - 2.925 miR_487b - 17.504 miR_143 - 6.288 miR_193b >= -9.1460
e02 - b - 10.939 miR_181a_1 - 14.359 let_7a_1 - 2.649 miR_487b - 17.505 miR_143 - 5.737 miR_193b >= -10.086
e03 - b - 12.051 miR_181a_1 - 13.189 let_7a_1 + 1.127 miR_487b - 15.325 miR_143 - 6.827 miR_193b >= -10.272
e04 - b - 13.410 miR_181a_1 - 14.244 let_7a_1 + 0.753 miR_487b - 16.150 miR_143 - 5.817 miR_193b >= -8.7370
e05 - b - 13.239 miR_181a_1 - 13.069 let_7a_1 + 0.280 miR_487b - 13.959 miR_143 - 5.694 miR_193b >= -10.146
e06 - b - 12.527 miR_181a_1 - 13.682 let_7a_1 - 2.294 miR_487b - 16.931 miR_143 - 6.243 miR_193b >= -10.173
e07 - b - 12.273 miR_181a_1 - 13.709 let_7a_1 + 0.659 miR_487b - 14.652 miR_143 - 6.394 miR_193b >= -8.7260
e08 - b - 12.346 miR_181a_1 - 13.919 let_7a_1 - 2.150 miR_487b - 15.747 miR_143 - 6.467 miR_193b >= -9.6450
e09 - b - 12.211 miR_181a_1 - 14.237 let_7a_1 + 0.666 miR_487b - 17.667 miR_143 - 5.692 miR_193b >= -8.5020
e10 - b - 12.656 miR_181a_1 - 13.607 let_7a_1 - 1.436 miR_487b - 15.776 miR_143 - 5.944 miR_193b >= -9.5560

Bounds

miR_181a_1 free
let_7a_1 free
miR_487b free
miR_143 free
miR_193b free

End
```

B) Gurobi output file format

```
# Objective value = 2.3587218089616080e+00
e01 0
e02 7.7933025597851469e-01
e03 0
e04 0
e05 0
e06 0
e07 6.6968598222521791e-01
e08 0
e09 5.3031928829273056e-01
e10 3.7938628246514483e-01
b 2.6688041826224627e+01
miR_181a_1 -3.1555227597285046e-02
let_7a_1 -1.4655227890979305e+00
miR_487b 1.3093536612889425e-01
miR_143 1.6623888375320484e-01
miR_193b 1.3079882101541807e-01
```

Fig. 1 An example of the Gurobi linear model file format and its output for gene expression modeling. (**a**) Gurobi input file format containing objective function in the **Minimize** section, mandatory constraints in the **Subject To** section, and optional constraints in the **Bounds** section. Gurobi estimates β coefficients for *b* (i.e., β_0 to capture additive effects) and *miR* or *let* prefixed variable names (β coefficients for miRNAs). Variables are set free in the bound section, which means they can have positive or negative coefficients. (**b**) Gurobi solution file format containing the value of the objective function (i.e., overall modeling error), followed by samples contributing to the objective function value (sample-wise errors), and then estimated β coefficients for β_0 and miRNAs, prefixed with *miR* or *let*. miRNAs with negative estimated β coefficients are potential regulators of the gene expression (i.e., positive or negative) from the measured expression. That's the reason there are two equations for each sample.

(b) Briefly, e01 to e10 represent errors for ten samples, then the parameter specification begins represented by β coefficients which will be estimated as part of the optimization solution. Except for β_0 (represented by the letter *b*), all other β coefficients are represented by the name of the miRNA and are preceded by the expression value of the corresponding miRNA. For example, "14.489 let_7a_1" equivalent to $x_{t,\ j} \times \beta_t$ represents the expression value of the let_7a_1 miRNA in the sample *j*, which is the first sample in Table 1 and Fig. 1, multiplied (denoted with a space) by its β coefficient (represented by the name of miRNA). Please *see* Eq. 1 or Eq. 4 for understanding.

(c) Every equation in this section ends with the number (preceded by >= operator) representing the measured expression of the gene, whose expression needs to be modeled. KRAS in this case.

6. The optional bounds section follows the mandatory constraint section. It begins with the word **Bounds**, on its own line, and is followed by a list of optional variable bounds, each on its own line.

   (a) The idea here is to put restrictions on the lower and upper possible value (ranges) on the parameters (coefficients) being estimated. By default, Gurobi estimates coefficients within a range of zero to positive infinite values, which Gurobi interprets as the high expression of a miRNA having a positive effect on the expression of its target gene, which biologically does not make sense, since miRNAs have a negative influence on the expression of their target gene. Hence, bounds for each of the miRNAs could be set between zero and negative infinite values in this section. This may however not work for many reasons, the most important ones of which are:

   - Biological systems are not rigid, in the sense that a gene could be a target of a certain miRNA in one tissue but not in others.
   - Restricting coefficients to negative could prevent the model from optimizing.

   (b) Therefore, bounds for each coefficient being estimated are set to free, so that they can be positive or negative. This is more natural and data-dependent.

7. The last line in an LP format file should be the word **End**, to conclude model formulation.

8. The file should be saved with the *lp* file extension so that Gurobi recognizes it; for example, *KRAS.lp*
9. As mentioned above, the whitespace between two variables or constants is treated as multiplication sign (resulting in their product being calculated). In addition, the backslash symbol starts a comment and the remainder of that line is ignored by Gurobi. Lastly, miRNA names are modified by substituting hyphen (-) to underscore (_) signs in the Gurobi model input file because Gurobi can treat hyphen as subtraction sign. Understanding the LP file format will take the user way ahead in formulating various optimization problems in Gurobi (please *see* user manual on Gurobi website).

3.4 Running Gurobi with Input Model File

1. The optimizer Gurobi can be executed by the simple command given below with default settings (Recommended, *see* **Note 3**).
2. This output file should be given the file extension *sol*, which stands for "solution information format" and is set using the *ResultFile* option, followed by the input file name, as shown below:

```
gurobi_cl ResultFile=output_filename.sol input_filename.lp
```

3. When the above command is executed, Gurobi should write an optimization solution to the output file. If the present model is an infeasible model, it writes an Irreducible Inconsistent Subsystem (IIS) instead (*see* **Note 4**).

3.5 Interpreting Gurobi Output

1. The Gurobi output file contains three components (Fig. 1b).
   (a) First, the objective function value, which is the sum of the errors in all samples. Since it is a minimization problem, a smaller value (close to zero) indicates a good optimal solution.
   (b) The second component lists how much each measurement or sample contributed to the objective function value on its own line. Higher values represent problematic samples. When the objective function value is too high for minimization problem, then the samples with high error rates may be removed.
   (c) The third and very important component represents the estimated β coefficients for each of the miRNAs, and the additive offset value of β_0.
2. Generally, it is not a good idea to interpret the β coefficients estimated by linear models. The model works reasonably well but it is not infallible. The β coefficients of miRNAs with negative values can be interpreted as representing inhibitory

effect on the expression of the gene. Therefore, those miRNAs could be selected as potential regulators of gene expression.

3. For example, the let-7a-1 miRNA has a negative estimated coefficient ($\beta = -1.4655$), consistent with its role in negative regulation of KRAS gene expression [18], as well as miR-181a-1($\beta = -0.0316$). The β coefficient of let-7a-1 is lower than that of miR-181a-1, suggesting that its inhibitory influence on KRAS gene expression is stronger. On the other hand, even though all remaining miRNAs in Table 1 have been reported to negatively regulate KRAS, they received positive β coefficients, suggesting they may not regulate KRAS expression at least in this biological context (skin cutaneous melanoma tissue).

3.6 Predicting Gene Expression

1. Gene expression can be predicted easily by plugging the estimated β coefficients in Eq. 4 and solve it for each sample. For example, the expression of KRAS gene in sample TCGA.FS.A1ZR.06 can be calculated by adding up the products of the expression values of miRNAs and their estimated β coefficients plus coefficient β_0 as follows:

$$\widetilde{y}_{\mathrm{TCGA.FS.A1ZR.06}} = \beta_0 + (\beta_{\mathrm{miR-181a-1}} \times 13.4) + (\beta_{\mathrm{let-7a-1}} \times 14.5) + (\beta_{\mathrm{miR-487b}} \times 2.925) + (\beta_{\mathrm{miR-143}} \times 17.5) + (\beta_{\mathrm{miR-193b}} \times 6.29)$$

2. The above equation can be solved by substituting the estimated β coefficients corresponding to each miRNA from the Gurobi output file. The solution, i.e., the predicted expression of KRAS in the TCGA.FS.A1ZR.06 sample ($\widetilde{y}_{\mathrm{TCGA.FS.A1ZR.06}}$) is 9.146, which is exactly the same as its measured expression. Table 2 shows a comparison of predicted and measured expression of KRAS for all samples. The predicted and measured expression differs slightly only in the samples TCGA.EE.A3JB.06, TCGA.BF.AAP0.06, TCGA.D3.A3MU.06, and TCGA.FS.A1Z4.06.

3. The estimated β coefficients can also be used with other expression datasets to predict KRAS expression in those and actually check how well the model performs on unseen data. This is a direct way to validate the model.

4. There are several ways to achieve better accuracy by tuning input data and the model when the solution of objective function has a very high error, and the estimated β coefficients do not reflect the expected relationship of corresponding miRNAs and the expression of the gene (*see* **Note 5**).

Table 2
A comparison between measured and predicted expression of KRAS gene using linear programming

Sample	Measured expression	Predicted expression
TCGA.FS.A1ZR.06	9.146	9.146
TCGA.EE.A3JB.06	10.086	9.3067
TCGA.D3.A3CF.06	10.272	10.272
TCGA.EE.A2GI.06	8.737	8.737
TCGA.D3.A1Q3.06	10.146	10.146
TCGA.EE.A2MC.06	10.173	10.173
TCGA.BF.AAP0.06	8.726	9.3957
TCGA.WE.A8K1.06	9.645	9.645
TCGA.D3.A3MU.06	8.502	9.0323
TCGA.FS.A1Z4.06	9.556	9.9354

4 Notes

1. After installing Gurobi, it may report an error that the *gurobi_cl* Gurobi "executable file not found." In this case, you need to set a global path for the Gurobi library. This can be done easily by following the installation guide provided by the Gurobi developers on their webpage.
2. Linear modeling tools assume that the predictor variables are independent, and hence fail to handle variables that are co-linear or strongly correlated at expression levels without an actual regulatory link. It is difficult to find such variables for removal, but there are a few methods available to do so. Gurobi handles the multicollinearity issue quite efficiently by its cutting plane algorithm and will find a solution. However, it can also underscore important predictor variables, and provide an unstable solution for the model, especially when the gene expression data has a small number of samples (personal experience). The variance inflation factor is a widely used diagnostics for multicollinearity. Another approach to handle such a situation is the clustering of miRNAs based on their expression similarity and keep one representative miRNA from each cluster for modeling.
3. Gurobi has deterministic and non-deterministic algorithms to solve models, which can be set by the *Method* parameter. The deterministic algorithms give the exact same result each time you run, while non-deterministic algorithms can produce different optimal solutions when running multiple times.

However, it is recommended not to change this parameter since there is no big gain over the default algorithm. One useful parameter is *TimeLimit*, for large sample and miRNA numbers, where model would take a lot of time for optimization, depending on the computational infrastructure. Gurobi will terminate optimization when the time expended exceeds the value specified in the *TimeLimit* parameter.

4. When the model shows the Irreducible Inconsistent Subsystem (IIS) message in the solution file, it is possible to identify the cause of the infeasibility. Users can run Gurobi again by setting one more output file, with an *ilp* extension, using the *ResultFile* option. Gurobi attempts to solve the model, and will automatically compute an IIS and write it to the requested file name (filename.ilp). This file contains the details on the subset of the constraints and variable bounds causing infeasibility, which when removed may change the model to be feasible. One model may have multiple IISs. The one first returned by Gurobi is not necessarily the one with minimum cardinality. Users need to play with the model and identify the problems.

5. In general, linear models work quite well. However, expression data is always subject to technical and biological variations. Genes and miRNAs can have correlated expression patterns if their expression is required at the same time point, even though these miRNAs have no role in regulating them. Some recommendations can be proposed to achieve better solutions.
   (a) Selection of expression samples derived from diverse physiological conditions in several replicates. It is better to use as many different samples as possible while excluding highly similar by clustering approach.
   (b) For modeling of one or few genes, it is possible to exclude one miRNA or a combination at a time and see whether the objective function value improves or deteriorates. When the objective function value deteriorates (i.e., increases for minimization problems), it means that the miRNA has a significant effect on the expression of the gene and is more likely a regulator.
   (c) Multicollinearity can hamper results and cause models to display entirely different behavior than expected (*see* **Note 2**). In such cases, removing multicollinear independent variables may improve the solution.

Acknowledgments

The author would like to sincerely thank Anne Hahn (Queensland Brain Institute, Australia) for critical reading of the manuscript.

This work was supported by DGAPA-UNAM research grant IA203920 to V.Y.M. The results shown here are based on data generated by the TCGA Research Network: https://www.cancer.gov/tcga.

References

1. Muley VY, Pathania A (2017) Gene expression. In: Vonk J, Shackelford T (eds) Encyclopedia of animal cognition and behavior. Springer, Cham
2. Mizuno T, Chou MY, Inouye M (1984) A unique mechanism regulating gene expression: translational inhibition by a complementary RNA transcript (micRNA). Proc Natl Acad Sci U S A 81(7):1966–1970
3. Lee RC, Feinbaum RL, Ambros V (1993) The C. elegans heterochronic gene lin-4 encodes small RNAs with antisense complementarity to lin-14. Cell 75:843–854
4. Wightman B, Ha I, Ruvkun G (1993) Post-transcriptional regulation of the heterochronic gene lin-14 by lin-4 mediates temporal pattern formation in C. elegans. Cell 75:855–862. http://www.ncbi.nlm.nih.gov/pubmed/8252622
5. Dougherty WG, Parks TD (1995) Transgenes and gene suppression: telling us something new? Curr Opin Cell Biol 7:399–405. http://www.ncbi.nlm.nih.gov/pubmed/7662371
6. Fire A, Xu S, Montgomery MK, Kostas SA, Driver SE, Mello CC (1998) Potent and specific genetic interference by double-stranded RNA in Caenorhabditis elegans. Nature 391:806–811
7. Plasterk RHA (2002) RNA silencing: the genome's immune system. Science 296:1263–1265. http://www.ncbi.nlm.nih.gov/pubmed/12016302
8. Lee RC, Ambros V (2001) An extensive class of small RNAs in Caenorhabditis elegans. Science 294:862–864. http://www.ncbi.nlm.nih.gov/pubmed/11679672
9. Murchison EP, Hannon GJ (2004) miRNAs on the move: miRNA biogenesis and the RNAi machinery. Curr Opin Cell Biol 163:223–229
10. Zamore PD, Haley B (2005) Ribo-gnome: the big world of small RNAs. Science 309 (5740):1519–1524
11. Meijer HA, Kong YW, Lu WT, Wilczynska A, Spriggs RV, Robinson SW et al (2013) Translational repression and eIF4A2 activity are critical for microRNA-mediated gene regulation. Science 340(6128):82–85
12. Friedman RC, Farh KKH, Burge CB, Bartel DP (2009) Most mammalian mRNAs are conserved targets of microRNAs. Genome Res 19 (1):92–105
13. Monga I, Kumar M (2019) Computational resources for prediction and analysis of functional miRNA and their targetome. Methods Mol Biol 1912:215–250. http://www.ncbi.nlm.nih.gov/pubmed/30635896
14. Mullokandov G, Baccarini A, Ruzo A, Jayaprakash AD, Tung N, Israelow B et al (2012) High-throughput assessment of microRNA activity and function using microRNA sensor and decoy libraries. Nat Methods 9:840–846. http://www.ncbi.nlm.nih.gov/pubmed/22751203
15. Gurobi Optimization LLC 2020 Gurobi optimizer reference manual. http://www.gurobi.com
16. Huang H-Y, Lin Y-C-D, Li J, Huang K-Y, Shrestha S, Hong H-C et al (2020) miRTarBase 2020: updates to the experimentally validated microRNA-target interaction database. Nucleic Acids Res 48:D148–D154. http://www.ncbi.nlm.nih.gov/pubmed/31647101
17. Hastie T, Tibshirani R, Narasimhan B, Chu G (2020) impute: impute: imputation for microarray data. R Packag version 1620
18. Johnson SM, Grosshans H, Shingara J, Byrom M, Jarvis R, Cheng A et al (2005) RAS is regulated by the let-7 microRNA family. Cell 120(5):635–647

Correction to: Inference of Gene Regulatory Network from Single-Cell Transcriptomic Data Using pySCENIC

Nilesh Kumar, Bharat Mishra, Mohammad Athar, and Shahid Mukhtar

Correction to:
Chapter 10 in: Shahid Mukhtar (ed.), *Modeling Transcriptional Regulation: Methods and Protocols*, Methods in Molecular Biology, vol. 2238, https://doi.org/10.1007/978-1-0716-1534-8_10

In the original version of this book, chapter 10 was published with incomplete list of citations. This has now been rectified in this revised version of the book.

The original version of chapter 10 was published without citing "SCENIC: single-cell regulatory network inference and clustering" by Aiber et al., *Nature Methods* 2017, its Python implementation "A scalable SCENIC workflow for single-cell gene regulatory network analysis" by Van de Sande et al., *Nature Protocols* 2020, GitHub repository (https://github.com/aertslab/pySCENIC), and the scRNA-seq dataset (GSE115978).

The updated online version of this chapter can be found at
https://doi.org/10.1007/978-1-0716-1534-8_10

Shahid Mukhtar (ed.), *Modeling Transcriptional Regulation: Methods and Protocols*, Methods in Molecular Biology, vol. 2328,
https://doi.org/10.1007/978-1-0716-1534-8_20, © Springer Science+Business Media, LLC, part of Springer Nature 2021

Index

Shahid Mukhtar (ed.), *Modeling Transcriptional Regulation: Methods and Protocols*, Methods in Molecular Biology, vol. 2328, https://doi.org/10.1007/978-1-0716-1534-8,

H

I

J

K

L

M

N

MIX
Papier aus verantwortungsvollen Quellen
Paper from responsible sources
FSC® C105338

If you have any concerns about our products,
you can contact us on
ProductSafety@springernature.com

In case Publisher is established outside the EU,
the EU authorized representative is:
Springer Nature Customer Service Center GmbH
Europaplatz 3, 69115 Heidelberg, Germany

Printed by Libri Plureos GmbH
in Hamburg, Germany